Structuring and Composition of Polyester

Structuring and Composition of Polyester

Edited by **Ken Lao**

NY RESEARCH PRESS

New York

Published by NY Research Press,
23 West, 55th Street, Suite 816,
New York, NY 10019, USA
www.nyresearchpress.com

Structuring and Composition of Polyester
Edited by Ken Lao

International Standard Book Number: 978-1-63238-428-7 (Hardback)

Printed in China.

Contents

Preface

Every book is initially just a concept; it takes months of research and hard work to give it the final shape in which the readers receive it. In its early stages, this book also went through rigorous reviewing. The notable contributions made by experts from across the globe were first molded into patterned chapters and then arranged in a sensibly sequential manner to bring out the best results.

This book primarily focuses on the structuring and composition of polyesters. Polyesters are being broadly applied in distinct fields because of their extraordinary characteristics. This book encompasses important aspects like characterization, applications and synthesis of polyester. Biological and chemical techniques for the synthesis of polyester are one of the most significant topics of research that have been added in this book. Physical, mechanical and chemical characterizations as well as enhancement of polyester have also been discussed in this book. Few aspects of applications of polyester in coating of aluminum alloys employed in aircraft, immobilization of radioactive wastes, formation of the novel composites and fabrics manufacturing have also been discussed, emphasizing on the extensive possibilities of its implementations. The aim of this book is to serve as a good source of reference for researchers, students and scientists as well as for other veterans, both in industry and academia.

It has been my immense pleasure to be a part of this project and to contribute my years of learning in such a meaningful form. I would like to take this opportunity to thank all the people who have been associated with the completion of this book at any step.

Editor

Section 1

Synthesis and Characterizations

Chapter 1

Whey Lactose as a Raw Material for Microbial Production of Biodegradable Polyesters

Martin Koller, Anna Salerno, Alexander Muhr, Angelika Reiterer, Emo Chiellini, Sergio Casella, Predrag Horvat and Gerhart Braunegg

Additional information is available at the end of the chapter

1. Introduction

Whey, a by-product of diary and cheese industry, constitutes the watery portion after the separation of fat and caseins from whole milk. Cheese whey is a surplus material produced in volumes almost equal to the milk processed in cheese manufactories, therefore its disposal as a waste causes serious pollution problems in the surrounding environment where it´s discarded. This is due to its enormous biochemical oxygen demand that is mainly caused by its high lactose content; as a consequence a large amount of industrial capital is requested for whey disposal. During the last years, the amounts of whey increased to such an extent that they cannot be simply used as animal feed as the most common application. To overcome these problems a sustainable alternative is to upgrade whey and its derivates to a resource for many value added industrial products, making whey not only a waste but also a valuable resource.

The article presents a future-oriented, alternative strategy to use surplus whey, namely its upgrading to the role of a raw material for cost-efficient production of polyhydroxyalkanoate (PHA) biopolyesters. PHAs are a group of bio-based, bio-compatible and compostable bio-plastics of increasing significance for numerous industrial applications. Data for PHA production from whey on different production scales and techniques by various microbial species are compared and embedded in the entire scientific field of biopolymers. The review shows how the smart solution of an industrial waste disposal problem can be combined with enhanced cost efficiency in production of "green plastics".

2. General: The need for sustainable utilization of whey

Whey is the liquid remaining after the coagulation of milk casein in cheese making or casein manufacture. Cheese whey, representing about the 85-95% (Guimarães et al., 2010) of the milk volume, constitutes a waste- and surplus material from dairy and cheese industries in many regions of the world (Illanes, 2011). The reported amounts of whey that are produced globally vary from $1.15*10^8$ tons (Peters, 2006) to $1.40*10^8$ tons (Audic, 2003) per year. OECD-FAO estimations for 2008 even report $1.60*10^8$ tons with annual increase of 1-2% (reviewed in Guimarães et al., 2010). Mainly in North America and Europe, huge quantities of whey are available; in 2008, the estimated accruing values are reported with $4*10^7$ tons for the USA, and $5*10^7$ tons for the European Union. Reliable data for Canada, another important whey producing country, are valid for the year 1997; here $2.2*10^5$ t are reported annually (Ghaly & El-Taweel, 1997).

Bovine whey is not only a cheap raw material, but also causes severe disposal problems because of the huge amounts generated and its high organic matter content. During cheese production, whey accrues in almost equal volumes to the processed milk. It can be estimated that, to make 1 kg of cheese, about 9L of whey are produced (Kosikowki, 1979). Whey shows a high biochemical oxygen demand (BOD, 40,000 –60,000 ppm) and chemical oxygen demand (COD, 50,000–80,000 ppm), making the disposal of surplus whey rather expensive (Kim et al., 1995; Viñas et al., 1994); lactose, the most representative compound of whey, is the main responsible component causing these high values for BOD and COD. The major part of whey is discarded as waste in the surrounding environments, causing dramatic pollution problems. Its disposal can be accomplished *via* different strategies: piping it into lagoons, rivers, lakes or oceans, funnelling in caves, feeding to ruminants and pigs, and spreading over fields. The negative impact of disposing whey in water bodies is visible by a quick reduction of aquatic life due to the depletion of the dissolved oxygen. In case of releasing on fields, the physical and chemical structures of soil are severely affected, with a consequent decrease in crop yield (Gonzales-Siso, 1996).

It has to be considered that the processing of one million litres of milk causes the challenging task of disposing of nearby one million litres of whey. This exemplary quantity contains up to 50 t of whey main carbon ingredient, namely lactose. The resulting problem becomes especially obvious considering the fact that today surplus whey is very often disposed of just by being poured into the sea. For example, in the Northern Italian Po-region, where a variety of well-known dairies is located, about 1 million litres of whey has to be disposed daily.

The market of whey products for human nutrition, e.g. for sweets, whey powder as nutritional supplement in body building formulations for increase of muscle size, whey beverages, additives for food processing (e.g. ice cream, meat products), baby food, or for application as pharmaceutical matrices, is also restricted. This is due to the high number of people suffering from lactose intolerance due to lacking activity of the enzyme lactase in their metabolism (Heyman, 2006). It is estimated that 75% of adults worldwide show some decrease in lactase activity (hypolactasia), resulting in lactose intolerance still during adulthood. The phenomenon "lactose intolerance" is mainly found in African, Asian, South American and South European regions (Bulhões et al., 2007). Unfortunately, this frequently occurring of lactase activity deficiency hampers a broader application of whey for nutrition of mankind. It has to be considered that whey is a material of substantial nutritional value due to its high contents of carbohydrates, proteins, lipids, and precious minerals like calcium (see Table 1). The consumption of 100 g of whey results in the considerable uptake of more than 100 kJ of nutritional energy.

The facts discussed above clearly underline the necessity to convert whey in a safe and, most favourably, value-adding way. During the last couple of years, the generated amounts of whey increased to such an extent that they cannot be simply used as porcine feed as it's nowadays most common application. It has to be emphasised that e.g in Italy this applications is not allowed any more for the production of *"Denominazione Origine Protetta"* (DOP, protected origin) porcine meet (ham) because the related disciplinary regulation does not admit the feeding of pigs with whey.

Considerable efforts are undertaken worldwide to upgrade surplus whey to a carbon feed stock for bioconversion towards various value-added products. The disaccharide lactose, as it´s major carbohydrate, can function as a carbon substrate for growth and product formation in numerous biotechnological processes. In literature, the production of bioethanol (Ghaly & El-Taweel, 1997; Zafar & Owais, 2006), vinegar (Parrondo et al., 2003), antibiotics, e.g. the bacteriocin Nisin, (Hickmann Flôres & Monte Alegre, 2001), yeasts for yeast extract production (de Palma Revillion et al., 2003), surface active compounds like sophorolipids (Daniel et al., 1999), single-cell protein (Schultz et al., 2006), "green bioplastics" like Polyhydroxyalkanoates, PHAs, (Ahn et al., 2000; Ahn et al., 2001; Kim, 2000; Povolo & Casella, 2003; Koller et al., 2007 a,b), and lactic acid that is of importance as food additive (E 270), for pharmaceutical matrices, and as monomer for the production of polylactic acid (PLA) is described (Kim et al., 1995). In addition, the induction of high-cell-density production of recombinant proteins can be accomplished by providing whey (Viitanen et al., 2003). Chemically, whey lactose can be converted to the artificial sweetener lactitol (E 966) or the laxative lactulose (Illanes 2011). Table 1 collects the value-added products that can be produced starting from whey.

Products from Whey	Production Mode	Strains	References
PHA	Biotechnological (direct conversion or upstream processing of substrate and conversion into PHA)	Recombinant *Escherichia coli, Hydrogenophaga psedoflava,* Azotobacter spp (direct conversion). Lactobacilli (from whey to lactic acid), (e.g.) *Cupriavidus necator* (from lactic acid to PHA). *Haloferax mediterranei, Pseudomonas hydrogenovora* (from hydrolysed lactose	Ahn et al., 2000; Ahn et al., 2001; Koller et al., 2007; Povolo & Casella, 2003

Products from Whey	Production Mode	Strains	References
		to PHA).	
Lactic Acid (PLA)	Biotechnological (anaerobic conversion) and chemical polymerization (PLA)	Lactobacilli (from whey to lactic acid)	Koller et al., 2010
Bioethanol	Biotechnological conversion	*Saccharomyces cerevisiae, Klyveromyces lactis, Klyveromyces marxianus, Candida pseudotropicalis*	Guimarães et al., 2010
SCP (single cell proteins)	Biotechnological conversion	*Klyveromyces lactis, K. fragilis, Torulopsis bovina, Candida intermedia*	Guimarães et al., 2010, Siso et al., 1996
Yeast	Biotechnological conversion	*K. marxianus*	de Palma Revillion, 2003
Vinegar	Biotechnological conversion (2 steps)	*K. marxianus* (alcoholic fermentation) *Acetobacter pasteurans* (acetic fermentation)	Parrondo et al., 2003
Artificial Sweeteners	Chemical reduction or enzymatic hydrolysis	*Aspergillus* and *Kluyveromyces* spp. (enzymes)	Gänzle et al. ,2008
Antibiotics	Biotechnological conversion	*Lactococcus lactis* (bacteriocin); *Listeria monocytogenes* and *Clostridium botulinum*	Hickmann, Flores & Monte Alegre, 2001
Sophorolipids	Biotechnological conversion (2 steps)	*Cryptococcus curvatus* (single cell oil, SCO), *Candida bombicola* (SCO converted in sophorolipids)	Daniel et al., 1999

Table 1. Products accessible from whey lactose

3. Polyhydroxyalkanoates: The only family of "Green plastics" completely synthesized by microbes

3.1. "Green plastics"

Nowadays, a variety of manufactures claim the fashionable labelling *"green plastic"* for polymeric materials they commercialize on the market. In most cases, the *de facto* properties of these products are not really in accordance with the generally valid definitions for classifying them as *"biobased"*, *"biodegradable"*, *"compostable"*, or *"biocompatible"* (CEN/TR

15391, CEN/TR 15932), thus making them "*green*". Such attributes do only apply to plastics if they come along with the strict requirements defined by standardized norms and certificates.

For instance, the norm EN-13432 that deals with biodegradation and composting of plastic packaging materials clearly postulates that a (plastic) material can be termed "*biodegradable*", if 90% of its carbon is metabolized within 180 days. According to the same norm, the material is "*compostable*", if not more than 10% of the material remains in a sieve of 2 mm pore size after 180 days of composting.

The classification „*biocompatible*" refers to the fact that, using standardized methods for assessing the ecotoxicity of the (plastic) material, it must not feature any negative impact on living organisms or the involved environment. This is described in the certification according to a standardized norm ISO 10993. Together with biodegradability, biocompatibility is of special significance for *in vivo* applications of polymers such as implants for surgery or other medical applications.

In addition, a polymer can be classified as „*biobased*", if the production of the building block monomeric units is based on renewable resources; afterwards, the polymerization of the monomers may occur chemically (e.g.: polymerization of biosynthesized lactic acid to PLA) or biologically (e.g.: in vivo polymerization of hydroxyacyl-CoAs by PHA synthases towards PHAs).

3.2. Particularities and significance of PHAs

PHAs constitute a family of biodegradable intracellular polyesters synthesized by a wide range of prokaryotic genera starting from renewable feedstocks (Koller et al., 2010). Among all known classes of bio-based polymers with plastic-like properties, PHAs are the only ones that are entirely produced and degraded by living cells. Figure 1 provides a Scanning Transmission Electron Microscopy (STEM) picture of *Cupriavidus necator* cells cultured on glucose as carbon source in a continuous cultivation process. The accumulated PHA inclusions are well visible as bright, refractive granules.

For the producing microbial cells, PHAs fulfil important biological functions; most prominently, they act as reserve and storage materials for energy and carbon (Steinbüchel & Hein, 2001). Under conditions of starvation due to lacking extracellular carbon source, these reserves can be mobilized and utilized as carbon and energy substrates. PHAs are key compounds for the regulation of balanced intracellular energy flow, e.g. for cell motility, and the target-oriented distribution and deviation of carbon reserves to different metabolic pathways. In addition to their role as reserve materials, a variety of important functions of PHAs in various ecosystems was elucidated especially during the last couple of years, such as cell protection against environmental stress conditions like osmotic shock, UV irradiation, desiccation, or thermal or oxidative stress. Further, they are involved in special metabolic on-goings in different microbial species, such as sporulation, cyste formation, germination, control of exopolysaccharide excretion, and, considering diazotrophic species, in the energy flow during nitrogen fixation (reviewed by Koller et al., 2011). In general, PHA accumulation is favoured by a sufficient availability of carbon source together with a restricted supply with macro-components like nitrogen, phosphate or dissolved oxygen, or micro-components like magnesium, sulphate, and various metals (Kim & Lenz, 2001; Helm et al., 2008, reviewed by Koller et al., 2010).

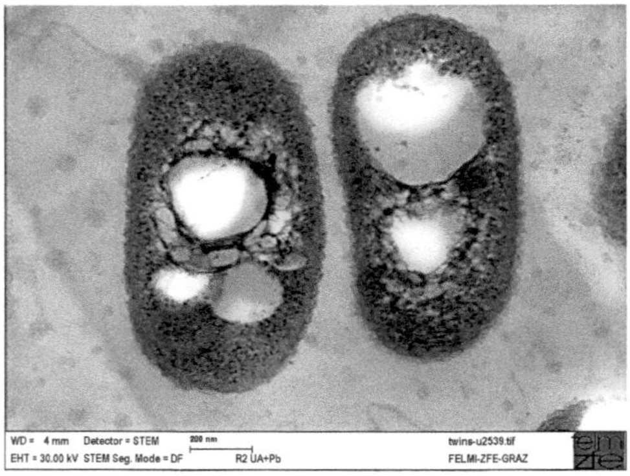

Figure 1. STEM picture of *Cupriavidus necator* cells harbouring about 37 wt.-% of PHA granules. Magnification: 1/65000.

PHAs mainly consist of 3-hydroxyalkanoates (3HAs) as monomeric building blocks. The general chemical structure is provided in Fig. 2.

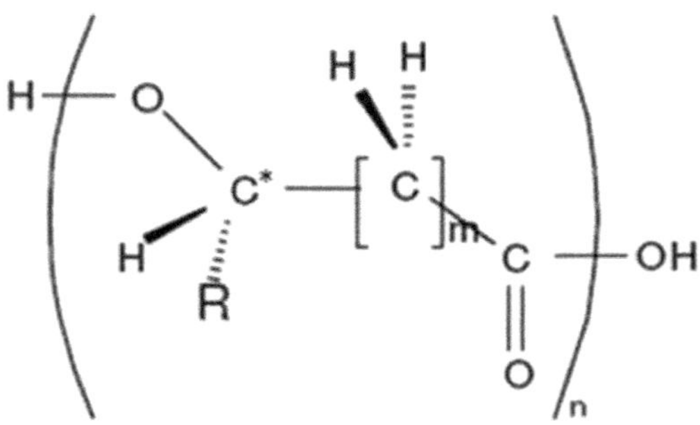

Figure 2. General chemical structure of Polyhydroxyalkanoates (PHAs). The chiral center is indicated by an asterisk (*).

3.3. Composition and production of PHAs

Depended on various factors like the type of microbial production strain, the feeding regime for nutrient supply, and the process parameters during the biosynthesis, PHA polymer chains can contain a magnitude of 10^2 to 10^5 3-hydroxyalkanoate (3HA) monomers. These

3HAs normally are enantiomerically pure, *R*-configurated chiral compounds. Among all known PHAs, Poly([*R*]-3-hydroxybutyrate) (PHB) is the most widely investigated and best characterized one. PHB is a homopolyester consisting merely of 3-hydroxybutyrate (3HB) building blocks; this material features a rather high degree of crystallinity and restricted processability. The low difference between the decomposition temperature (typically around 270°C) and the high melting point (typically around 180°C) provides a too small window of processability for many processing techniques, e.g. in melt extrusion technology or production of polymeric films. This drawback can be overcome by interrupting the crystalline PHB matrix by incorporation of additional building blocks like 3-hydroxyvalerate (3HV) or the achiral building blocks 4-hydroxybutyrate (4HB) and 5-hydroxyvalerate (5HV). This results in co-polyesters with enhanced material properties, opening the door for a broader range of applications. The exact material properties strongly depend on the monomeric composition of the co-polyesters; this composition can be triggered during the PHA biosynthesis by co-feeding of precursor substrates in order to obtain the desired monomeric building blocks. For example, 3HV building blocks are produced by many PHA producing strains if they are supplied with 3HV related precursors, namely odd-numbered fatty acids, such as propanoic or pentanoic acid (Braunegg et al., 1998).

By methods of industrial (white) biotechnology, PHAs can be biosynthesized starting from renewable resources. For this purpose, monosaccharides, lipids, methanol, agroindustrial wastes like the hydrolyzates of various (ligno)cellulosics, starch, beet sugar, cane sugar, maltose, glycerol from biodiesel production, or, as it is the focus of the review at hands, whey lactose from dairy industry, are available in sufficient quantities (Lee, 1996; Braunegg et al., 2007; Koller et al., 2010).

After production of PHA-rich bacterial biomass, appropriate methods of downstream processing are needed, encompassing cell harvest, product isolation and product refining. The biopolymers can be recovered from the microbial cells by extraction, chemical or enzymatic digestion of the cell wall, or by mechanical cell disruption (Kunasundari & Sudesh, 2011). After the release from cells, they can be processed to a marketable form, e.g. to granulates, and used as sustainable biodegradable substitutes for a variety of "classical" petrochemical plastics such as poly(ethylene) (PE), poly(propylene) (PP), poly(ethlyleneterephtalate) (PET) and many others (Reddy et al., 2003; Khanna & Srivastava, 2005; Ren et al., 2005; Chen, 2009). According to calculations accomplished by Akiyama et al. (2003), Harding et al. (2007), Pietrini et al. (2007), and Titz et al. (2012), the production of PHAs should be more beneficial if considering full cradle-to-gate life cycle analysis (LCA) than the production of the petrochemical plastics mentioned before.

4. The potential of Life Cycle Assessment (LCA) in the field of biopolymers

Life cycle analysis (LCA) is an excellent method to quantify the sustainability of a product or a process, encompassing all factors from raw materials, transportation, product manufacture, end use, to disposal (Gonzalez et al., 2010; Champrateep, 2010).

The energy use and water pollution, as well as the hazards of the chemicals are assessed and compared to alternativee products or solutions. The total impact on the environment is quantified for each step (Patel et al., 2005).

As indicated in the ISO-14040 series, LCA methodology distinguishes four phases: goal and scope definition, inventory analysis, impact assessment, and interpretation as reported in Figure 3.

The Goal and Scope Definition involves LCA application, type, reason, and audience together with the geographical and temporal scope. The system boundaries and functional units (FU) are also determined.

The Inventory Analysis (LCI) creates a flow chart for the process where the inputs (energy, materials) and the environmental releases of each process phase are determined.

The Impact Assessment (LCIA) is a consequence of LCI and it represents the environmental impacts related to the functional unit. The results can be grouped and weighted.

The Interpretation is the discussion of the data obtained in previous steps of LCA.

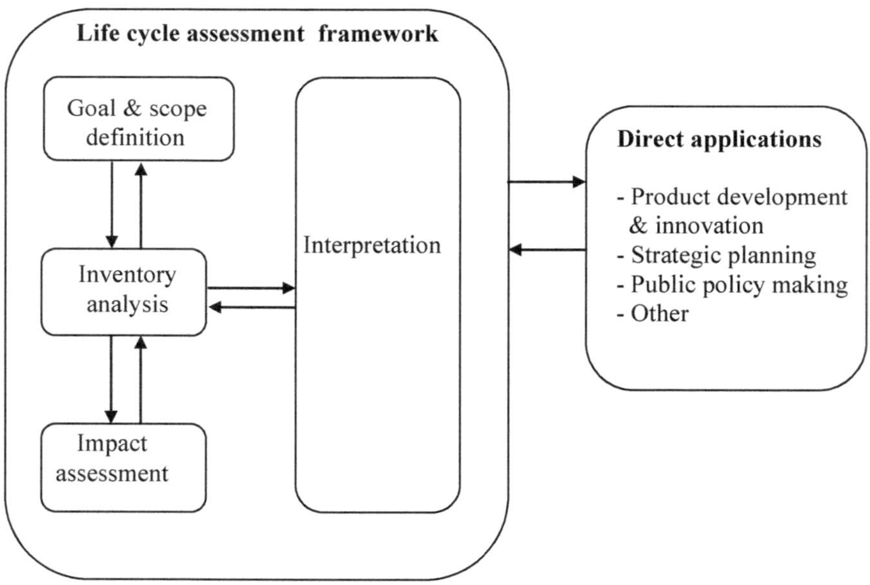

Figure 3. General Scheme for Life Cycle Assessment Studies.

The goal of LCA is the prediction of the environmental benefits that could be reached by the replacing conventional petrochemical polymers with bio-based polymer composites, such as

PHB or PLA. The evaluation of the indicators NREU and GWP100 permits to understand how each step of the LCA and the material properties can influence the final environmental performance of the end products.

In full-scale LCA studies, the impact categories taken into account are eutrophication (EP), acidification (AP), ozone layer depletion (ODP), ecotoxicity (ETP), and human toxicity (HTP).

The functional units are typical commodities made of petrochemical polymers taken into account to perform a comparison with the same items made of materials from renewable resources.

The investigation of all material properties is useful to collect the information on which property has to be improved to maximize the material performance in the application.

The system boundary is a simplified flow chart describing the commodity production in terms of processing technique and production process of all components included in the composite.

The polymer production process considers all the steps from the agricultural cultivation up to down-stream processing.

After the product's life span, the products post-consumer have to submit to a waste management method where the NREU for transport and waste treatment has been considered 1% of NREU of polymers production. The GWP100 impacts have been calculated by assuming that all the carbon fixed in the petrochemical polymers is converted to carbon dioxide during the combustion process.

In order to reflect energy recovery during incineration of the composites after their useful life, credits are introduced in the calculation. One can assume that for each joule of incinerated waste there is a credit of 0.12 J of electricity and 0.12 J of heat (Patel et al., 2002; Pietrini et al., 2007).

The inventory analysis involves the energy used for the polymer production. For example, when bagasse is used as filler for a PHA based material, the final impact of PHA production must consider the obtainment of the biomass from the renewable resource, as well as a large amount of this can be combusted producing electricity (Pietrini et al., 2007).

For the utilization phase of the commodity, we can consider the correlation between the lightness of the PHB based product for the substitution of the interior car panel, and the car economy in terms of kilometres driven per kilogram of fuels.

The impact assessment is correlated with the indicators NREU and GWP100, from which values is possible to make a comparison between the new products and the conventional ones.

The interpretation is a collection of the previous data, where their complete analysis permits to have an improved application due to the PHB based material (Table 2).

Usually the production of materials from natural resources requires a relatively small amount of fossil energy, because the main contribution comes from solar energy. As a consequence, also GWP emissions are lower if compared with materials coming from non-renewable resources. In addition the score of other categories could be greatly affected by the cultivation of the renewable resources and complicated processes, possibly involving toxic compounds. Pesticides, needed for the cultivation of sugar cane and corn, can release in the environment a considerable amount of phosphates and nitrates, increasing the final score in HTP, ETP, and EP indicators.

PHA Type or Product	Renewable Resource	Compared Commodity or Process	Petro-Based Polymer	Balance Energy (NREU, GWP100)	Reference
PHAs	Corn grain, corn stover	Film	Polystyrene (PS)	Favorable	Kim et al., 2005
[P3HB-*co*-5mol% 3HHx)]	Soybean oil, glucose	-	LDPE, HDPE, PP, PS, PET	Favorable	Akiyama et al., 2003
-	Genetically engineered corn	-	PE	-	Kurdikar et al., 2001
PHB, PHV	Corn plant, bacterial fermentation	-	HDPE, PET, PS	Favorable	Patel et al., 2002; Shen & Patel, 2008
PHB	Sugar beet, starch, fossil methane, fossil-based methanol	Film	PE, PS	Favorable	Harding et al., 2007; Heyde, 1998
-	Corn crop	-	LDPE, HDPE, PP, PVC,PET, PC	-	Tabone et al., 2010
Mixed culture PHA	Waste stream	-	PS	-	Guerrif et al., 2007
-	Corn grain, corn stover	Ethanol production system	-	Favorable	Anastas et al., 2000
-	Corn plant, bacterial fermentation	bottles	HDPE, PET, PS	Favorable	Kim & Dale, 2004
PHB-SCB PHB-OMMT composites	Sugar cane, corn starch	Cathode ray tube monitor housing, internal panels of an average car	High-impact PS, glass-fibers-filled polypropylene	Favorable	Patel et al., 2002
PHB	Sugar cane crop and waste	Ethanol and sugar production	PE, PP	-	Pietrini et al., 2007
PHB	Beet or cane molasses, plant oils, plant derived fatty acids, alkanes, steam, sucrose, corn	-	PE and PP production	Favorable	Nonato et al., 2001;
PHB	Corn-derived glucose	-	-	Unfavorable	Kendall, 2010
PHB	Corn grain	-	-	Favorable	Kim & Dale, 2008

Table 2. Types of PHA, Originating Renewable Resource, Compared Petrochemical Commodity, the balance energy (NREU, GWP100) and the source reference.

Another very important aspect is represented by the feasibility of using biodegradable materials for durable products. The lifetime of products such as monitor housings or car panels is expected to be in the order of some years. In this view, it is of fundamental importance that the biodegradation of the product does not take place during this useful period.

The life cycle interpretation considers all contributions in terms of energy for each production step for the new PHB based commodity to NREU.

Then it becomes important to know the renewable resource used for PHA production and what the item made of petrochemical polymers it has to compete with in order to evaluate a comparison with the new PHA based material.

4.1. Life Cycle Assessment of PHB Based Composites

A cradle-to-grave environmental life cycle assessment (LCA) of some relevant PHB based composites was performed, with the purpose of assessing the potential environmental benefits that could be reached with the application of these new biodegradable materials instead of petrochemical polymers.

The investigation was pointed out on two functional units, specifically the cathode ray tube (CRT) monitor housing that actually is made of high impact polystyrene (HIPS) with an average weight of 2.2 kg and on the internal panels of an average car with a total distance travelled of 150000 km, that is made of glass fibres filled polypropylene and compatibilizers in the percent of 30/63/7. The used key environmental parameters were the non-renewable energy use (NREU) and global warming potential over a 100 year's time horizon (GWP100).

In particular the assessment was carried out on composites reinforced with 10 and 20 wt-% of BNTf and 5 and 10 wt-% of OMMTSi, because of their enhanced mechanical properties.

Figure 4 shows a simplifying flow chart describing a conventional production of the CRT monitor housing and the internal car panels (Pietrini et al., 2007).

For the collection of cradle-to-factory gate data, all the steps involved in the production of the components mixed in our composites have to be taken into account. So clay production included extraction, ion exchange, hydro-cycloning, spray drying, organic modification, filter pressing, heating and milling (Roes et al., 2007).

The PHB production process involved agricultural cultivation and sugar production, the bio-fermentation and the downstream processing. The sugar used to feed the micro-organisms can be extracted from sugar cane or corn starch. The residual biomass produced along with the sugar was considered to be combusted for electricity production, with an assumed power generation efficiency of about 35%.

The sugar cane milling gives 1-3% of SCB that can be used as PHB filler in correlation with the composition of PHB composite.

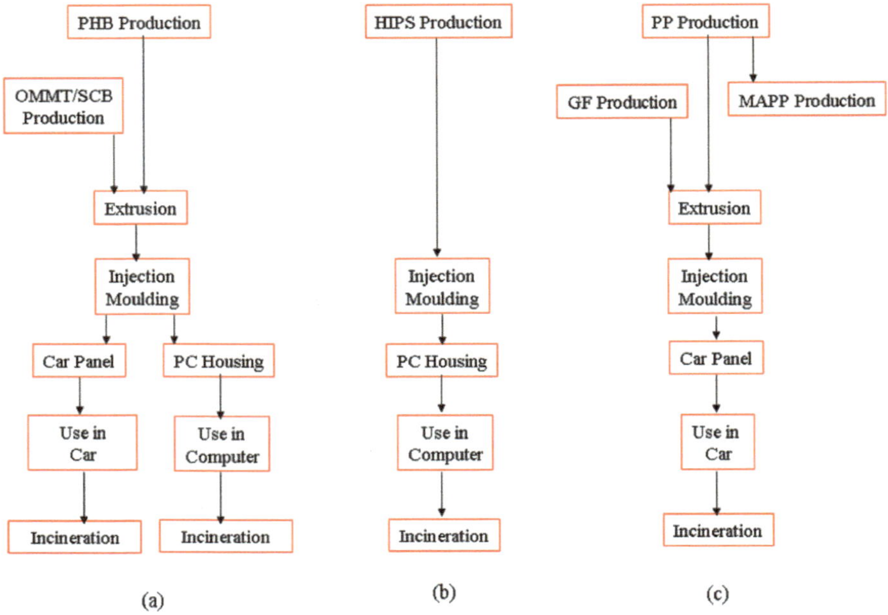

Figure 4. Simplified flow chart for the production process of both functional units: PHB based composites CRT monitor housing and internal car panels (a), HIPS CRT monitor housing (b), PP-GF internal car panels (c).

The maleic acid anhydride (MAPP), a product obtained by the PP and maleic anhydride extrusion in presence of a peroxidic agent, transfers a better adhesion between PP and the fillers inside the PP-GF formulation.

The use phase of internal car panels was represented by the fuel consumption of the car, that is strongly affected by the type of car and the car panels weights.

At the end-life, for both products, an energy recovery from the incineration of the municipal solid waste was assumed.

The NREU for transport and waste treatment was considered 1% of NREU of polymer production, while the GWP100 impacts were calculated assuming that all the carbon fixed in the petrochemical polymers is converted to CO_2 during the combustion process.

For the materials originated from renewable resources (PHB, SCB), this contribution was not considered, because the CO_2 originating from the incineration of the composites is equal to that extracted from the atmosphere during plant growth, creating a closed cycle.

The energy and CO_2 bonus coming from the incineration was calculated on the basis of the calorific value of the polymers and SCB and then deducted from the total NREU and GWP100 impacts.

For the inorganic materials (GF, OMMT), these parameters are neutral, if correlated with the energy and greenhouse gases (GHG) emissions in the waste incineration end stage.

4.2. Inventory analysis

The sugar used for the PHB fermentation process is produced from sugar cane or corn starch.

Both process create also biomass, combusted for the electricity production: a large amount of bagasse can be used for power generation in the first case, while the amount of biomass is smaller if the material used for the PHB fermentation is corn starch.

The indicators NREU and GWP100 have negative values calculated for the final impacts of PHB production in the first process and positive ones for the second case.

For injection moulding, data published by PlasticEurope were used (Boustead et al., 2005).

According to PlasticEurope, the NREU for PP production pellets and injection moulded PP items are 73.0 GJ/t and 113.2 GJ/t respectively.

During the use phase of the car, the share of fuel that can be assigned to the weight of the internal car panels was calculated following these assumptions:

Weight of the car without the panels: 1222 kg

Car weight: 1110 kg and passengers weight:112 kg. The second value is obtained considering an average of 1.6 passengers of 70 kg for each car.

Car life distance: 150000 km

Fuel economy for a conventional car: 15.13 km/kg of fuel, value calculated taking into account that for 1 driven km, 0.0125 kg of diesel and 0.0536 kg of petrol are combusted on average.

Table 3 shows the results of the calculations.

Composite	Panels weight (Kg)	Car fuel consumption (Kg)	Panels fuel use (Kg)
PP-GF	20.0	15.13	115.2
5-OMSi	25.5	15.08	147.0
10-OMSi	25.7	15.08	147.8
10-BNTf	24.6	15.09	141.6
20-BNTf	23.4	15.10	134.9

Table 3. Data used in the calculation of the use phase for internal car panels.

The fuel consumption of the car decreases with increasing weight of panels: the best result for PHB composites is observed for 20BNTf (17% more than PP-GF panels), that is the lightest PHB based product, whereas the larger value of fuel consumption is attributed to 10-OMSi composite (28% more than the conventional panel).

4.3. Impact assessment

Table 4 reports the NREU and GWP100 impacts for all selected products (Pietrini et al., 2007).

	CRT housing		Internal car panels	
	NREU (MJ/FU)	**GWP100 ($kgCO_2eq/FU$)**	**NREU (MJ/FU)**	**GWP100 ($kgCO_2eq/FU$)**
HIPS	200.0	15.1	-	-
PP-GF	-	-	8.2	569.9
PHB1-5-OMSi	8.3	0.5	8.4	602.0
PHB1-10-OMSi	22.4	1.2	8.6	612.1
PHB1-10BNTf	1.3	0.1	8.1	576.3
PHB1-20BNTf	6.8	0.5	7.7	552.6
PHB2-5-OMSi	160.2	8.2	9.9	677.2
PHB2-10-OMSi	167.0	8.6	10.0	683.7
PHB2-10BNTf	139.8	7.2	9.4	645.0
PHB2-20BNTf	124.9	6.4	8.9	610.7

Table 4. Estimation of NREU and GWP100 for the cradle-to-grave system of conventional and PHB based CRT monitor housings and internal car panels.

For CRT monitor housing, all PHB relevant composites scored better if compared to their conventional counterpart made of HIPS. Both indicators were lower both for PHB produced from sugar cane (PHB1) and PHB derived from corn starch (PHB2).

PHB1-10BNTf and PHB1-20BNTf composites showed a reduction for both indicators by about 99% and 97% respect to conventional HIPS housing.

On the other hand, clay filled composites showed savings ranging from 97% (GWP100 of PHb1-5-OMSi) to 89% (NREU of PHB-10-OMSi).

For internal car panels, best results were found for PHB produced from sugar cane (PHB1), even if there were not relevant savings for both indicators, in comparison with the conventional PP-GF panels.

The composite PHB1-20BNTf showed a lower impact for both indicators, with a reduction of around 5% and 3% for NREU and GWP100 respectively.

4.4. Life cycle interpretation

For HIPS, the total NREU was 200 MJ/housing, while the value associated with PHB production is -22.7 GJ/t for PHB produced from sugar cane (PHB1) and 38.6 GJ/t for PHB produced from corn starch (PHB2)(Pietrini et al., 2007).

For PHB based composites, the contribution of injection moulding was higher than for HIPS (from 60 to 66 MJ/housing against 54 MJ/housing for HIPS) because of the higher weight of PHB based monitor housing.

In this process, also the extrusion and the production of the filler must be taken into account.

The energy credit from post-consumer waste incineration was higher for HIPS (44 MJ/housing) compared to PHB composites (from 26 to28 MJ/housing) because of the higher calorific content of HIPS. The use phase for this end product had no contribution, because the electricity use of the monitor did not depend on the weight, the shape or the material of the housing. These properties could influence the transportation of the monitors (to and from retail and to waste management), but this parameter was not included in the calculation, because it was typically negliable.

So the life cycle stage that makes PHB composites so environmentally competitive in this application is the polymer production.

In fact for the HIPS monitor housing, polymer production accounted for 190 MJ/housing, that represents about 80% of the total positive contribution to NREU.

This indicator for the production of PHB1 was negative so the overall NREU connected to PHB1 production decreases by around 43 MJ/housing (PHB1-20BNTf composite) to 56 MJ/housing (PHB1-5-OMSi).

On a qualitative base, analogous considerations could be made in the case of GWP100 for PHB1 based CRT monitor housings. The only substantial difference between the contributions to the two indicators was represented by the incineration step.

CO_2 emissions originated from PHB and SCB incineration stem from renewable resources, so they do not pollute the environment, while the HIPS emissions coming from of fossil polymers were considered polluting.

The contribution to GWP100 due to the incineration was calculated assuming the complete conversion of the carbon contained in the composite to CO_2: this consideration gives CO_2 emissions of around 7 and 5 kg for HIPS and PHB based monitor housings respectively, due to the higher carbon content calculated for HIPS respect to PHB and SCB.

However, an equivalent CO_2 amount was decreased from the GWP100 emissions related to PHB and SCB production, in order to reflect the neutrality of bio-based materials with regard to CO_2 emissions.

The internal car panels showed very different results: the main difference compared to the display housing was the large contribution of the use phase in the case of the car panels.

Considering a cradle-to-factory gate system for the LCA data collection related to the car panels, the results would be quite similar to the display housing (Pietrini et al., 2007).

The NREU of PHB based internal car panels represented only approximately 15% (PHB1-10BNTf) to 25% (PHB1-10-OMSi) of NREU value of PP-GF panels.

When the use phase was included the NREU for the end products from PHB composites scored much worse than the product made from conventional polymer. The reason is the higher weight of the composites that leads to an higher fuel consumption.

While for the CRT monitor the housing have not any influence on electricity consumption of the monitor, in this case the fuel consumption of the car is related to the weight of the internal panels.

The energy recovery without incineration represented 84% of total NREU for PP-GF service phase, while is 100% of the total impact for PHB based composites.

Hence, the low mechanical properties and high densities of the composites nullify all the savings provided by the environmentally favourable PHB production process. The same considerations are valid for the GWP100 impact of the panels, although the contribution of the use phase to the final value of this indicator was higher if compared to NREU.

The GWP100 value was influenced by the large amount of carbon dioxide produced during fuel combustion.

4.5. Sensitivity analysis on internal car panels

The relation between the relative impacts of both NREU and GWP100 and PHB1 based composite for car internal panels is represented by an exponential trend (Pietrini et al., 2007).

So it becomes possible to evaluate the critical values of Young modulus that should be reached by PHB based composites in order to render these materials environmentally attractive in the case of the car panels application.

PHB1-20BNTf was the only composite that showed lower values for both relative indicators at the measured Young modulus (1.73 GPa) if compared with the conventional panels.

The composite PHB1-10BNTf showed values of relative NREU and GWP100 of 0.99 and 1.01, indicating that the environmental performances of the panels made with this composite can be considered equivalent to those of the panels made with PP-GF composite.

The higher values of relative impacts were found for clay filled composites: these impacts were found to be from 3% (NREU of PHB1-5-OMSi) to 7% (GWP100of PHB1-10-OMSi) higher that the impact calculated for the conventional PP-GF panels.

So PHB1-5-OMSi and PHB1-10-OMSi composites should reach the values of Young modulus of 2.1 and 2.3 GPa respectively.

5. The broad field of potential applications of PHAs

Concerning the potential applications, PHAs constitute very versatile materials that raise the attention of different industrial branches. As the best-known and most simple application, these biopolymers are of interest for packaging purposes, especially in such areas where compostable packaging is wanted, e.g. in the food producing industry. Especially in the field of packaging of easily spoiling food, the high oxygen barrier of PHA films is very beneficial. In addition, bottles for shampoos (Wella, Germany) made of PHAs were commercially available in the past. PHAs can be used for paper coating, production of daily commodity items like razors, diapers, hygiene products, or cups and dishes (Metabolix, USA; BASF, Germany). For these applications, PHAs can be processed by techniques of injection moulding or film blowing using the same equipment as known from the well-established processing of petrochemical plastics.

In the medical field, PHAs were already investigated as bone implant materials, for tissue engineering, for *in-vivo* application as implants, surgical pins, screws, meshes and sutures, and as carrier matrices for controlled drug release. Also the production of highly sophisticated surgical articles such as artificial blood vessels and vein valves, spinal fusion cages, bone marrow scaffolds, and meniscus regeneration devices is reported (Chen and Wu, 2005). Especially the possibility to change the composition of PHA allows the manufacture of materials with tailor-made mechanical properties and a fine-tuned degradation rate under *in-vivo* conditions. In fact, one can expect the fast increase of the number of medical applications for PHAs and its composites; a first step in this direction was done recently by approval of Poly-4-hydroxybutyrate (P4HB) as implant material (http://www.tepha.com).

Hydrolysis of PHA to the monomers generally results in a rich source of chiral synthons that can be used as starting materials for synthesis of fine chemicals and marketable products such as pheromons, aromatics, vitamins or antibiotics, or can even be used as pharmaceutically active compounds (Ren et al., 2005). It was demonstrated that 3HB and its oligomers reveal therapeutic effects. They promote cell proliferation and prevent necrotic cell death. Some of these chiral acids also display biological activity against pathogenic bacteria or viruses (Ruth et al., 2007).

Further, PHAs harbouring special building blocks can be applied as so-called "functional materials" for different niche applications. Here, they can act as heat sensitive adhesives, latex materials, or smart gels (reviewed by Chen et al., 2010a)

A completely new field of application for PHA is their conversion towards alkyl esters by means of transesterification. Here, the conversion leads to 3-hydroxyalkanoate methyl esters (3HAME) (Zhang et al., 2009). Chemically, 3HAME have a composition very similar to esters stemming from the alkaline transesterification of vegetable oils or tallow, the so called "biodiesel". In fact, PHA-stemming alkyl esters of 3HAs were successfully tested as engine fuels; similar combustion heats were determined for 3HAME if compared with gasoline. The conversion of PHA to "biofuels" seems to be reasonable for such biopolymer fractions that show modest material properties, e.g. PHA blends stemming from mixed cultures grown in sewage water. PHA produced by such microbial consortia often contain building blocks with a rather long carbon side chain, so called medium chain length (mcl) PHAs. Examples for such mcl-building blocks are 3-hydroxyhexanoate (3HHx), 3-hydroxyoctanoate (3HO), 3-hydroxydecanoate (3HD), and 3-hydroxydodecanoate (3HDD). The longer the carbon chain, the higher the expected combustion heat for the deriving 3HAMEs should be.

6. Economics challenges in the production of PHAs

Despite the huge efforts globally devoted to biopolymer research, PHAs are still not really competitive to petrochemical plastics mainly considering production costs and, to a certain extent, also regarding the material properties (Choi & Lee, 1999; Sudesh & Iwata, 2008; Koller et al., 2010). A major share of up to half of the entire production costs is related to the carbon substrates. Hence, the reduction of these expenses by utilizing cheap carbon-rich raw materials available in huge quantities is the first pre-condition to make the PHA production process economically competitive. Considering the increasing amounts of carbon-rich whey as discussed above, it is easily understandable that this surplus material has attracted and still attracts the attention of several research groups active in the field of PHA biopolymers.

In addition to the substrate expenses, costs have to be saved by optimizing the downstream processing for PHA recovery and refining after cell harvest. As intracellular products, PHAs have to be separated from the surrounding non-PHA cell mass, mainly consisting of proteins, lipids, nucleic acids and special polysaccharides. Here, high input of (often hazardous!) solvents and enormous energy demand still constitute the state-of-the art in PHA recovery, compromising the demanding claims of these bio-plastics to be ecologically sound materials (Koller et al., 2010).

Apart from the selected raw materials and the downstream processing, the increasing of productivity by designing of the optimal engineering set-up is indispensable for the final break-through of these biopolymers on the market. Batch and fed-batch discontinuous fermentation mode are up to date the most common techniques for microbial PHA production (Kim et al., 1994; Ryu et al., 1997; Ahn et al., 2000; Nonato et al., 2001; Atlić et al., 2011). In contrast, continuous biotechnological production mode is well known as a precious tool for achieving high productivities, lower costs and constant product quality. Due to these facts, a growing number of research activities were accomplished during the last couple of years, investigating and assessing the potential of continuous PHA production processes (Zinn et al. 2003; Sun et al. 2007; Atlić et al., 2011). Recently, the continuous production of PHB in a five-stage cascade of stirred bioreactors was investigated. Here, the biopolymer was produced on glucose by eubacterial producing strains belonging to *Cupriavidus necator* species. As a main results, the authors report high productivities of 1,85 g/L h for PHB and a constant and satisfying product quality (Atlić et al., 2011).

Figure 4 illustrates schematically the principle process line for PHA biopolymer production from the feedstock to the final commercial bioplastic product (Chen et al., 2010b).

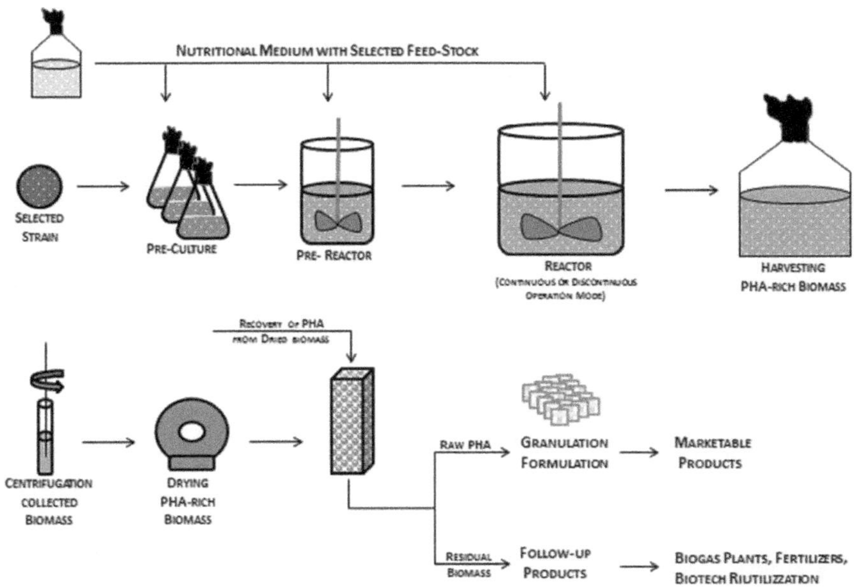

Figure 5. General process scheme for PHA biopolymer production

7. Current routes of march in PHA research

Nowadays, research in the field of PHAs focuses on several key topics. The application of growth additives that shorten the time for production of catalytically active biomass is a pre-requisite to enhance the entire volumetric productivity of the process. Such cheap growth additives can be found in agriculture, e.g. side streams from the cultivation of green grass land, and were already tested successfully on laboratory scale (Koller et al., 2005b). Concerning the raw materials, simple "unrelated" carbon sources that are available at low prices, or even constitute waste streams, should act as sole feedstocks for production of high-value PHA co-polyesters. This provides the possibility to safe costs for precursor compounds normally needed for co-polyester production.

Efforts done in the field of genetic engineering mainly aim at the increase of volumetric PHA productivity and at higher molecular masses of the biopolymers. This can be accomplished *via* the knock-out of the genes encoding for the enzymes responsible for intracellular PHA degradation, so called depolymerases. Such a strategy could also face other metabolic bottle necks that can hamper a fast and complete substrate conversion by the selected strain. For instance, *Cupriavidus necator*, a well-known PHA producer, but unable to grow on lactose, has been genetically modified in order to construct a recombinant strain that can use lactose-containing waste material such as cheese whey, and one of the intracellular PHA depolymerases (*phaZ1*) was chosen to insert the *lacZ, lacI* and *lacO* genes of *Escherichia coli* (Povolo et al., 2010)

This would have the effect to allow polymer production on lactose and, at the same time, to remove part of the PHA intracellular degradation system.

The enhancement of the microbial oxygen uptake by inserting genes encoding catalase or peroxidase was also attempted. This should enhance microbial growth kinetics and higher final biomass concentrations (Ouyang, 2007).

In the field of downstream processing, environmentally safe and efficient solvents are investigated for enhanced recovery of PHA from the cells. This goes in parallel with the examination of novel biological lysis methods and enhanced strategies for mechanical cell disruption. In any case, enhanced downstream processing has to feature lower energy demands if compared to the contemporary methods. For efficient polymer recovery, the increase of the intracellular polymer content as well as the increase of the PHA granule size is of importance; these factors are determined during the PHA bio-production (Chen, 2010a). Of course, also the remaining non-PHA cell mass (NPCM) has to be converted in a sustainable, value-adding way. Research in this direction is devoted to the anaerobic digestion of NPCM in biogas plants, or to the chemical or enzymatic hydrolysis of NPCM to a rich carbon- and nitrogen source for subsequent microbial cultivations. As an alternative, NCPM can be applied in agriculture as "green fertilizer". Also downstream processing can be facilitated by genetic modification; an excretion of high quantities of nuclease enzymes after cell disruption results in decreased amounts of nucleic acids in the medium, leading to lower viscosities that facilitate the separation of PHA granules from the surrounding liquid phase by centrifugation or flocculation. With this aim, a nuclease-encoding gene from *Staphylococcus aureus* was integrated into the genomes of a number of PHA-producing bacterial species, including *Ralstonia eutropha* (now renamed as *C. necator*). Here, the staphylococcal nuclease was functionally integrated into the chromosome and readily expressed in *Pseudomonas* strains, directed to the periplasm and occasionally to the culture medium, without affecting PHA production or strain stability (Boynton et al., 1999).

The technological drawbacks of the bio-production itself can be handled by the application of robust microbial production strains that remain genetically stable for a long time period under continuous cultivation conditions, and, at the same time, can resist the risk of contamination by microbial competitors. Here, extremophilic species like the highly salt requiring archaeon *Haloferax mediterranei* might be a viable solution in order to minimize the normally indispensible, highly energy demanding sterility requirements for PHA production set-ups (Koller et al., 2007a,b). In future, continuous PHA production should not only aim at the increase of volumetric productivity, but should also open the door for tailor-made material properties by fine-tuning the polyester composition. This can be accomplished by the formation of block-copolymers, where the sequential arrangement of softer and harder polymer parts can result in well-adjusted novel polymeric materials. Here, a multistage bioreactor cascade for PHA production as presented by Atlić and colleagues (2011) might be the adequate process engineering equipment.

During the last two to three decades, the preparation of composites and blends has become one of the key research fields in biopolymer science. For enhancement of the material properties, PHAs can be processed together with a variety of compatible matters, resulting in the creation of novel PHA-based blends and composites. For this purpose, the utilization of polymeric materials like Poly(vinyl alcohol) (PVA), poly-ε-caprolactone (PCL) etc., including synthetic analogues of PHA (e.g. atactic PHB), inorganic fillers (clays, sepiolites, Montmorillonite, or calcium carbonate), and organic fillers of agricultural origin was already tested (Chiellini et al., 2004; Pietrini et al., 2007). Concerning fillers from agriculture, the application of surplus materials such as lignocelluloses like sugar cane bagasse, wheat flour, fruit peels, crop fruit fibres, saw dust and wheat straw is reported in literature (Chiellini et al., 2004). In general, nanocomposites and natural fibers composites can be distinguished. Nanocomposites have the potential to improve special polymer properties, such as gas permeability and thermal and mechanical characteristics. For creation of nanocomposites, rather small amounts of filler, commonly an organophilic modified clay, are needed for efficient enhancement of the properties. Natural fibers composites often display excellent mechanical properties, and, as desired for many applications, they lower the density of the final product. Due to the fact that in most cases fibres constituting agricultural residues are used as fillers, the biodegradability of the final product is enhanced. Because such fibers do not feature a considerable price, this normally goes in parallel with a reduction of the entire production cost of the marketable product (reviewed by Pietrini et al., 2007).

8. From the feedstock milk to whey to PHA biopolyesters

Cheese whey is a surplus product in the dairy industry. Figure 6 illustrates the process line from the feed stock milk via whey towards PHA biopolyester production. From the feed stock milk, casein is precipitated enzymatically or by acidification. This so called "transformation" results in the generation of solid curd cheese (predominately consisting of caseins), and liquid full fat whey. After removing the major part of lipids from full fat whey by skimming, skimmed whey remains. Sweet skimmed whey is subjected to a concentration step, removing 80% of its water content. This whey concentrate is separated via ultra-filtration in whey permeate (carbohydrate fraction) and whey retentate (protein fraction with considerable lactose residues).

Whereas whey permeate that contains about 80% of the lactose originally included in milk can be used as carbon source for fermentative production of e.g. PHA, special proteins of the retentate like lactoferrin and lactoferricin are of significance for pharmaceutical application. The predominant proteins in whey retentate, namely α-lactalbumin and β-lactoglobulin, are potential candidates for food- and feed supplements. Biotechnologically, they can be applied as nitrogen source for enhanced cultivation of microbial PHA production strains. More recently, plastic films coated by whey proteins are developed; also here, classical polymers are replaced by recyclable, bio-based materials featuring low oxygen and moisture permeability, making them especially interesting for food packaging (http://www.wheylayer.eu/project.html). Table 5 summarizes the composition of sweet whey, fermented whey, whey permeate and retentate.

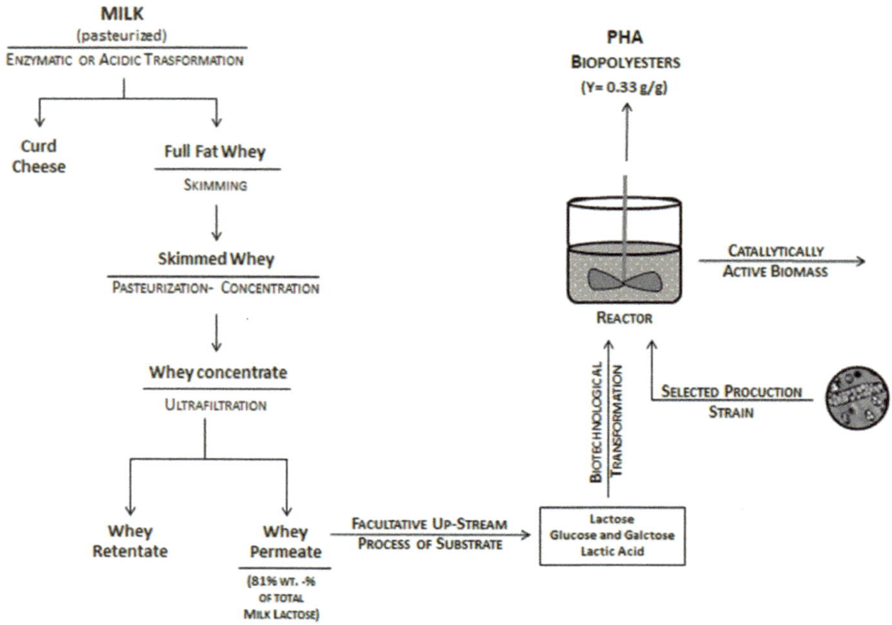

Figure 6. From the feedstock milk to whey to the biopolyester including a rough estimation of the material balances. The necessity for hydrolysis and/or desalting of the raw material whey depends on the PHA production strain (see text)

Compound [% (w/w)]	Sweet Whey	Fermented Whey	Whey Permeate	Whey Retentate
Lactose	4.7–4.9	4.5–4.9	23	14
Lactic acid	traces	0.5	-	-
Proteins	0.75–1.1	0.45	0.75	13
Lipids	0.15–0.2	traces	-	3-4
Inorganic compounds (minerals like e.g. calcium)	ca. 7	6-7	ca. 27	ca. 7

Table 5. Composition of different types of whey (Braunegg et al., 2007)

9. Reported PHA production from whey by different microbial strains

Converting the pollutant whey into valuable products combines the creation of economic benefit with abating of inherent ecological concerns. Biotechnological production of PHAs from different sugars via condensation of Acetyl-CoA units stemming from the catabolic breakdown of simple sugars such as hexoses is well known and exhaustively described in literature (Braunegg et al., 1998; Lee, 1996; Steinbüchel, 1991; Steinbüchel & Valentin, 1995; Sudesh et al., 2008). Nevertheless, only a rather restricted number of prokaryotic wild-type microorganisms directly convert the disaccharide lactose into PHAs (Young et al., 1994; Koller et al., 2008a,b). Reports for PHA biopolyester production starting from whey lactose as the main or even sole carbon source is reported in literature for microbial wild-type species as well as for genetically modified strains, mainly for engineered *Escherichia coli*. Volumetric productivities for PHA of more than 4 g/L h can be obtained by optimized experimental set-ups (Ahn et al., 2001).

In 1994, Young and co-workers delivered the very first report for PHB production from the whey-related carbon source lactose by a wild-type strain, namely *Pseudomonas cepacia* (now renamed to *Burkholderia cepacia*). Here, the authors asses the strain's potential for growth and PHB production on shaking flask scale on the sugars lactose, glucose and xylose. Focusing on lactose, the organism was able to accumulate up to 56% of PHB in cell dry mass that amounted to nearby 5 g/L. Some years later, an Austrian research group reports further experiments dealing with PHA production by this promising organism on an artificial carbon substrate consisting of an equimolar mixture of glucose and galactose, corresponding to a hydrolysed lactose solution. This experiment was carried out under controlled conditions on bioreactor scale; the results obtained exceeded by far the data obtained from lactose on shaking flask scale (volumetric productivity for PHB: 0.18 g/Lh, final PHB concentration 4.0 g/L). The authors concluded that the application of hydrolysed lactose (mixture of the monomeric sugars) for this organism is more beneficial in terms of specific growth rates, and propose the application of enzymatically hydrolysed whey lactose for cost efficient PHA production based on the surplus material whey (Wallner et al., 2001).

From a *Bacillus megaterium* strain isolated from the sludge of a sewage treatment plant, the production of a rather low amount of 26% PHB in cell dry mass from lactose was reported on shaking flask scale (Omar et al., 2001).

The principal possibility of direct conversion of whey lactose towards PHA using different wild type bacterial strains was first investigated on shaking flask scale. Yellore and Desai (1998) isolated a *Methylobacterium* sp. ZP24 from a local pond in India. This organism grew on pure lactose and, on shaking flask scale, produced PHB at a final concentration of 3.1 g/L polymer, corresponding to a PHA content in biomass of about 59%. Using whey as substrate and optimization of the nitrogen supply, the PHB yield amounted to up to 2.6 g/L, corresponding to a final percentage of PHB in biomass of 44%. These findings were extended later on bioreactor scale, where Nath and colleagues (2008) cultivated this organism on cheese whey. In batch mode, final PHB concentrations of 2.07 g/L with 67% of PHB in biomass and 0.06 g/L h of volumetric PHB productivity are reported.

Povolo & Casella (2003) reported the production of PHA from lactose by *Paracoccus denitrificans* DSM 413, *Sinorhizobium meliloti* 41 and *Hydrogenophaga pseudoflava* DSM1034. The latter two strains were also able to produce the polymer directly from cheese whey permeate and especially *Hydrogenophaga pseudoflava* turned out to be a promising candidate for PHA production from whey. The same research group constructed a genetically engineered *Cupriavidus necator* strain harbouring the *lacZ, lacI* and *lacO* genes of *Escherichia coli* for lactose conversion. The recombinant strains obtained here were able to produce the polymer directly from lactose and from whey permeate. In addition, as reported above, the insertion of the lac operon within *phaZ* gene may reduce the amount of PHA depolymerised by the cell, thus improving the final polymer yield (Povolo et al., 2010). Catalán and co-workers (2007) constructed a modified strain of *Herbaspirillum seropedicae* Z69 carrying the *lacZ, lacI* and *lacO* genes of *Escherichia coli* for lactose conversion in plasmid pHM3. They obtained production of 1.8 g/L PHB with 5 g/L biomass in shaking flasks experiments.

Recombinant *Escherichia coli* strains harbouring PHA synthesis genes were also well studied for directly converting lactose to PHAs. In contrast to the work presented by Povolo et al., (2010), these activities start from a direct lactose converter (*E. coli*) that is later equipped with the required genes for PHA production. These works are of special interest due to high volumetric productivities found, opening a route for industrial scale PHA production from lactose (Lee, 1997; Wong & Lee, 1998). These high productivities are mainly due to the fact that only the genetic information for PHA production is inserted into the cells, but not the information for PHA degradation. Moreover, also the direct utilization of whey permeate as carbon source for growth and PHA production by recombinant *E. coli* strains is well investigated (Ahn et al., 2000; Ahn et al., 2001; Kim, 2000). The until today most promising results for PHA biosynthesis on whey by recombinant *E. coli,* namely a final PHA concentration of 168 g/L and a volumetric PHA productivity of 4.6 g/L h are reported by Ahn and colleagues (2001). In this study, the authors employed a cell-recycle system in order to overcome the typical problems arising from the continuous addition of the whey feed in fed batch cultures, i.e. a rapidly increasing volume of fermentation broth in the bioreactor.

If the natural *ß*-galactosidase activity featured by an investigated production strain is not sufficiently high for efficient conversion of the substrate lactose, additional processing of lactose prior to the application as substrate is needed. Here, the disaccharide can be hydrolyzed by enzymatic or chemical means to equimolar mixtures of the monomeric sugars D-(+)-glucose and D-(+)-galactose. Compared to lactose these monosaccharides are converted by a much higher number of organisms together with typically higher conversion rates (Koller et al., 2008b).

A third, more complex possibility arises from the anaerobic conversion of lactose to lactic acid in a first process step using lactobacilli capable of producing lactic acid with high yields (more than 0.9 g of lactic acid per gram of carbon source). In a subsequent aerobic cultivation lactic acid is metabolized to acetyl-CoA and further to PHAs by numerous strains, e.g. most common PHA producers like *Cupriavidus necator* or *Alcaligenes latus* or *Azotobacter vinelandii.* Alternatively, lactic acid can be converted to polylactic acid (PLA), if wanted.

Based on the above, for PHA production from whey, the decision for applying whey lactose, hydrolysed whey lactose or first-step fermentation towards lactic acid mainly depends on the production strain (Fig.7).

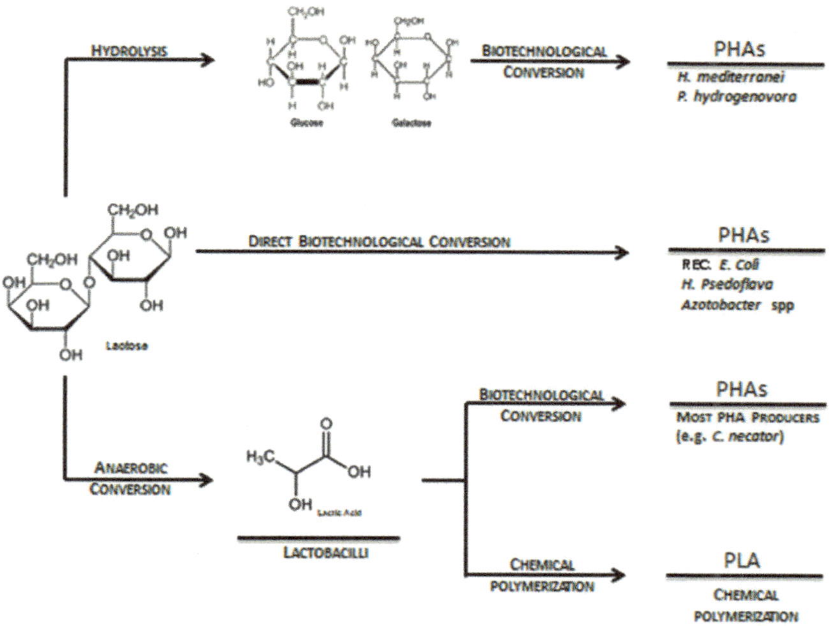

Figure 7. Three different routes from whey lactose to PHA; alternative products like PLA and follow-up products of PHA are included

In a recent study, Koller and colleagues (2007a) compare the potential of three prokaryotic wild type strains for utilization of whey as carbon source for PHA production. In this work, the archaeon *Haloferax mediterranei*, and the eubacterial strains *Pseudomonas hydrogenovora* and *Hydrogenophaga pseudoflava* were investigated in laboratory scale bioreactors. Among these organisms, *H. mediterranei* turned out to be the most promising candidate for eventual industrial scale PHA production starting from whey. This is due to the strain's high robustness and stability; the risk of microbial contamination during cultivation is restricted to an absolute minimum, thus a lot of energy can be saved by the lower sterility demands. Additionally the strain produces a P(3HB-*co*-8%-3HV) copolyester directly from the 3HV-unrelated carbon source whereby the normally high costs for propanoic acid or pentanoic acid as precursors can be saved. The strain grew well on hydrolyzed whey permeate with a maximum specific growth rate μ_{max} of 0.11 1/h. PHA was accumulated at a maximum specific production rate of 0.08 g/g h. The conversion yield for whey sugars to PHA was calculated with 0.33 g/g. After further optimizing of the production conditions, the productivity for this strain on hydrolysed whey permeate was increased to 0.09 g/L h (specific rate 0.15 g/g h); 16.8 g/L biomass containing 73% PHA were obtained (Koller et al., 2007b). By co-feeding of precursors for 3HV and 4HB production (pentanoic acid and γ-butyrolactone, respectively) together with hydrolysed whey permeate as main carbon source, a high value P-(3HB-*co*-21.8%-3HV-*co*-5.1%-4HB) terpolyester was produced by *H. mediterranei*. Also in this case, the polymer was recovered from the cells and underwent a detailed characterization of thermal properties and molecular mass distribution. The promising results for polymer characterization indicate that the material might be of special interest for application in the medical field (Koller et al., 2007b). The partial conversion of whey sugars to 3-hydroxyvalerate (3HV) units and the excellent polymer characteristics (low melting temperature, high molecular masses within narrow distribution) together with a viable cheap and simple downstream processing (Munoz et al., 1994) make the strain especially interesting. The estimated production price amounted to € 2.82 per kg PHA. For further improvement, the recycling of the highly saline side streams has to be tested and optimized. Additionally, high salinity requires special material demands for the bioreactor equipment and the probes (Hezayen et al., 2000).

P. hydrogenovora features the disadvantage of low final polymer contents, low productivities and product yields due to redirection of the carbon flux towards unwanted by-products such as organic acid. Using this organism, the final PHB homopolyester content amounted to 12 wt.-% (q_P: 2.9 mg/g h). By co-feeding of pentanoic acid, the strain accumulated 12 wt.-% of poly-3(HB-*co*-21%-HV) (q_P: 2.0 mg/g h) (Koller et al., 2007a; Koller et al., 2008a, b).

H. pseudoflava produces biopolyesters of rather good quality (high molecular masses and low polydispersities) directly from whey lactose at acceptable specific production rates and yields, but is not competitive with *H. mediterranei* in terms of strain stability and robustness. In details, using this strain, 40 wt.-% of poly-3(HB-*co*-5%-HV) in cells with addition of pentanoic acid (q_P: 12.5 mg/g h) were obtained. Without pentanoic acid, the strain accumulated 30 wt.-% of the homopolyester PHB in cells (q_P: 16.0 mg/g h)(Koller et al., 2007a).

A number of very recent reports describe the production of PHA by other bacterial species, suggesting that the occurrence of these traits in new isolates from various environmental samples is underestimated and has still to be deeply investigated.

For instance, the thermophilic bacterium *Thermus thermophilus* HB8 (DSM 579) was found to be able to utilize lactose from whey-based media, using both glucose and galactose, for the biosynthesis of polyhydroxyalkanoates under nitrogen limitation (Pantazaki et al., 2009) PHA was accumulated up to 35% (w/w) of its biomass and revealed a novel heteropolymer consisting of the short chain length 3-hydroxyvalerate (3HV; 38 mol%) and the medium chain length 3-hydroxyheptanoate (3HHp; 9.89 mol%), 3-hydroxynonanoate (3HN; 16.59 mol%) and 3-hydroxyundecanoate (3HU; 35.42 mol%).

Moreover, *Azohydromonas lata* DSM 1123 was reported to produce poly-3(HB-*co*-27%-HV) from whey hydrolysate (Baei et al., 2010), *Bacillus megaterium* CCM 2032 was shown to accumulate more than 50% of its biomass (w/w) in optimized whey media (Obruca et al., 2011).

The subsequent Table 2 collects the data for PHA production from lactose, whey and whey-derived substrates available in literature as discussed above. Here, the applied microbial production strains, the pre-treatment of the raw material whey, PHA productivities, final concentrations of cell mass and PHA biopolymer, production scale and information about post synthetic polymer characterization are provided.

Production strain	Applied carbon source	Type of PHA produced	Vol. productivity for PHA [g/L h]	CDM final [g/L]	PHA final [g/L]	Information about PHA quality available	Production scale	Reference
Pseudomonas cepacia ATCC 17759	Lactose	PHB	0.03	4.9	2.75	Yes: Molecular mass characterization	Shaking flask scale	Young et al., 1994
Pseudomonas cepacia ATCC 17759	Equimolar mixtures of glucose and galactose	PHB	0.18	16.7	4.0	no	Shaking flask scale	Young et al., 1994
rec. E. coli GCSC 4401 (pSYL107)	bovine whey powder solution	PHB	n.r.	6.4	5.2	no	Shaking flask scale	Lee et al., 1997
rec. E. coli GCSC 6576 (pSYL107)	bovine whey powder solution	PHB	n.r.	5.7	4.5	no	Shaking flask scale	Lee et al., 1997
rec. Cupriavidus necator mRePT	lactose	PHB	0.04	8.1	0.21	no	Shaking flask scale	Povolo et al., 2010
rec. Cupriavidus necator mRePT	Hydrolyzed whey permeate	PHB	0.05	8.0	2.4	no	Shaking flask scale	Povolo et al., 2010
rec. Cupriavidus necator mRePT	Not hydrolyzed whey permeate	PHB	0.03	6.5	1.4	no	Shaking flask scale	Povolo et al., 2010
Sinorhizobium melioti 41	Whey permeate	PHB	0.00018	0.483	0.017	no	Shaking flask scale	Povolo & Casella, 2003
Hydrogenophaga pseudoflava DSM 1034	Whey permeate	PHB	0.00018	0.375	0.017	no	Shaking flask scale	Povolo & Casella, 2003
rec. Escherichia coli CGSC 4401 harbouring pJC4	Processed bovine whey powder solution; controlled DOC	PHB	1.15	83.1	46.8	no	6.6-liter jar fermentor; fed-Batch	Ahn et al., 2000
rec. Escherichia coli CGSC 4401 harbouring pJC4	Processed bovine whey powder solution; higher concentration of whey feed	PHB	1.42	102.9	59.6	no	6.6-liter jar fermentor; fed-Batch	Ahn et al., 2000

Production strain	Applied carbon source	Type of PHA produced	Vol. productivity for PHA [g/L h]	CDM final [g/L]	PHA final [g/L]	Information about PHA quality available	Production scale	Reference
rec. Escherichia coli CGSC 4401 harbouring pJC4	Processed bovine whey powder solution; controlled DOC	PHB	2.57	119.5	96.2	no	6.6-liter jar fermentor; fed-Batch	Ahn et al., 2000
rec. Escherichia coli CGSC 4401 harbouring pJC4	Processed bovine whey powder solution;	PHB	4.6	194	168	no	6.6-liter jar fermentor; Fed-Batch with cell recycle membrane module system	Ahn et al., 2001
Hydrogenophaga pseudoflava DSM 1034	Whey permeate + 3HV precursor pentanoic acid	P(3HB-*co*-4,6%-3HV)	0.05	6.7	2.7	yes: Thermal analysis characterization, Molecular mass characterization	Bioreactor, 10L; Fed-Batch	Koller et al., 2007a
Pseudomonas hydrogenovora DSM 1749	Enzymatically hydrolyzed whey permeate	PHB	0.03	10.6	1.27	yes: Thermal analysis characterization, Molecular mass characterization	Bioreactor, 2L; Fed-Batch	Koller et al., 2007a, Koller et al., 2008
Pseudomonas hydrogenovora DSM 1749	Enzymatically hydrolyzed whey permeate + 3HV precursor pentanoic acid	P(3HB-*co*-21%-3HV)	0.05	11.7	1.44	yes: Thermal analysis characterization, Molecular mass characterization	Bioreactor, 2L; Fed-Batch	Koller et al., 2008
Haloferax mediterranei DSM 1411	Enzymatically hydrolyzed whey permeate	P(3HB-*co*-8%-3HV)	0.05	11.0	5.5	yes: Thermal analysis characterization, Molecular mass characterization	Bioreactor, 42L; Fed-Batch	Koller et al., 2007a
Haloferax mediterranei DSM 1411	Enzymatically hydrolyzed whey permeate	P(3HB-*co*-6%-3HV)	0.09	16.8	12.2	yes: Thermal analysis characterization, Molecular mass characterization	Bioreactor, 42L; Fed-Batch	Koller et al., 2007b
Haloferax mediterranei DSM 1411	Enzymatically hydrolyzed whey permeate + 3HV precursor pentanoic acid + 4HB precursor γ-butyrolactone	P-(3HB-*co*-21.8%-3HV-*co*-5.1%-4HB)	0.14	16.8	14.7	yes: Thermal analysis characterization, Molecular mass characterization	Bioreactor, 10L; Fed-Batch	Koller et al., 2007b
Methylobacterium sp. ZP24	Lactose	PHB	0.08	5.25	3.1	no	Shaking flask scale	Yellore & Desai, 1998

Production strain	Applied carbon source	Type of PHA produced	Vol. productivity for PHA [g/L h]	CDM final [g/L]	PHA final [g/L]	Information about PHA quality available	Production scale	Reference
Methylobacterium sp. ZP24	Whey supernatant	PHB	0.017	3.8	0.8	no	Shaking flask scale	Yellore & Desai, 1998
Methylobacterium sp. ZP24	Whey supernatant + nitrogen	PHB	0.054	5.9	2.6	no	Shaking flask scale	Yellore & Desai, 1998
Methylobacterium sp. ZP24	Whole whey	PHB	0.023	5.1	1.1	no	Shaking flask scale	Yellore & Desai, 1998
Methylobacterium sp. ZP24	Whole whey + nitrogen	PHB	0.008	7.1	0.4	no	Shaking flask scale	Yellore & Desai, 1998
Methylobacterium sp. ZP24	Processed cheese whey	PHB	0.06	3.34	2.07	no	semiautomatic jar fermenter, 2 L Batch	Nath et al., 2008
Methylobacterium sp. ZP24	Processed cheese whey	PHB	0.09	5.53	3.54	no	semiautomatic jar fermenter, 2 L Fed-Batch	Nath et al., 2008
Methylobacterium sp. ZP24	Processed cheese whey	PHB	0.056		3.91	no	semiautomatic jar fermenter, 30 L Fed-Batch	Nath et al., 2008
Bacillus megaterium	lactose	PHB	0.01	1.8	0.47	no	Shaking flask scale	Omar et al., 2001
Bacillus megaterium SRKP-3	Dairy waste	PHB	0.31	n.r.	11.32	no	Bioreactor, 3 L; Fed-Batch	Pandian et al., 2010
Bacillus megaterium CCM2037	whey	PHB		2.87	1.46	no	Shaking flask scale	Obruca et al., 2011
Azohydromonas lata DSM 1123	whey	P(3HB-co-12.7%-3HV)		9.2	1.66	no	Shaking flask scale	Baei et al, 2010
Herbaspirillum seropedicae Z69 (pHM3)	lactose	PHB		5	1.8		Shaking flask scale	Catalan et al., 2007

Table 6. PHA production from lactose, whey and whey-derived substrates: an overview about literature data

10. Mathematical models for value-added conversion of whey

Due to the complex nature of the regulatory mechanism in PHA production, mathematical modelling of the processes, based on experimental results, becomes more and more interesting. This is especially true for multi-substrate carbon sources like whey. Such models provide a precious tool to shorten the number of needed experiments during biotechnological process development. The first, early developed mathematical models for PHA synthesis (Sonnleitner et al., 1979; Heinzle & Lafferty, 1980) have postulated a "two compartment" strategy which is applicable and valid till today. Biomass (X) was structured as having two components:

1. The catalytically active fraction consisting of proteins, glycolipids, glycoproteins and nucleic acids (known as residual biomass);
2. The intracellular, catalytically inert fraction which is the biopolymer (PHB).

In addition, it was experimentally proved that:

3. Nitrogen (N) or phosphorus (P), on the one hand, act as limiting substrates affecting the growth kinetics, but N or P starvation are inductivity factors provoking PHA synthesis;
4. TIn some strains, the production of PHB is uncoupled from the biomass growth (not growth associated PHA production), whereas in some other strains PHB synthesis can be evidenced in both, growth and stationary phase respectively. In the last case, PHB synthesis occurs with significantly different rates (partially associated PHA production to biomass growth).

In above citied works, the specific growth rate was defined as function of C-source and N-source concentration respectively. Limitation by N-source was expressed as linear sum of two different kinetics: Monod saturation type and sigmoidal Hill type. Above postulates are also useful tools for modelling of processes based on whey as C-source. In literature a lot of different types of mathematical models (formal kinetic, low and high structured models) are available arranged for microbia1 PHA production, but predominantly dealing with glucose, fructose, sucrose and glycerol as the C sources (Sonnleitner et al., 1979; Mulchadani et al., 1988; Raje & Srivastava, 1998; Leaf & Srienc, 1998; Katoh et al., 1999; Patwardhan & Srivastava, 2004; Franz et al., 2011). In contrast, mathematical models of processes with whey itself or with whey sugars are still rare.

Different model types can be considered as suitable for this purpose, depending on chemical composition of whey (raw native whey or whey permeate, partially fermented whey, hydrolysed whey or permeate) and on PHA accumulating microorganism (growth associated or partially growth associated synthesis of PHA). Native whey contains lactose (C-source), casein (complex N-source), phosphorus compounds, and, due to partial fermentation, lactic acid and some other metabolites (see also Table 5). Due to its high N- and P-content, this substrate is not suitable for microbial PHA synthesis, especially not for partially growth associated PHA producers with week lactase expression (able to activate PHA synthesis under N/P starvation). In contrast, hydrolysed casein-free whey permeate containing hydrolysed lactose (i.e. glucose and galactose) is more promising for this purpose. However, if any type of whey will be used as the C-source, the multisubstrate kinetic models will be necessary for successful modelling of biomass growth. Recently, unstructured mathematical models for production of PHB from whey and sucrose by *Azotobacter vinelandii* were developed using kinetic patterns from batch experiments (Dhanasekar & Viruthagiri, 2005). The mentioned organism is able to use sucrose, cane molasses and cheese whey as a C-source. It is a growth associated PHB producer which synthetizes product also in the non-growth stationary phase. The authors have applied Logistic equation and modified Logistic equation (Mulchadani et al., 1988) for the modelling of sigmoidal growth curves (to compensate a certain lag-phase and to reduce it to the desired length) combined with Leudeking-Piret specific growth rate/specific production rate relation (Leudeking & Piret, 1959).

For the biosynthesis of P(3HB-*co*-3HV-*co*-4HB) terpolyesters from whey plus cosubstrates by *H. mediterranei*, Koller et al. (2006) compiled a formal kinetic model as well as a low structured model for PHB synthesis from whey by *P. hydrogenovora*. Batch and feed-batch processes with *H. mediteranei* were modelled by these authors using the subsequent assumptions:

a. Residual biomass (non-PHA biomass) is synthesized from all three main carbon sources i.e. glucose, galactose and γ-butyrolactone (independent growth on each substrate according Monod relation was introduced!) and from yeast extract as an obligate complex nitrogen source.
b. There is no direct influence of the consumption of one C-source on the consumption rate of two other C-sources (simultaneous independent consumption of C-sources).
c. Both 3HB and 3HV were supposed to be synthesized from both glucose and galactose, 4HB is synthesized only from γ -butyrolactone.
d. PHA synthesis is inversely influenced by the increase of the intracellular mass fraction of PHA (known as steric disturbing effect). The equation according to Luong (1985) was applied for this purpose.

Metabolic pathways for the applied microorganism were not well known (except a few basic metabolic routes!). This is especially valid for the strain's galactose degradation pathway. An insufficient metabolic knowledge was the reason why authors have chosen a formal-kinetic modelling instead of low structured or high structured metabolic flux model as available for *Escherichia coli* (Van Aalst et al., 1997; Van Wegen et al., 2001). Simulated values of process variables were in good but not in excellent agreement with the experimental data. This is especially true for PHB concentration in the time period of late exponential phase and for glucose concentration in the transient time period before stationary phase of growth.

Additionally, based on experiences and results of above formal-kinetic model a low structured metabolic mathematical model for fed-batch cultivation of *Pseudomonas hydrogenovora* was established. *P. hydrogenovora* is a typical strain having 3HB synthesis provoked by nitrogen limitation of growth. The total uncoupling of growth phase and 3HB synthesis phase ("non-growth-associated PHA production") was evidenced. Here, the production of PHB from glucose and galactose (C-sources) using ammonia and casein hydrolyzate as N-sources was studied.

The following assumptions were necessary to establish a useful mathematical model:

Ac-CoA (originated from the central metabolic pathways - that mean from the metabolism of both sugars) is chosen as the ubiquitary precursor for PHB. This was based on the fact that all sugar degrading pathways in microorganisms (Entner–Doudoroff, pentose phosphate, Leloir, DeLey–Doudoroff, and EMP pathway) lead to EMP substrates (i.e. pyruvate, consecutively transformed to Ac-CoA!). Ac-CoA was chosen because it constitutes the substrate for the thiolase reaction (first step in the PHB synthesis), and for the citrate synthase reaction (substrates for biomass synthesis). Furthermore, the breakdown of the complex nitrogen source (deamination and degradation of casamino acids) leads to the Ac-CoA. An independent Monod kinetic of Ac-CoA synthesis from both sugars was applied as the modelling strategy.

i. Negative feedback control mechanism of Ac-CoA synthesis is built in as strategy for its own regulation /Luong type of inhibition pattern; Luong, (1985)/.
ii. Ac-CoA is foreseen for consumption toward biomass formation, energy supply including maintenance energy (NADPH generation in TCA cycle), 3HB accumulation (thiolase substrate), α-ketoglutarate excretion, for one chemically unknown excreted compound production and for PHB-polymerase synthesis.
iii. A small quantity of biomass is assumed to be synthesized directly from the complex nitrogen source (casamino acids could be used as a separate C/N source).
iv. PHB synthesis rate is proportional to the intracellular biocatalyst concentration (PHB-polymerase complex). Its synthesis rate from Ac-CoA and nitrogen sources (multiple Michaelis-Menten kinetic) with desired degradation rate (protein turnover) was assumed.
v. Ac-CoA synthesis from sugars is inhibited (but not stopped) by complex nitrogen (Jerusalimsky type of inhibitory influence applied).
vi. The PHB-polymerase complex synthesis is inhibited by the presence of high level of complex nitrogen source (Jerusalimsky type of kinetic equation), its activity started after complex nitrogen sources was almost completely depleted.
vii. Inorganic nitrogen source consumption is inhibited by complex nitrogen source (no consumption of ammonia in the first part of growth phase when casamino acids were provided at sufficient amounts).
viii. The excreted metabolite α-ketoglutarate is assumed to be a negative feedback controller of own synthesis as well as for unknown metabolite. Ammonia and PHB polymerase were adopted as inhibitory agents of metabolites production (Jerusalimsky type of equation).

The authors emphasise that the developed models provide feasible tools for better understanding the complex intracellular on-goings during PHA production from multi-substrate carbon sources without a specific analysis of the involved enzymatic reactions.

Despite of the complex intracellular regulatory mechanisms of PHA metabolism and despite of complexity of whey as the substrate, mathematical modelling of PHA synthesis could be a powerful tool to predict different situations and to reduce the number of needed experimental set ups. Significant progress has been achieved during the last two decades in the field of metabolic flux and metabolic control modelling (Ghadkar et al., 2003;

Thilakavathi et al., 2007; Franz et al., 2011). Cited cybernetic, kinetic, dynamic and structured metabolic modelling methods have not been applied yet in the analysis of PHA production from whey. The developed models are structured kinetic models, simulating the effects of enzyme and metabolite concentrations on the rate of PHB synthesis in microorganisms (Franz et al., 2011). They suggest how the physiological state of the cell, intracellular concentrations of metabolites, and enzyme levels affect the rate of PHB synthesis.

Extensive experimental tools, suitable computational methods, a detailed kinetically-metabolic knowledge of metabolic fluxes are necessary for the development of high structured models. That's why for practical laboratory and industrial purposes respectively, a certain degree of simplification of the model structure is desired and welcome. A typical example for this type of simplification was done by Visser and colleagues (2004) dealing with the metabolic modelling of *Saccharomyces cerevisiae*. Compared to more detailed models, the simulations with simplified models are less accurate. This simplified approach in modelling is perhaps a promising tool in the case when metabolic pathways and/or rate control mechanisms are insufficiently known, and therefore the building of metabolic flux models is not possible.

11. Conclusion

The study presents a strategy how industrial waste streams like surplus whey can be upgraded to the role of substrates for production of high-value bio-products. The selection of waste materials like surplus whey from dairy industry can be regarded as the most promising route to make the entire PHA biopolymer production process economically competitive to the petrochemical competitors like PE and PP. Improvements in the fermentation strategy by switching from discontinuous to continuous mode can be considered as another decisive step for achieving enhanced productivities and to obtain tailor-made bioplastics under constant qualities. Beside the raw material costs and the fermentation process itself, downstream processing for polymer recovery from the surrounding cells is a decisive cost-determining factor in biopolymer production.

Uniting the potential enhancements of each process step, one can definitely make substantial progress towards a cost-efficient technology. In any case, the development of really efficient biopolymer production processes needs the narrow cooperation of experts from different scientific fields. Chemical engineers, microbiologists, enzymologists, polymer scientists, genetic engineers and experts in the fields of mathematical modelling of bio-processes, LCA and Cleaner Production studies have to concentrate their special expertise and know-how in order to close the existing gaps between promising data from the laboratory scale and industrial realization.

Independent from the selected production strain, an industrial plant for PHA production from whey should be integrated into existing process lines of large dairies, were the raw material whey directly accrues. This can be considered as a viable strategy to minimize production costs by taking profit of synergistic effects. Regarding the already broad amount of available data from literature, one can conclude that important progress has been achieved in terms of combining the environmental benefit of future-oriented biopolyesters with a cost efficiency that makes them interesting for potential industrial partners.

Author details

Martin Koller*
Graz University of Technology; Institute of Biotechnology & Biochemical Engineering, Graz, Austria
ARENA Arbeitsgemeinschaft für Ressourcenschonende & Nachhaltige Technologien, Graz, Austria

Anna Salerno, Alexander Muhr and Angelika Reiterer
Graz University of Technology; Institute of Biotechnology & Biochemical Engineering, Graz, Austria

Emo Chiellini
Laboratory of Bioactive Polymeric Materials for Biomedical and Environmental Applications via Vecchia Livornese, Loc. S. Piero a Grado (PI), University of Pisa, Department of Chemistry and Industrial Chemistry, Pisa, Italy

Sergio Casella
DAFNAE, Department of Agronomy Food Natural Resources Animals and Environment, Università degli Studi di Padova, Legnaro, Padova, Italy

Predrag Horvat
Department of Biochemical Engineering, Faculty of Food Technology and Biotechnology, University of Zagreb, Croatia

Gerhart Braunegg
ARENA Arbeitsgemeinschaft für Ressourcenschonende & Nachhaltige Technologien, Graz, Austria

Acknowledgement

The authors thank for financial support by the European Commission, granting the FP5 and FP7 projects WHEYPOL (GRD2-2000-30385) and ANIMPOL (Contract No: 245084) and the Austrian project FFG-Laura Bassi, centre of expertise "BRIC – BioResobable Implats for Children". In addition, the authors are thankful for the electron microscopic picture of *C. necator* cells that was provided by Dr. Elisabeth Ingolić, FELMI-ZFE Graz.

12. References

Ahn, W.S., Park, S.J. & Lee, S.Y. (2000). Production of Poly(3-Hydroxybutyrate) by Fed-Batch Culture of Recombinant *Escherichia coli* with a Highly Concentrated Whey Solution, *Applied Environmental Microbiology*. 66(8): 3624–3627.

* Corresponding Author

Ahn, W.S., Park, S.J. & Lee, S.Y. (2001). Production of poly(3-hydroxybutyrate) from whey by cell recycle fed-batch culture of recombinant *Escherichia coli*, *Biotechnological Letters* 23: 235-240.

Akiyama, M., Tsuge, T. & Doi, Y. (2003). Environmental Life Cycle Comparison of Polyhydroxyalkanoates Produced from Renewable Carbon Resources by Bacterial Fermentation, *Polymer Degradation and Stability* 80: 183-194.

Anastas, P.T. & Lankey, R.L. (2000). Life Cycle Assessment and Green Chemistry: the Yin and the Yang of Industrial Ecology, *Green Chemistry* 2: 289-295.

Atlic, A., Koller, M., Scherzer, D., Kutschera, C., Grillo Fernandes, E., Horvat, P., Chiellini, E. & Braunegg, G. (2011). Continuous Production of Poly([(R]-3-hydroxybutyrate) by *Cupriavidus necator* in a Multistage Bioreactor Cascade, *Applied microbiology and biotechnology* 91: 295-304.

Audic, J.L., Chaufer, B. & Daufin, G. (2003). Non-food applications of milk components and dairy co-products: A review, *De Lait* 83(6): 417-438.

Baei, M.S., Najafpour, G.D., Lasemi, Z., Tabandeh, F., Younesi, H., Issazadeh, H. & Khodabandeh, M. (2010). Optimization PHAs production from dairy industry wastewater (cheese whey) by *Azohydromonas lata* DSMZ 1123, *Iranica Journal of Energy and Environment* 1(2): 132-136.

Boustead, I. (2005). Eco-Profiles of the European Plastic Industry: Polypropylene (PP). *Plastics Europe*, Brussels.

Boynton, Z.L., Koon, J.J., Brennan, E.M., Clouart, J.D., Horowitz, D.M., Gerngross, T.U. & Huisman, G.W. (1999). Reduction of Cell Lysate Viscosity during Processing of Poly(3-Hydroxyalkanoates) by Chromosomal Integration of the Staphylococcal Nuclease Gene in *Pseudomonas putida*, *Applied and Environmental Microbiology* 65(4): 1524–1529.

Braunegg, G., Koller, M., Hesse, P.J., Kutschera, C., Bona, R., Hermann, C., Horvat, P., Neto, J. & Dos Santos Pereira, L. (2007). Production of Plastics from Waste Derived from Agrofood Industry, *in*: Graziani, M. & Fornasiero, P. (eds.), *Renewable resources and renewable energy: a global challenge,* Taylor and Francis Group, Boca Raton, New York, London, pp. 119 – 135.

Braunegg, G., Lefebvre, G. & Genser, K.F. (1998). Polyhydroxyalkanoates, biopolyesters from renewable resources: physiological and engineering aspects, *Journal of Biotechnology* 65(2,3):127-161.

Bulhões, A.C., Goldani, H.A.S., Oliveira, F.S., Matte, U.S., Mazzuca, R.B. & Silveira, T.R. (2007). Correlation between lactose absorption and the C/T-13910 and G/A-22018 mutations of the lactase-phlorizin hydrolase (LCT) gene in adult-type hypolactasia, *Brazilian Journal of Medical and Biological Research* 40 (11): 1441–6.

Catalán, A.I., Ferreira, F., Gill, P.R. & Batista, S. (2007). Production of polyhydroxyalkanoates by *Herbaspirillum seropedicae* grown with different sole carbon sources and on lactose when engineered to express the lacZlacY genes, *Enzyme and Microbial Technology* 40: 1352–1357.

Champrateep, S. (2010). Current trends in biodegradable Polyhydroxyalkanoates, *Journal of Bioscience and Bioengineering* 110 (6): 621-632.

Chen, G.Q. & Wu, Q. (2005). Polyhydroxyalkanoates as tissue engineering materials, *Biomaterials* 26:6565-6578.

Chen, C.Q. (2009) A microbial polyhydroxyalkanoates (PHA) based bio- and materials industry, *Chemical Society Review* 38:2434–2446.

Chen, G.Q. (2010) Plastics completely synthesized by bacteria: Polyhydroxyalkanoates, *in*: G.-Q. Chen (ed.), *Plastics from Bacteria.*, Alexander Steinbüchel, Series (ed.) *Microbiology Monographs*, Springer-Verlag, Berlin, Heidelberg, pp. 17-37.

Chen, G.Q. (2010), Industrial Production of PHA. *in*: G.-Q. Chen (ed.), *Plastics from Bacteria. Microbiology Monographs,* Alexander Steinbüchel, Series (ed.) Springer-Verlag, Berlin, Heidelberg, pp. 121-132.

Chiellini, E., Cinelli, P., Chiellini, F. & S.H. Imam (2004). Environmentally Degradable Bio-Based Polymeric Blends and Composites, *Macromolecular Biosciences* 4 (3): 218-231.

Choi, J. & Lee, S.Y. (1999). Factors affecting the economics of polyhydroxyalkanoate production by bacterial fermentation, *Applied Microbiological Biotechnology* 51:13-21.

Dhanasekar, R. & Viruthagiri, T. (2005). Batch kinetics and modeling of poly-β-hydroxy butyrate synthesis from *Azotobacter vinelandii* using different carbon sources, *Indian Journal of Chemical Technology* 12(3): 322-326.

Daniel Ferrari, M., Bianco, R., Froche, C. & Loperena, M.L. (2001). Baker's yeast production from molasses/cheese whey mixtures, *Biotechnology Letters* 23: 1–4.

Daniel, H.J., Otto, R.T., Binder, M., Reuss, M. & Syldatk, C. (1999). Production of sophorolipids from whey: development of a two-stage process with *Cryptococcus curvatus* ATCC 20509 and *Candida bombicola* ATCC 22214 using deproteinized whey concentrates as substrates, *Applied Microbiology and Biotechnology* 51(1):40-45.

De Palma Revillion, J.P., Brandelli, A. & Záchia Ayub, M.A. (2003). Production of yeast extract from whey using *Kluyveromyces marxianus, Brasilian Archives of Biology and Technology* 46(1): 121-127.

Franz, A., Song, H., Ramkrishna, D. & Kienle, A. (2011). Experimental and Theorethical analysiys of Poly (b-hydroxybutyrate) formation and consumptions in *Ralstonia eutropha, Biochem Engineering Journal* 55:49-58.

Gadkar, K.G., Doyle, F.J., Crowley, J.T. & Varner, J.G. (2003). Cybernetic Model Control of a continuous Bioreactor with Cell Recycle, *Biotechnology Progress* 19: 1487-1497.

Gänzle, M.G., Haase, G. & Jelen, P. (2008). Lactose: crystallization, hydrolysis and value-added derivatives, *International Dairy Journal* 18:685–94.

Ghaley, A.E. & El-Taweel, A.A. (1997). Kinetic modelling of continuous production of ethanol from cheese whey, *Biomass and Bioenergy* 12(6), 461-472.

Gonzales Siso, M.I. (1996) The biotechnological utilization of cheese whey. A review. Bioresearch Technology, 57: 1-11.

Gonzalez, C.J. & Woodley, J.M. (2010). Bioprocesses: Modelling needs for process evaluation and sustainability assessment, *Computers and Chemical Engineering* 34: 1009-1017.

Gueriff, N. & Lant, P. (2007). Comparative Life Cycle Assessment and Financial Analysis of Mixed Culture Polyhydroxyalkanoate Production, *Bioresource Technology* 98: 3393-3403.

Guimarães, P.M.R., Teixeira, J.A. & Domingues, L. (2010). Fermentation of lactose to bio-ethanol by yeasts as part of integrated solutions for the valorisation of cheese whey, *Biotechnology Advances* 28:375–384.

Guo-Qiang, C. & Qiong, W. (2005). The application of polyhydroxyalkanoates as tissue engineering materials, *Biomaterials* 26 : 6565–6578.

Harding, K.G., Dennis, J.S., Blottnitz, H.V. & Harrison, S.T.L. (2007). Environmental Analysis of plastic Production Processes: Comparing Petroleum-Based Polypropylene and Polyethylene with Biologically-Based Poly-β-Hydroxybutyric Acid Using Life Cycle Analysis, *Journal of Biothechnology* 130: 57-66.

Heinzle, E. & Lafferty, R.M. (1980). A mathematical model for growth and synthesis of poly-b-hydroxybutyric acid (PHB) in *Alcalygenes eutrophus* H16, *European Journal of Applied Microbiology and Biotechnology* 11: 8-16.

Helm, J., Wendlandt, K.D., Jechorek, M. & Stottmeister, U. (2008). Potassium deficiency results in accumulation of ultra-high molecular weight poly-beta-hydroxybutyrate in a methane-utilizing mixed culture, *Journal of applied microbiology* 105(4):1054-61.

Heyde, M. (1998). Ecological Considerations on the Use and Production of Biosynthetic and Synthetic Biodegradable Polymers, *Polymer Degradation & Stability* 59: 3-6.

Heyman, M.B., (2006). Lactose Intolerance in Infants, Children, and Adolescents, *Pediatrics* 118(3): 1279–1286.

Hezayen, F.F., Rehm, B.H., Eberhardt, R. & Steinbuchel, A. (2000). Polymer production by two newly isolated extremely halophilic archaea: application of a novel corrosion-resistant bioreactor, *Applied Microbiology and Biotechnology* 54(3): 319-25.

Hickmann Flôres, S. & Monte Alegre, R. (2001). Nisin production from *Lactococcus lactis* A.T.C.C. 7962 using supplemented whey permeate, *Biotechnology and Applied Biochemistry* 34: 103–107.

Illanes, A. (2011). Whey upgrading by enzyme biocatalysis, *Electronic Journal of Biotechnology* 14(6): 15.

Katoh, T., Yuguchi, H., Shi, H. & Shimizu, K. (1999) Dynamics and modeling on fermentative production of poly (β-hydroxybutyric acid) from sugars via lactate by a mixed culture of *Lactobacillus delbrueckii* and *Alcaligenes eutrophus*, *Journal of Biotechnology* 67(2-3): 113-134.

Khanna, S., Srivastava, A.K. (2005). Recent advances in microbial Polyhydroxyalkanoates, *Process Biochemistry*. 40, 607–619

Kendall, A. (2010). A life cycle assessment of biopolymer production from material recovery facility residuals, *Resources, Conservation and Recycling* 61: 69-74.

Kim, B.S., Lee, S.C., Lee, S.Y., Chang, H.N., Chang, Y.K. & Woo S.I. (1994). Production of poly(3-hydroxybutyric acid) by fedbatch culture of *Alcaligenes eutrophus* with glucose concentration control. *Biotechnology and Bioengineering* 43:892-898.

Kim, H.O., Wee, Y.J., Kim, J.N., Yun, J.S., Ryu, H.W. (1995). Production of lactic acid from cheese whey by batch and repeated batch cultures of *Lactobacillus* sp. RKY2. Applied Biochemestry and Biotechnology 129-132:694-704

Kim, B.S. (2000) Production of poly(3-hydroxybutyrate) from inexpensive substrates, *Enzyme and Microbial Technology* 27: 774–777.

Kim, S. & Dale, B.E. (2005)Life cycle assessment of integrated biorefinery-cropping systems: all biomass is local," in *Agriculture as a Producer and Consumer of Energy*, JL Oulaw, KJ Collins, and JA Duffield (Eds.), CAB International, Wallingford, Oxfordshire, UK, 2005, conference proceedings of Farm Foundation and USDA Office of Energy Policy and New Uses, Arlington, VA, June 24-25, 2004.

Kim, S. & Dale, B.E. (2005). Life Cycle Assessment Study of Biopolymers (Polyhydroxyalkanoates) Derived from No-Tilled Corn, *International Journal of Life Cycle Assessment* 10(3): 200-210.

Kim, S. & Dale B.E (2008). Energy and greenhouse gas profiles of polyhydroxyalkanoates derived from corn grain: a life cycle perspective, *Environmental Science and Technology* 42: 7690-7695.

Kim, Y.B. & Lenz, R.W. (2001). Polyesters from microorganisms, *Advances in Biochemical Engineering/Biotechnology* 71:51-79.

Koller, M., Bona, R., Braunegg, G., Hermann, C., Horvat, P., Kroutil, M., Martinz, J., Neto, J., Varila, P. & Pereira, L. (2005a). Production of Polyhydroxyalkanoates from Agricultural Waste and Surplus Materials, *Biomacromolecules* 6: 561-565.

Koller, M., Bona, R., Hermann, C., Horvat, P., Martinz, J., Neto, , Pereira, L., J., Varila, P. & Braunegg, G. (2005b).Biotechnological production of poly(3-hydroxybutyrate) with *Wautersia eutropha* by application of green grass juice and silage juice as additional complex substrates, *Biocatalysis and Biotransformation* 23(5): 329–337.

Koller, M., Hesse, P.J., Bona, R., Kutschera, C., Atlic, A. & Braunegg, G. (2007a). Various Archae- and Eubacterial Strains as Potential Polyhydroxyalkanoate Producers from Whey Lactose, *Macromolecular bioscience* 7: 218 – 2266.

Koller, M., Hesse, P.J., Bona, R., Kutschera, C., Atlic, A. & Braunegg, G. (2007b). Biosynthesis of High Quality Polyhydroxyalkanoate Co- and Terpolyesters for Potential Medical Application by the Archaeon *Haloferax mediterranei*, *Macromolecular symposia* 253: 33-39.

Koller, M., Atlic, A., Gonzalez-Garcia, Y., Kutschera, C. & Braunegg, G. (2008a). Polyhydroxyalkanoate (PHA) Biosynthesis from Whey Lactose, *Macromolecular symposia* 27: 287-92.

Koller, M., Bona, R., Chiellini, E., Grillo Fernandes, E., Horvat, P., Kutschera, C., Hesse P.J. & Braunegg, G. (2008b). Polyhydroxyalkanoate production from whey by *Pseudomonas hydrogenovora*, *Bioresource technology* 99: 4854-4863.

Koller, M., Salerno, A., Miranda de Sousa Dias, M., Reiterer, A. & Braunegg, G. (2010). Modern Biotechnological Polymer Synthesis: A Review, *Food technology and biotechnology* 48 (3), 255 – 269.

Koller, M., Gasser, I., Schmid, F. & Berg, G. (2011). Linking ecology with economy: insights into polyhydroxyalkanoate-producing microorganisms, *Engineering in life sciences* 11(3):222-237.

Kosikowski , F.V. (1979) Whey utilisation and whey products, *Journal of Dairy Science* 62: 1149-60

Kunasundari, B. & Sudesh, K. (2011). Isolation and recovery of microbial polyhydroxyalkanoates, *eXPRESS Polymer Letters* 5(7): 620–634.

Kurdikar, D., Fournet, L., Slater, S.C., Paster, M., Gerngross, T.U. & Coulon, R. (2001). Greenhouse Gas Profile of a Plastic Material Derived from a Genetically Modified Plant, *Journal of Industrial Ecology* 4(3): 107-122.

Leaf, T.E.A. & Srienc, F. (1998). Metabolic modeling of polyhydroxybutyrate biosynthesis, *Biotechnology and Bioengineering* 57(5): 557.

Lee, S.Y., Middelberg, A.P.J. & Lee, Y.K. (1997). Poly(3-hydroxybutyrate) production from whey using recombinant Escherichia coli, *Biotechnology Letters* 19(10):1033–1035.

Lee, S.Y. (1996). Bacterial Poly- b-hydroxyalkanoates, *Biotechnology and Bioengineering* 49: 1-14.

Luedeking, R. & Piret E.L. (1959). Transient and steady states in continuous fermentation. Theory and experiment, *Biotechnology and Bioengineering* 1(4): 431-459.

Luong, J.H.T. (1985). Kinetics of ethanol inhibition in alcohol fermentation, *Biotechnology and Bioengineering* 27(3):

Mulchandani ALuong, J.H.T. & Groom, C. (1988). Substrate inhibition kinetics for microbial growth and synthesis of PHB by *Alcaligenes eutrophus* ATCC 17697, *Applied Microbiology and Biotechnology* 32:639–646.

Munoz, E.A., Rodriguez-Valera, F. & Marcilla Gomis, A. (1994). Procedure for the extraction of polyhydroxyalkanoates from halophilic bacteria which contain them, European Patent 0622462.

Nath, A., Dixit, M., Bandiya, A., Chavda, S. & Desai, A.J. (2008). Enhanced PHB production and scale up studies using cheese whey in fed batch cultures of *Methylobacterium* sp. ZP24. *Bioressource Technology* 99(13):5749-5755.

Nonato, R.V., Mantelatto, P.E. & Rossell, C.E. (2001). Integrated Production of Biodegradable Plastic, Sugar and Ethanol, *Applied Microbiology and Biotechnology* 57: 1-5.

Obruca, S., Marova, I., Melusova, S. & Mravcova, M. (2011). Production of polyhydroxyalkanoates from cheese whey employing *Bacillus megaterium* CCM 2037, *Annals Microbiology* 61: 947–953.

Omar, S., Rayes, A., Eqaab, A., Viss, I. & Steinbuchel, A. (2001). Optimization of cell growth and poly(3-hydroxybutyrate) accumulation on date syrup by a *Bacillus megaterium* strain, *Biotechnology Letters* 23:1119–1123.

Ouyang, S.P., Liu, Q., Fang, L. & Chen, G.Q. (2007). Construction of pha-Operon-Defined Knockout Mutants of *Pseudomonas putida* KT2442 and their Applications in Poly(hydroxyalkanoate) Production, *Macromolecular Bioscience* 7(2): 227–233.

Pandian S.RK., Deepak, V., Kalishwaralal, K., Rameshkumar, N., Jeyaraj, M. & Gurunathan, S. (2010). Optimization and fed-batch production of PHB utilizing dairy waste and sea water as nutrient sources by *Bacillus megatrium* SRKP-3, Bioresource Technology 101: 705-711.

Pantazaki, A.A., Papaneophytou, C.P., Pritsa, A.G., Liakopoulou-Kyriakides, M. & Kyriakidis, D.A. (2009). Production of polyhydroxyalkanoates from whey by *Thermus thermophilus* HB8, *Process Biochemistry* 44: 847-853.

Parrondo, J., Herrero, M., García, L.A. & Díaz, M. (2003). A Note Production of Vinegar from Whey, *The Journal of the The Institute of Brewing&Distilling* 109(4):356–358.

Shen, L. & Patel, M.K. (2008). Life Cycle Assessment of Polysaccharide Materials: A Review, *Journal of Polymers and the Environment*, 16(2): 154-167.

Patel, M.K., Bastioli, C., Marini, L. & Wurdinger, G.E. (2005). Life-cycle Assessment of Bio-based Polymers and Natural Fiber Composites Environmental Assessment of Bio-based Polymers and Natural Fibres, *Biopolymers online* 21-24.

Patwardhan, P.R., Srivastava, A.K. & Luong J.H.T. (1985). Kinetics of ethanol inhibition in alcohol fermentation, *Biochemical Engineering Journal* 20:21–28.

Peters, D. (2006). Carbohydrates for fermentation, *Biotechnology Journal* 1(7-8): 806-814.

Phylipsen, D., Kerssemeeckers, K., Blok, K., Patel, M.K. & de Beer, J. (2002). Clean Technologies in the Materials Sector- Current and Future Environmental Performance of Material Technologies, Report Commissioned by European Commission's Institute for Prospective Technological Studies (IPTS), Ecofys, Department of Science, Technology and Society, Utrecht University, Utrecht, 2002.

Pietrini, M., Roes, L., Patel, M.K. & Chiellini, E. (2007). Comparative Life Cycle Studies on Poly(3-hydroxyalkanoate)-Based Composites as Potential Replacement for Conventional Petrochemical Plastics, *Biomacromolecules* 8: 2210-2218.

Povolo, S. & Casella, S. (2003). Bacterial production of PHA from Lactose and cheese whey permeate, *Macromolecular symposia* 197:1-9.

Povolo, S., Toffano, P., Basaglia, M., Casella, S. (2010). Polyhydroxyalkanoates production by engineered *Cupriavidus necator* from waste material containing lactose, *Bioresource Technology* 101: 7902–7907

Reddy, C.S.K., Ghai, R., Rashmi, K.V.C. (2003). Polyhydroxyalkanoates: an overview, *Bioresource Technology* 87:137–146.

Ren, Q., Grubelnik, A., Hoerler, M., Ruth K., Hartmann, R., Felber, H., Zinn M (2005). Bacterial poly(hydroxyalkanoates) as a source of chiral hydroxyalkanoic acids, *Biomacromolecules* 6:2290–2298.

Roes, L., Marsili, E., Nieuwlaar, E. & Patel M.K. (2007). Environmental and Cost Assessment of a Polypropylene Nanocomposite, *Journal of Polymers and the Environment* 15(3):212-226.

Ruth, K., Grubelnik, A., Hartmann, R., Egli, T., Zinn, M. & Ren, Q. (2007). Efficient Production of (R)-3-Hydroxycarboxylic Acids by Biotechnological Conversion of Polyhydroxyalkanoates and Their Purification, *Biomacromolecules* 8:279-286.

Ryu, H.W., Hahn, S.K., Chang, Y.K. & Chang, H.N. (1997). Production of poly(3-hydroxybutyrate) by high cell density fed-batch culture of *Alcaligenes eutrophus* with phosphate limitation, *Biotechnology and Bioengineering* 55:28-32.

Schultz, N., Chang, L., Hauck, A., Reuss, M. & Syldatk, C. (2006). Microbial production of single-cell protein from deproteinized whey concentrates, *Applied Microbiology and Biotechnology* 69(5): 515-520.

Siso, M.I.G., Ramil, E., Cerdan, M.E. & Freire Picos, M.A. (1996). Respirofermentative metabolism in *Kluyveromyces lactis*: ethanol production and the Crabtree effect, *Enzyme and Microbial Technology* 18:585–91.

Sonnleitner, E., Heinzle, E., Braunegg, G. & Lafferty, R.M. (1979). Formal kinetics of poly-hydroxy butyric acid (PHB) production in *Alcaligenes eutrophus* H16 and Mycolplasma rubra R14 with respect to dissolved oxygen tension in ammonia limited batch cultures, *European Journal of Applied Microbiology and Biotechnology* 7:1–10.

Spielmann, M., Kägi, T., Stadler, P. and Tietje, O. (2004). Life Cycle Inventories of Transport Services. Final report, ecoinvent 2000 No. 14. Swiss Centre for Life Cycle Inventories, Dübendorf, 246.

Steinbuchel, A. (1991). Recent advances in the knowledge of the metabolism of bacterial polyhydroxyalkanoic acids and potential impacts on the production of biodegradable thermoplastics, *Acta Biotechnologica* 11: 419-427.

Steinbüchel, A. & Hein, S. (2001). Biochemical and molecular basis of microbial synthesis of polyhydroxyalkanoates in microorganisms, *Advances in Biochemical Engineering/ Biotechnology* 71:81-123.

Steinbüchel, A. & Valentin, H.E. (1995). Diversity of bacterial polyhydroxyalkanoic acids, *FEMS Microbiology Letters* 128: 219-228.

Sudesh, K. & Iwata, T. (2008). Sustainability of biobased and biodegradable plastics, *Clean* 36(5-6):433-442.

Sun, Z., Ramsay, J.A., Guay, M. & Ramsay, B.A. (2007). Fermentation process development for the production of medium-chain-length poly-3-hydroxyalkanoates, *Applied Microbiology and Biotechnology* 75:475-485.

Tabone, M.D., Cregg, J.J., Beckman, E.J. & Landis, A.E. (2010). Sustainability Metrics: Life Cycle Assessment and Green design in Polymers, *Environmental Science & Technology* 30(20): 8264–8269.

Thilakavathi, M., Basak, T. & Panda, T. (2007). Modelling of enzyme production kinetics. *Applied Microbiology and Biotechnology* 73: 991-1007.

Titz, M., Kettl, K.H., Shahzad, K., Koller, M., Schnitzer, H. & Narodoslawsky, M. (2012). Process Optimization for Efficient Biomediated PHA Production from Animal-Based Waste Streams, *Clean technologies and environmental policy* DOI: 10.1007/s10098-012-0464-7.

Van Aalst-Van Leeuwen, M.A., Pot, M.A., Van Loosdrecht, M.C.M. & Heijnen, J.J. (1997). Kinetic modeling of poly(b-hydroxybutyrate) production and consumption by *Paracoccus pantotrophus* under dynamic substrate supply, *Biotechnology and Bioengineering* 55:773–782.

Van Wegen, R.J., Lee, S.Y. & Middelberg, A.P.J. (2001). Metabolic and kinetic analysis of poly(3-hydroxybutyrate) production by recombinant *Escherichia coli, Biotechnology and Bioengineering* 74:70–80.

Viitanen, M.I., Vasala, A., Neubauer, P. & Alatossava, T. (2003). Cheese whey-induced high-cell-density production of recombinant proteins in *Escherichia coli, Microbial Cell Factories* 2:2.

Viñas, M., Borzacconi, L. & Martínez, J. (1994). Anaerobic treatment of yeast manufacturing wastewater in UASB reactors, *Environmental technology* 15: 79–85.

Viñas, M., Borzacconi, L., Martínez, J., Mallo, M. & Galisteo, M. (1994). Simultaneous degradation of carbohydrates and proteins in anaerobic treatment, *VII International Symposium on Anaerobic Digestion, January 23–27*, Cape Town, South Africa, pp. 345–348.

Visser, D., van Zuvlen, G.A., van Dam, J.C., Eman, M.R., Pröl, A., Ras, C., Wu, L., van Gulik, W.M. & Heijnen, J.J. (2004). Analysis in vivo kinetics in aerobic *Saccharomyces cerevisiae* by application of glucose and ethanol pulses, *Biotechnology and Bioengineering* 20: 157-167.

Wong, H.H., & Lee, S.Y. (1998). Poly (3-hydroxybutyrate) production from whey by high-density cultivation of recombinant *Escherichia coli, Applied Microbiology and Biotechnology* 50 (*1*), 30-33.

Wallner, E., Haage, G., Bona, R., Schellauf, F. & Brauengg, G. (2001). The production of poly-3-hydroxybutyrate-co-3-hydroxyvalerate with *Pseudomona cepacia* ATCC 17759 on various carbon sources, in E. Chiellini et al (ed.), *Biorelated polymers Sustainable Polymer science and Technology* Kluwer Academic / Plenum Publishers, New York, Boston, Dordrecht, London, Moscow 2001, pp. 139-145.

Yellore, V. & Desai, A. (1998). Production of poly-3-hydroxybutyrate from lactose and whey by *Methylobacterium* sp. ZP24, *Letters in Applied Microbiology* 25:391–394.

Young, F.K., James, R., Sheldon, K. & May, W. (1994). Microbial Production of Poly-β-Hydroxybutyric Acid from d-Xylose and Lactose by *Pseudomonas cepacia*. Applied and Environmental Microbiology 60(11): 4195-4198.

Zafar, S. & Owais, M. (2006). Ethanol production from crude whey by *Kluyveromyces marxianus, Biochemical Engineering Journal* 27(3): 295-298.

Zhang, X., Luo, R., Wang, Z., Deng, Y. & Chen, G.Q. (2009). Application of (R)-3-Hydroxyalkanoate Methyl Esters Derived from Microbial Polyhydroxyalkanoates as Novel Biofuels, *Biomacromolecules* 10(4): 707-711.

Zinn, M., Weilenmann, H.U., Hany, R., Schmid, M. & Egli, T. (2003). Tailored synthesis of poly([R]-3-hydroxybutyrate-co-3-hydroxyvalerate) (PHB/HV) in *Ralstonia eutropha* DSM 428, *Acta Biotechnolica* 23:309-316.

Chapter 2

Synthesis of Thermally Stable Polyesters

Hossein Mighani

Additional information is available at the end of the chapter

1. Introduction (Roger, 2003)

Polyesters are defined as polymers containing at least one ester linking group per repeating unit. They can be obtained by a wide range of reactions, the most important being polyesterifications between dibasic acids and diols or their derivatives(Scheme 1).

$$nHOOC - R^1 - COOH + nHO - R^2 - OH \xrightarrow[0c\ Bulk]{160-290} H - (OOC - R^1 - COO - R^2)_n OH + (2n-1)H_2O \uparrow$$

Scheme 1.

Many other reactions have been reported for the synthesis of polyesters, such as reactions between dicarboxylic acid salts and dialkylhalides, reactions between chlorocarbonyloxy-terminated monomers and diacids, or reactions between bisketenes and diols. These reactions, however, cannot be applied to the synthesis of high molar mass polyesters under economically viable conditions and are limited to very specific laboratoryscale syntheses. Two notable exceptions are the ringopening polymerization of lactones and lactides for the production of degradable polyesters and the biosynthesis of aliphatic polyesters by bacteria or genetically modified plants during the last few years, a number of companies have put biodegradable polymers on the market. Almost all these polymers are polyesters or copolyesters: aliphatic polyesters such as poly (ε- caprolactone), poly (butylene-co-ethylene succinate), poly (lactic acid) and microbial poly (hydroxyalkanoic acid), copolyesters of terephthalic acid with various aliphatic diols and diacids and polyesteramides .

1.1. Structure property relationships (Roger, 2003)

The nature- aliphatic or aromatic- of the bivalent - R^1- and - R^2- radicals in polyester chains (Scheme 1) exerts a prefound influence on the properties of polyesters and define four main classes of linear polyesters.

1.1.1. Aliphatic polyesters are low melting (40 - 80 °C) semicrystalline polymers or viscous fluids and present inferior mechanical properties Notable exceptions are poly (α-hydroxyacid)s and poly (β-hydroxyacid)s.

1.1.2. Aromatic–aliphatic polyesters, in which R^1 or R^2 is aromatic, are generally high melting (150-270 °C) semicrystalline materials that find applications as engineering thermoplastics, films, or fibers.

1.1.3. Wholly aromatic polyesters, in which both R^1 and R^2 are aromatic, are either high-T_g amorphous polymers or very high melting semicrystalline polymers that often exhibit liquid crystalline properties.

1.1.4. Polyester thermoplastic elastomers, which are obtained by replacing a part of the R^2 diol by dihydroxy polyether macro monomer, present biphasic morphology and rubberlike properties.

2. History of thermally stable polyesters

Polyester resin is attractive in different ways. This material is being used efficiently as folding resin, blend forming, film, fiber, surface covering, rubber and plasticizer. The common factor in these different materials is that all of them contain some joint in their main chain. Different methods for preparation of thermally stable polyester are shown below.

2.1. Flame retardant polyesters

Kim & Yang (Kim & Yang, 2001, 2003, 2008) used phosphorous compounds and copolymerized together to produce polyesters which are resistant to flame and because of their suitable properties (Scheme 2).

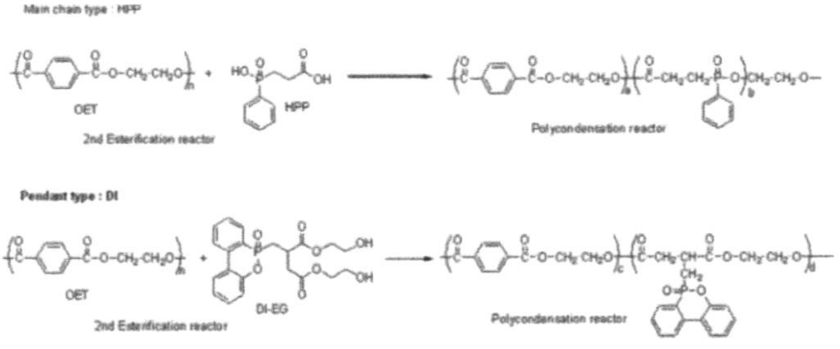

Scheme 2.

2.2. Polyester with alkyl group in the addition chain

Wang et al. (Wang, 2008) produced polyesters in which substituents have many alkyl groups. These polyesters have better solubility than polyesters with one alkyl group; however they are less thermo resistant. Polymer mit C_5 - C_{10} are amorph and more than C_{10} are crystaline, its film is totally sleek (Scheme 3).

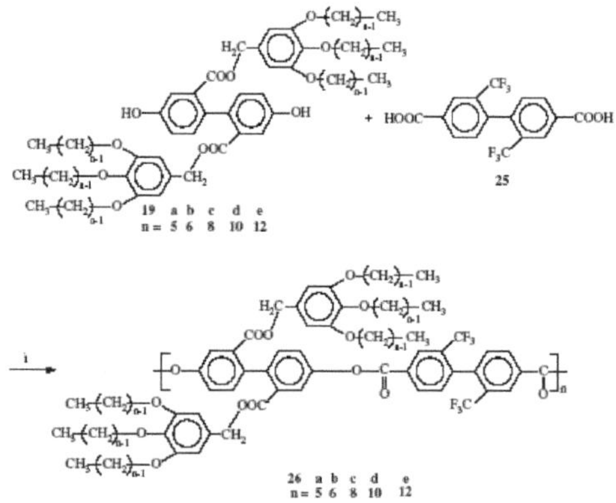

Scheme 3.

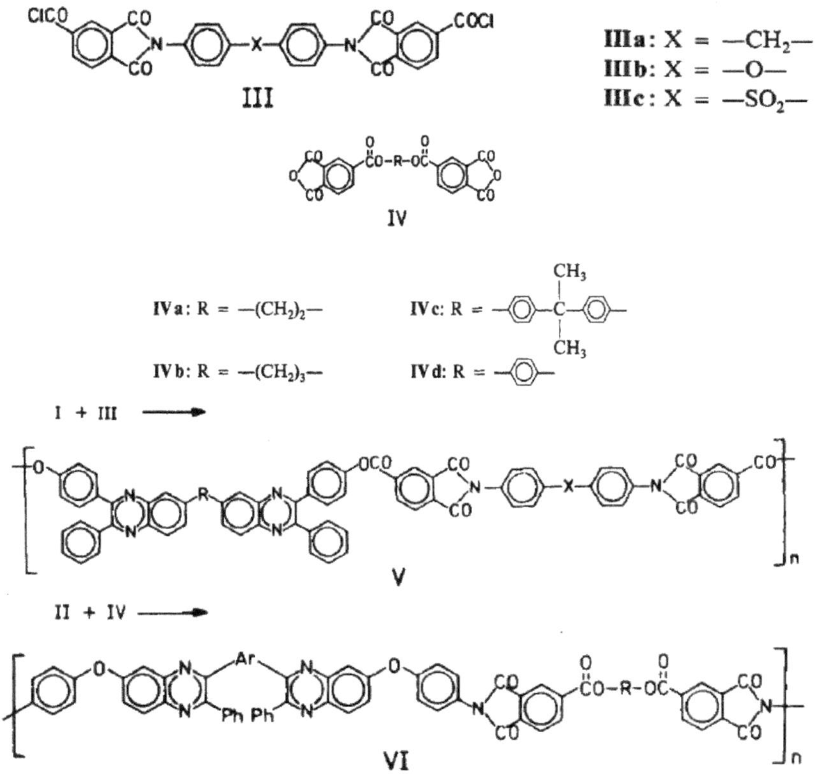

Scheme 4.

2.3. Polyester with quinoxaline ring in the main chain

This type of polyesters (Akutsu et al., 1985, 1987, 1990, 1998, 1999) are produced with heterocyclic groups in their main chain by addition of quinoxaline cycle into the polymer main chain and reaction of quinoxaline hydroxyl compound with diacids and quinoxaline amine compound with dianhydrides. Existence of these groups causes high thermoresistance of the polymer where they will not show any mass depreciatory until 300 °C. They have suitable electrical properties and have capability to form film. These groups are also used in production of pigments (Scheme 4).

2.4. Polyester-imide

This kind of polyester is formed by solution (dilution) condensation polymerization. This polyester has been produced from aromatic diols with aromatic and aliphatic diacidchlorides in high efficiency (Asadi & Shadpour, 2010). These polymers are capable of forming film and semi-cristalline aromatic polyester-imides. The aromatic-aliphatic polyester-imides are amorphous due to their structure (Scheme 5).

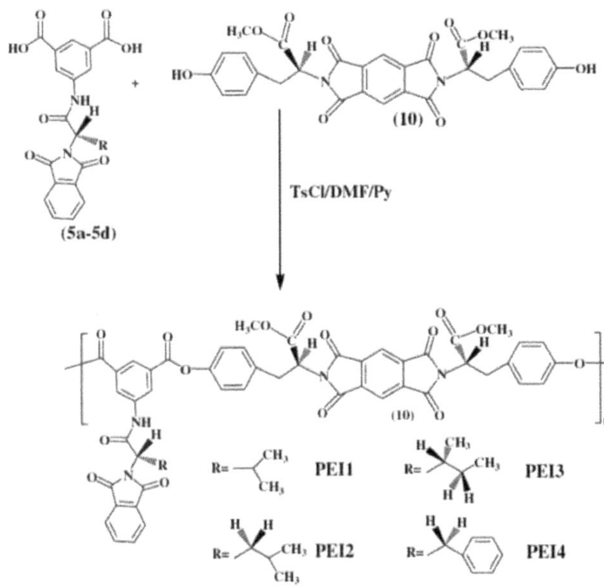

Scheme 5.

2.5. Polyester –Amide – Ester

Polyester-amide-esters are produced by reaction of diols which contain ester-amide-ester groups with aromatic and aliphatic diacidchlorides (Einollahi & Mehdipour, 2004). Totally, aromatic polymers have a minimum thermoresistance and maximum solubility. Films from these polymers are brittle (Scheme 6).

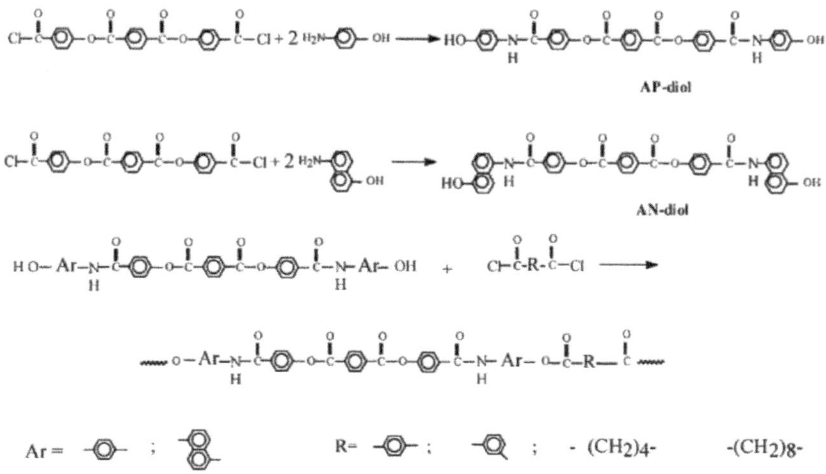

Scheme 6.

2.6. Unsaturated polyester

Aromatic unsaturated polyester is produced by solid state polymerization of glycolized diol (Boehme et al., 2006). Continuous glycolization is done by an extruder that unsaturated bonds appears as conjugated (Scheme 7).

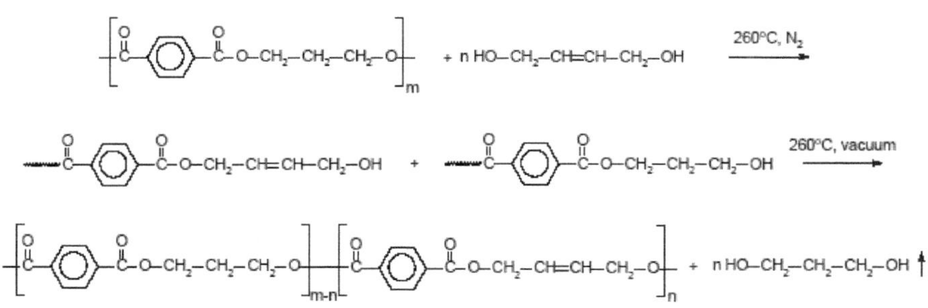

Scheme 7.

2.7. Aromatic polyester-imide

Aromatic polyester-imide is produced by soluble condensation polymerization in high temperature (Behniafar et al., 2005). Pyridine is used as absorbent of produced hydrogen chloride. High solubility in polar solvent such as dimethyformamide and tetrahydrofuran, T_g about 300 °C and mass reduction about 10 % in 450 °C have been proven (Scheme 8).

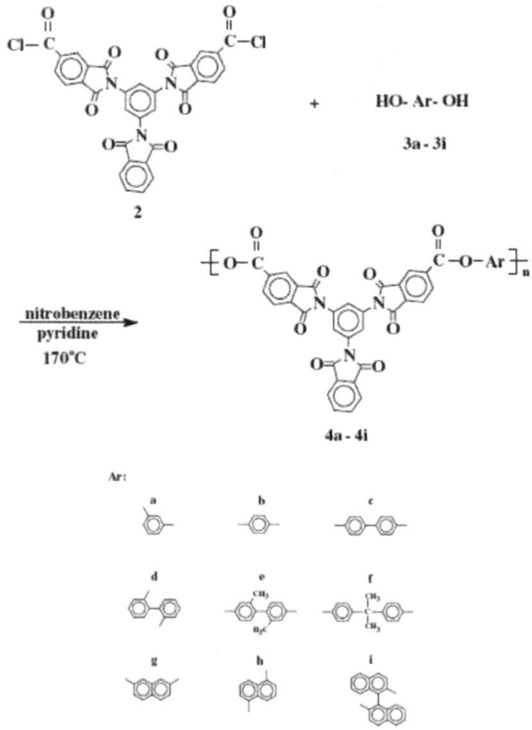

Scheme 8.

2.8. Polyester with sulfur in the main chain

This type of polyesters are produced by from sulfonic diols with sulfure in the main chain with aliphatic diacidchlorides possessing methylene groups from 2 to 10 groups (Hirano, 2004). They have high molecular weight and crystalline structure and also high thermo resistance (Scheme 9).

2.9. Polyester with aryl-ether group

This polyester produced (Chern & Huang, 1998) with solution polycondensation in high temperature. These polyesters have good solubility and high molecular weight. Their 5 % mass reduction temperature is about 400 °C in N_2 atmosphere. Their Young's modulus is about 40 MPa and their elongation at break is about 4 % (Scheme 10).

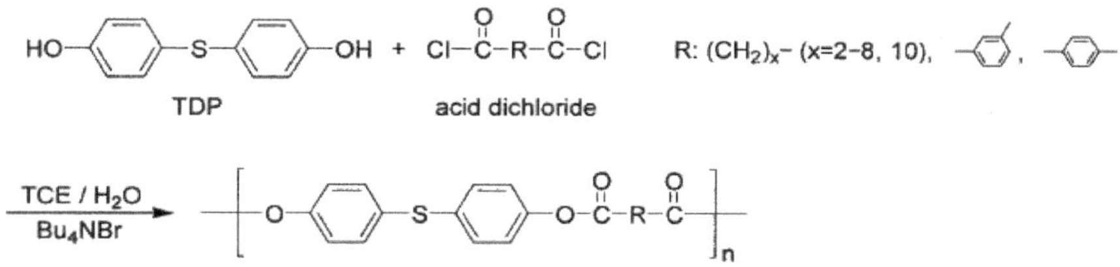

Scheme 9.

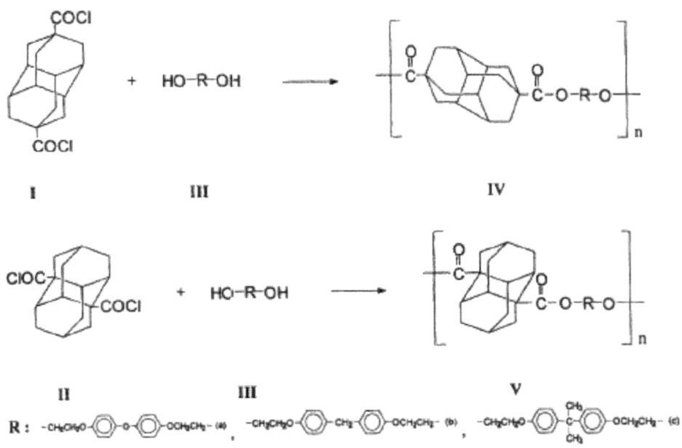

Scheme 10.

2.10. Aromatic fluoro polyester

This polymer produced by Hahn & Zhu (Hahn & Zhu, 2007). This polymer is not soluble in organic solvents while soluble in polar solvents such as dimethylsulfoxide (DMSO), dimethyformamide (DMF) and dimethyacetamide (DMAC). This polymer is crystalline (Scheme 11).

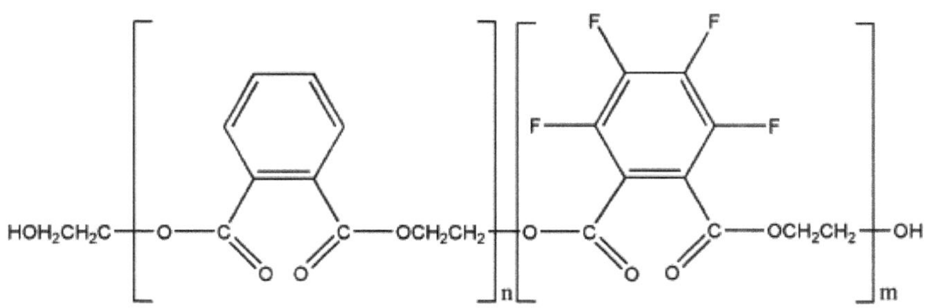

Scheme 11.

2.11. Aliphatic polyester with high molecular weight

This polyester is produced in two steps in bulk polycondensation. These polyesters are among biopolymer and specifically polymers which are among green chemistry family (Fu & Liu, 2008). They have high molecular weight and to achieve this high molecular weight they have to be purified by washing (Scheme 12).

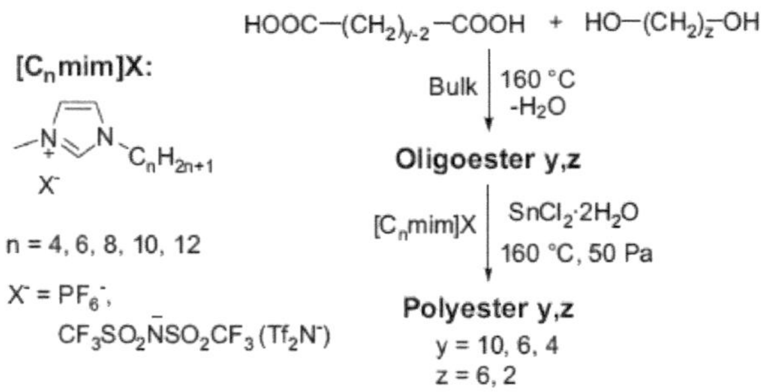

Scheme 12.

3. Experiments

3.1. Synthesis of polyesters with azaquinoxaline ring (Ghaemy et al., 2009)

3.1.1. Synthesis of the monomer

Synthesis of the monomer 2,3-Bis(4-hydroxy phenyl)-5-azaquinoxaline diol (DIOL) took place in two steps: first, the synthesis of 4,4'-dihydroxybenzil (DHB), and second, the reaction of DHB with 2,3 diamino pyridine. Both steps are described in Scheme 13.

3.1.1.1. 4, 4'-Dihydroxybenzil (Moylan et al., 1993)

In a 1 L, three-necked, round-bottom flask equipped with a reflux condenser, a magnetic stirrer bar, and a nitrogen inlet were placed 6 g (22 mmol) of 4,4'- dimethoxybenzil, 60 ml of acetic acid (HOAc), and 300 ml HBr (68%). The suspension was heated at reflux with stirring for 4 h to become a homogeneous yellow solution which was cooled to room temperature, during which a yellow solid was formed. The solid was filtered, washed with cold water several times, recrystallized in HOAc, and dried in vacuum oven. The yellow needles obtained in a yield of 82% (3.3 g), which starts to melt at 244 °C.

3.1.1.2. 2,3-Bis(4-hydroxy phenyl)-5-azaquinoxaline

A 500 ml, round-bottom flask equipped with a magnetic stirrer bar, a reflux condenser, a Dean–Stark trap and a nitrogen inlet was charged with 0.48 g (2 mmol) of 4,4'-dihydroxybenzil, 0.22 g (2 mmol) of 2,3-diaminopyridine, 50 ml of toluene, and 36 ml of deoxygenated acetic acid. The reaction mixture was stirred at reflux for 12 h, during which the generated water was collected in the Dean-Stark trap. The reaction vessel was let to cool to room temperature, and then poured into 500 ml slurry of ice and water containing 7.2 ml of HCl (37%). The precipitate was filtered and recrystallized in ethanol. The dark yellow powder was obtained with a yield of 85% and melting at 324 °C.

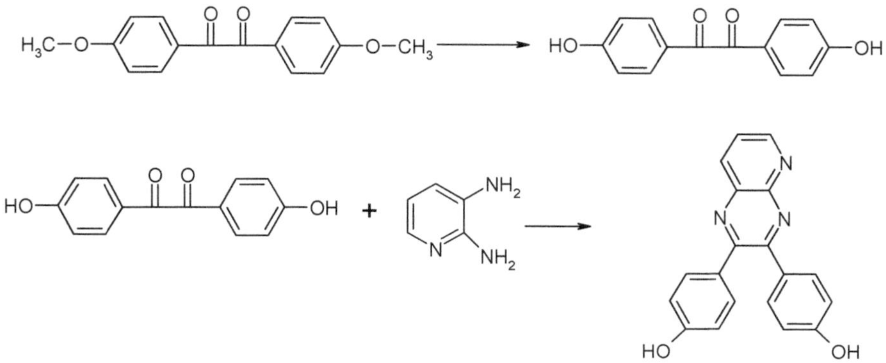

Scheme 13.

3.1.1.3. Synthesis of the model compound

Model compound was prepared from 2,3-bis(4- hydroxyphenyl)-5-azaquinoxaline and benzoyle chloride by using conventional method, Scheme 14. A two-necked round-bottom flask equipped with a magnetic stirrer bar, a reflux condenser and a nitrogen inlet/outlet tube was charged with DIOL (0.314 g, 1 mmol) in 20 ml DMF and 0.8 ml triethylamine. A solution of benzoylechloride (0.244 g, 2 mmol) in 10 ml DMF was added drop wise at 0 °C. The reaction was stirred for 5 h at room temperature. The solution was then poured into water, and the precipitate was filtered and washed several times with the solution of $NaHCO_3$. The solid product was then dried in a vacuum oven at 60 °C (Scheme 14).

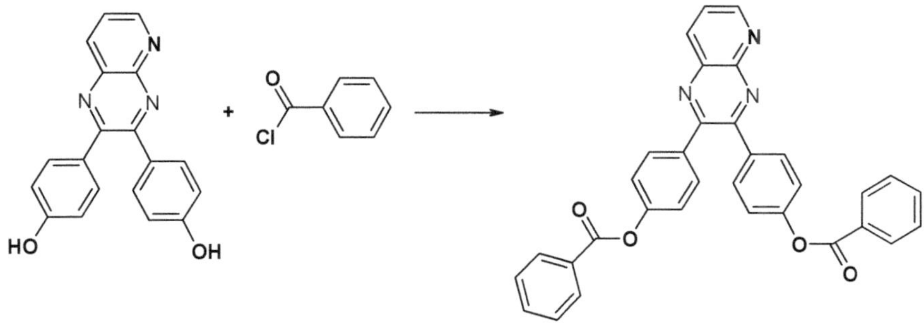

Scheme 14.

3.1.1.4. Synthesis of polyesters

Polyesters were prepared from polycondensation of 2,3-bis(4-hydroxyphenyl)-5-azaquinoxaline with aromatic and aliphatic dicarboxylic dichlorides such as terephthaloyle dichloride, isophthaloyle dichloride, sebacoyle dichloride and adipoyle dichloride. A typical synthesis procedure for the preparation of polyesters was conducted in three-necked, round-bottom flask equipped with a nitrogen inlet, a condenser and a magnetic stirrer bar. The flask was charged with the diol (0.628 g, 2 mmol) in 20 ml DMF and 0.8 ml triethylamine. A solution of terephthaloyle dichloride (0.406 g, 2 mmol) in 10 ml DMF was added drop wise at 0 °C. The reaction mixture was stirred for 5 h at room temperature. The solution was then poured into water and the precipitate was filtered and washed several times with the solution of $NaHCO_3$. The solid product was then dried in a vacuum oven at 60 °C. The synthesis procedure and the polymer designations are shown in Scheme 15.

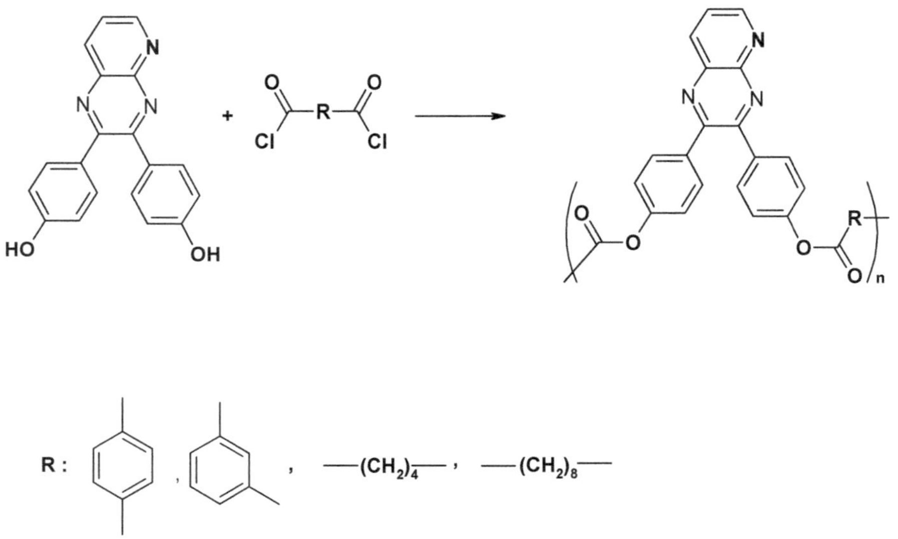

Scheme 15.

3.2. Synthesis of polyesters with urazole ring (Mighani et al., 2011)

3.2.1. Preparation of monomers

3.2.1.1. 4-Nitrobenzoylchoride (Mallakpour & Nasr, 2002)

A 100 ml flask was charged with a mixture of 4-nitrobenzoic acid. (5.00 g, 29,9 mmol), 5 ml thionyl chloride and 20 ml ethylacetate. Subsequently the mixture was stirred at reflux temperature for 2 hours to obtain a transparent solution. Next the additional content of thionyl chloride and solvent was extracted out of solution by using distillation and was poured into cold water. The yellow powder produced was separated by filtration. Then the product dried in vacuum oven at 50 °C. A purified sample was obtained by recrystallization form carbontetrachloride and obtained in 95% yield (5.436 g) with the melting point of 71-72 °C.

3.2.1.2. 4.Nitrobenzoylazide (Mallakpour & Nasr, 2002)

A 100 ml flask was charged with 4-introbenzoylchloride (5.43 g, 29.2 mmol) and 10 ml acetone. The mixture was subsequently stirred at 5 ^{0}C temperature and then added dropwise of a sodiumazide solution (1.69 g, 30.1 mmol) in 7 ml water for 30 minutes. The solution was stirred for additional 1 hour and the white precipitate was filtered and dried at air. A purified sample was obtained in a 90% yield (4.95 g) with the melting point of 73-75 ^{0}C.

3.2.1.3. 1-Ethoxycarbonly-4-(4-introphenyl) semicarbazide (Mallakpour & Nasr, 2002)

A 250 ml flask was charged with 4-nitrobenzoylazide (4.00 g, 21 mmol) and 75 ml dried toluene. The mixture was subsequently stirred under N_2 at reflux temperature for 6 hours. The solutions was cooled and filtered. The solution was cooled at 5 ^{0}C and then charged dropwise with a mixture of ethylhydrazin carboxilate (ethyl carbazate) (2.70 g, 21 mmol) and 40 ml dried toluene in 15 minutes. The solution was stirred for 30 minutes in ice bed and for 1 hour at room temperature. The solution refluxed for 3 hours, cooled, filtered and dried at 70 ^{0}C for one day. A purified sample was obtained in a 93% yield (4.8 g) with the melting point of 219-220 ^{0}C.

3.2.1.4. 1-Ethoxycarbonyl-4-(4-aminophenyl)semicarbazide (Mallakpour & Nasr, 2002)

A 250 ml flask was charged with 1-ethoxycarbonyl- 4- (4-nitrophenyl) semicarbazide (3.04 g, 11.3 mmol), $SnCl_2.2H_2O$(9.37 g, 4.3 mmol) and 15 ml ethanol. The mixture was subsequently stirred under N_2 at reflux temperature for 4.5 hourse to obtain a transparent solution. The solution was cooled with ice bed and water and stirred for 15 minutes. The pH of solution was raised to 10 with a solution of sodium hydroxide 30%. The mixture was put in 30 ml ethylacetate. Afterward the white powder was filtered and dried at 70 ^{0}C for one day. A purified sample was obtained in a 82% yield (2.2 g) with the m.p. >340 ^{0}C.

3.2.1.5. 4-(4-aminophenyl)-1,2,4-triazolidyne -3.5-dione (Urazole) (Mallakpour & Nasr, 2002)

A 100 ml round flask was charged with pure sodium (0.23 g, 10 mmol) dissolved in 14 ml absolute ethanol. The solution was under N_2 and added 1-ethoxycarbonyl -4-(4-aminophenyl) semicarbazide (2 g, 8.4 mmol). The mixture was subsequently stirred under N_2 at reflux temperature for 4.5 hours and cooled with ice bed. The solution was neutralized with HCl 30%. The with product was filtered (1.38 g, 86%), recrystallized with hot water and dried. A purified sample was obtained in a 86% yield (1.38 g) with the melting point 270-273 ^{0}C.

3.2.1.6. 4-(4- phthalimidophenyl) -1, 2, 4-triazolidyne -3, 5-dione (Mallakpour & Nasr, 2002)

A 25 ml round flask was charged with 4-(4-aminophenyl) urazole (0.2 g, 1.04 mmol) and a mixture of solvents acetic acid/ pyridine (3/2) and anhydride phthalic (0.15 g, 1.04 mmol). The mixture was subsequently stirred at room temperature for 2 days till the amic acid was produced as a white precipitate. The solution was stirred at reflux temperature for 8 hours and then filtered and the white precipitate washed with ethanol, filtered and dried. The white purified sample (1.38 g) was obtained in a 76% yield (0.46 g) with the melting point of 363-365 ^{0}C (Scheme 16).

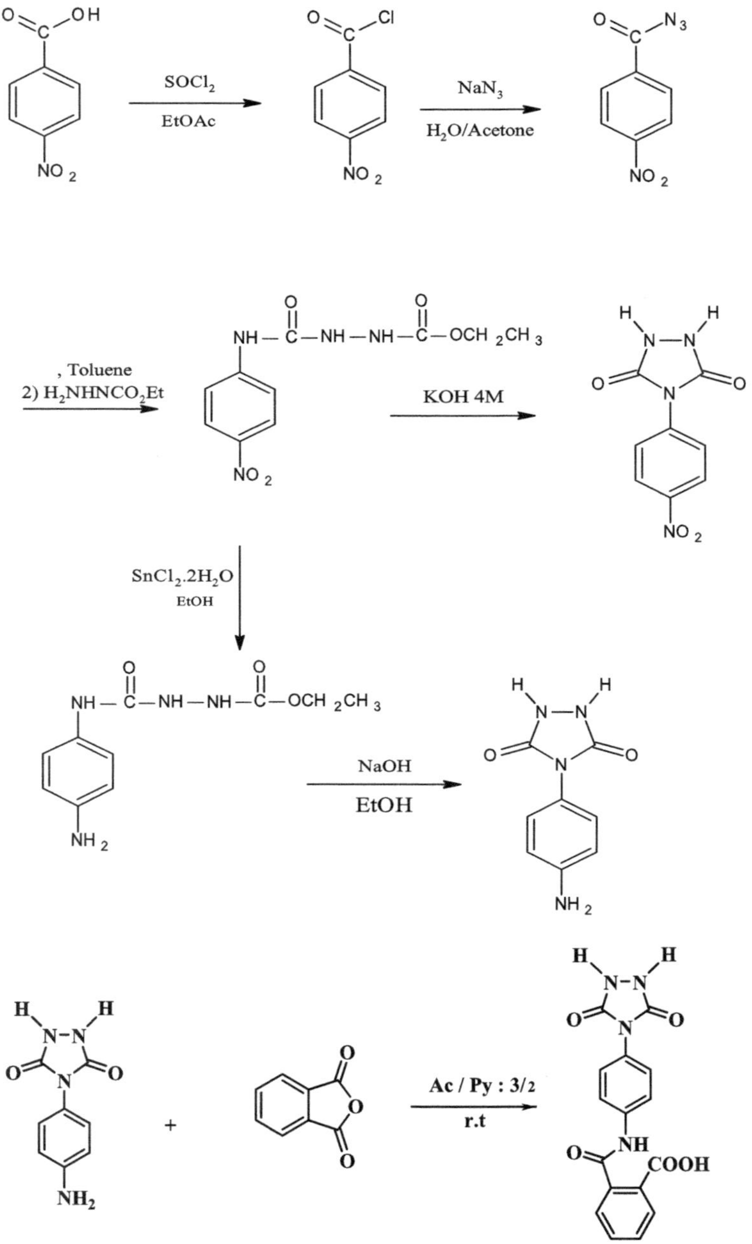
SOCl2
EtOAc
NaN3
H2O/Acetone
NO 2
, Toluene
2) H2NHNCO2Et
KOH 4M
SnCl2.2H2O
EtOH
NaOH
EtOH
NH 2
Ac / Py : 3/2
r.t
COOH

Scheme 16.

3.2.1.7. 4)-4-phthalimidophenyl)-1,2,4-triazolidyne-3,5-diamine

A 10 ml round flask was charged with 4-(4-phthalimidophenyl)-1,2,4-triazolidyne-3,5dione(2.5 g, 7.7737 mmol), and hexamethylenediisocyanate (0.2615 g, 1.555 mmol) and 3 ml dimethlformamide. The mixture was subsequently stirred at room temperature for 2 days and the precipitate was obtained in a Solution with low concentration of water/acid and filtered and dried. The purified sample was obtained in a 85% yield with the melting point 124 – 125 °C (Scheme 17).

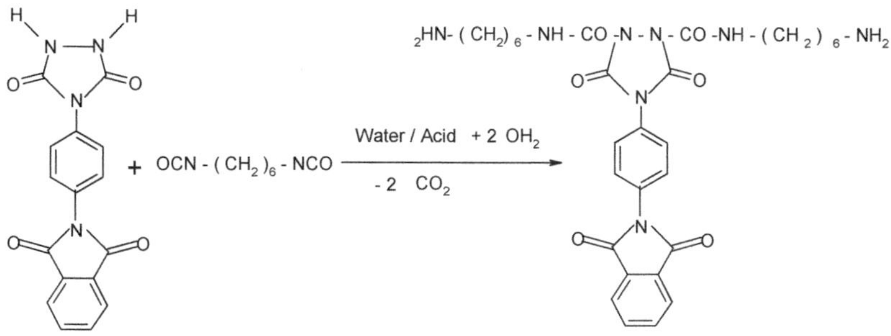

Scheme 17.

3.2.1.8. 4)-4-phthalimidophenyl)-1,2,4-triazolidyne-3,5-diol

To a 250 ml. three –necked flask equipped with a mechanical stirrer and an addition funnel were added 4-(4-phthalimidophenyl)-1,2,4-triazolidyne-3,5-diamine (6.06 g, 0.0100 mol), concentrated hydrochloric acid (22 ml), and water (100 ml). The solution was cooled to 0 °C in an ice bath. A solution of sodium nitrite (1.45 g, 0.0210 mmol) in water (10 ml) was added at 0 - 5 °C. the resulting solution was poured into a cold solution of phosphoric acid (20 ml) in water (1.8 1 l). After stirring for 5 min. The organe mixture was heated at the boiling point for 10 min. The mixture was cooled and extracted with diethyl ether (3 x 200 ml). The combined organic phase was extracted with 2 N sodium hydroxide solution (2 x 50 ml).The combined aqueous phase was acidified with concentrated hydrochloric acid, extracted with diethyl ether (3 x 100 ml) and dried with magnesium sulfate (Scheme 18).

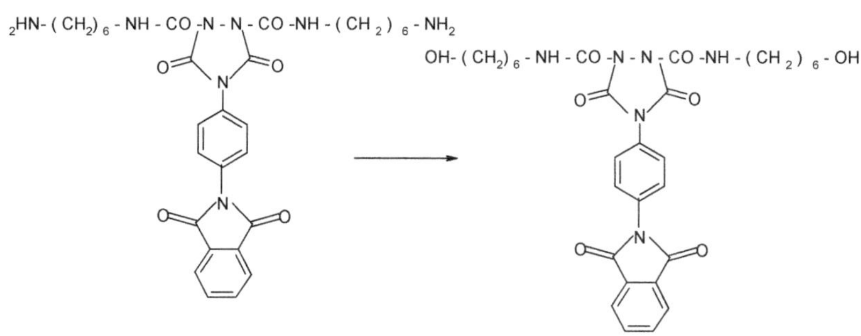

Scheme 18.

3.3. Synthesis of polyesters with thiosemicarbazid ring (Ghaemy et al., 2008)

3.3.1. Preparation of monomers

3.3.1.1. Benilbisthisosemicarbazone diamine LH6 (Arquero et al., 1998)

The thiosemicabazide (3.64 g, 40.30 mmol) was dissolved in 40 ml of 2N HCl and 1ml of conc. HCl and then added to a suspension of benzyl (4,24 g, 20.20 mmol) in 50 ml of methanol and a few drops of conc. HCl. The mixture was stirred for 6 h at room temperature. The yellow solid was filtered off, washed with methanol and dried in a vacuum oven at 70 °C for 2 h. A yellow solid product was obtained in a 75% yield which starts to melt and decompose at 240 ⁰C (Scheme 19).

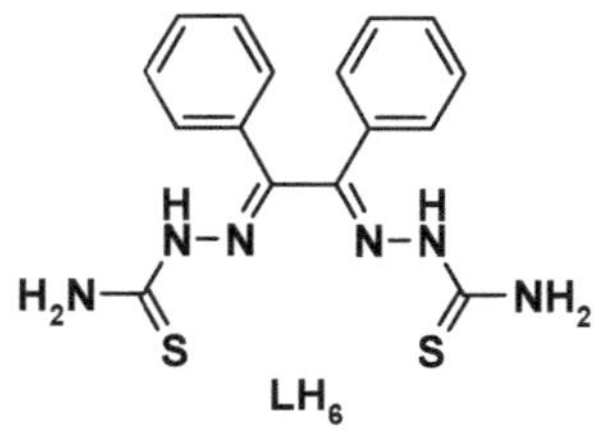

Scheme 19.

3.3.1.2. Benilbisthisosemicarbazone diol

To a 250 ml. three –necked flask equipped with a mechanical stirrer and an addition funnel were added Benilbisthisosemicarbazone LH6 (3.56 g, 0.0100 mol), concentrated hydrochloric acid (22 ml) and water (100 ml). The solution was cooled to 0 °C in an ice bath. A solution of sodium nitrite (1.45 g, 0.0210 mmol) in water (10 ml) was added at 0-5 °C. The resulting solution was poured into a cold solution of phosphoric acid (20 ml) in water (1.8 l). After stirring for 5 min. The organe mixture was heated at the boiling point for 10 min. the mixture was cooled and extracted with diethyl ether (3 x 200 ml). The combined organic phase was extracted with 2 N sodium hydroxide solution (2 x 50 ml). The combined aqueous phase was acidified with concentrated hydrochloric acid , extracted with diethyl ether (3 x 100 ml) and dried with magnesium sulfate (Scheme 20).

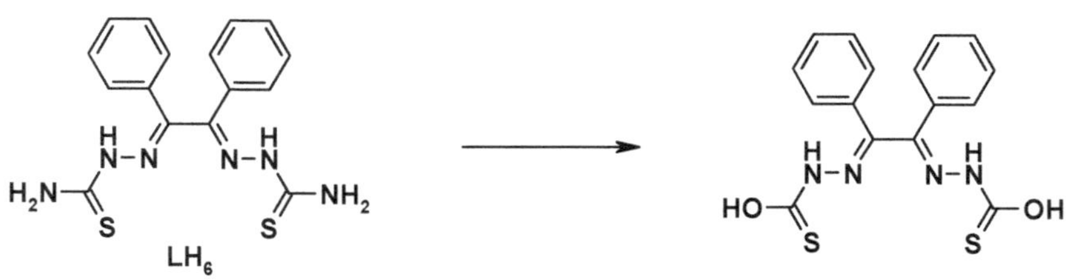

Scheme 20.

Author details

Hossein Mighani
Uni of Golestan, Iran

4. References

Aciernoa, D.; Fresab, R.;Iannellib, P. & Vacca, P. (2000) segmented liquid-cristalline polyesters with allyl groups as lateral substituents, synthesis and characterization, *Polymer*, 41, pp. 4179-4187, ISSN 0032-3861

Akutsu, F.; Inoki, M.; Sunouchi, K.; Sugama, Y.; Kasashima, Y.; Naruchi K. & Miura, M. (1998) synthesis and properties of novel polyamides containing 2-methyl-4,5-oxazoldiyl structure, *Polymer*, 39, pp. 1637-1641, ISSN 0032-3861

Akutsu, F.; Hirata, H.; Hayashi, H.; Miura, M. & Nagakubo K. (1985) Synthesis of polyamides containing α-diketone moieties and 2,3-quinoxalinediyl units in the main chain, *Macromol.chem.rapid commun* 6,pp. 215-218, ISSN 1521-3927

Akutsu, F.; Inoki, M.; Inagawa, T.; Kasashima, Y.; Sonoda, Y. & Marushima, K. (1999) Metal adhesive properties of polyamides having the 2,3-bis(1,4-phenylene)quinoxalinediyl structure, *Journal of Appl Polym Science*, 74, pp. 1366-1370, ISSN 1097-4628

Akutsu, F.; Kuze, S.; Matsuo, K.; Naruchi, K. & Miura, M. (1990) Preparation of polyimides from 2,3-bis(4-aminophenyl)quinoxalines and aromatic tetracarboxylic dianhydrides, *Macromol. chem. Rapid commun.*, 11, pp. 673-677, ISSN 1521-3927

Akutsu, F.; Hayashi, H.; Miura, M. & Nagakubo, K. (1985) synthesis of polyamides having 2,3-benzo[g]quinoxalindiyl units in the main chain, *Macromol. Chem. rapid commun.*, 6, pp. 475-479, ISSN 1521-3927

Akutsu, F.; Narochi, K.; Miura, M. & K. Nagakubo, K. (1987) polyesters with 2,3-quinoxalinediyl units in the main chain, *Macromol. Chem. rapid commun.*,8, pp. 113-117, ISSN 1521-3927

Arquero, A.; Cafiadas, M.; Martinez-Ripoll, M.; Mendiola, A. & Rodriguez, A. (1998) Selective access to new semicarbazone and thiosemicarbazone drived from benzyl. Study of their conversion reactions, *Tetrahedron*, 54, pp. 11271-11284, ISSN 0040-4020

Bruma, M.; Schulz, B.; Koepnick, T.; Stiller, B.; Belomoina, N. & Mercer, F. (1999) Synthesis and study of aromatic polyamides containing silicon and phenylquinoxaline ring in the main chain, *Europ.polymer journal*, 35, pp. 1253-1260, ISSN 0014-3057

Behniafar, H.; Akhlaghinia, B. & habibian, S. (2005) synthesis and characterization of new soluble and thermally stable poly(ester-imide)s drived from N-[3,5-bis(trimellitoylphenyl]phthalimide and various bisphenols, *European polymer journal*, 41, pp. 1071-1078, ISSN 0014-3057

Boehme, F.; Komber, H. & Jafari, S.H. (2006) synthesis and characterization of a novel unsaturated polyester based on poly(trimethylene trephthalate), *polymer*, 47, pp. 1892-1898, ISSN 0032-3861

Bruma. M.; Hamciuc, S.E.; Hamciuc, C.; Belomoina, N.M. & Krongiuz, E.S. (1992) Thermostable polyesters containing imide and phenylquinoxaline units , *Die Angewandte Makromolekulare Chemie*, 194, pp. 179-187, ISSN 1522-9505

Chern, Y.T. & Huang, C.M. (1998) synthesis and characterization of new polyesters derived from 1,6- or 4,9-diamantanedicarboxylic acyl chlorides with aryl ether, *polymer*, 39, pp. 2325-2329, ISSN 0032-3861

Fu, C. & Liu, Z. (2008) synthesis of high molecular weight aliphatic polyesters in 1-alkyl-3-methylimidazolium ionic liquids, *polymer*, 49, pp. 461-466, ISSN 0032-3861

Ghaemy, M.; Mighani, H. & Ziaei, P. (2009) Synthesis and characterization of novel organosoluble polyesters based on a DIOL with azaquinoxaline ring, *Journal of Applied Polymer Science*, 114, pp. 3458-3463, ISSN 1097-4628

Ghaemy, M.; Mighani, H. & Behmadi H. (2008) synthesis and characterization of novel Schiff-base polyamides from copper/benzilbisthiosemicarbazone complexes, *Journal of Applied Polymer Science*, 109, pp. 2388-2394, ISSN 1097-4628

Hirano, H.; Watase, S. & Tanaka, M. (2004) Linear polymers with sulfur in the main chain. II. Synthesis of polyesters by interfacial polycondensation of bis(4,4'-hydroxyphenyl)sulfide with several aliphatic acid dichlorides and their properties , *Journal of applied polymer science*, 91, pp. 1865-1872, ISSN 1097-4628

Mallakpour, S. & Asadi, P. (2010) Novel chiral poly(ester-imide)s with different natural amino acids in the main chain as well as in the side chain: synthesis and characterization, *Colloid polymer science*, 288, pp. 1341-1349, ISSN 0303-402X

Mehdipour-Ataei, S. & Einollahy, P. (2004) Novel diols containing ester and amide groups and resulting poly9ester-amide-ester)s, *Journal of applied Polymer Science*, 93, pp. 2699-2703, ISSN 1097-4628

Mighani, H.; Nasr Isfahani, H. & Mighani, S. (2011) Novel organosoluble polyamides with phthalimide pendant group:Synthesis, physical and thermal characterization, *macromolecules , an Indian journal*,7, pp. 78-84, ISSN 0974-7478

Moylan, C.R.; Miller, R.D.; Twieg, R.J.; Betterton, K.M.; Lee, V.Y.; Matray, T.J. & Nguyen, C. (1993) Synthesis and nonlinear optical properties of donor-acceptor substituted triaryl azole derivatives, *Chem Mater*, 5, pp. 1499-1508, ISSN 0897-4756

Rogers, M.E. & Long, T.E. (2003) synthetic Methods in Step-Growth polymers, John Wiley & Sons. Inc, ISBN:0.471-38769-X, New Jersey, USA

Shadpour-Mallakpour, E. & Nasr-Isfahani, H. (2002) Synthesis of Novel Azo dyes derived from 4-Phenylurazole, *Indian Journal of Chemistry*, 41B, pp. 169-174, ISSN: 0376-4699

Wang, D.H.; Cheng, S.Z.D. & Harris, F.W. (2008) synthesis and characterization of aromatic polyesters containing multiple n-alkyl side chain, *Polymer*, 49, pp. 3020-3028 , ISSN 0032-3861

Yang, C.P.; Chen, R.S. & Yu, C.W. (2001) Preparation and characterization of organosoluble polyimides based on 1,1-bis[4-(3,4-aminophenoxy)phenyl]cyclohexane and commercial aromatic dianhydrides, *Journal of Applied Polymer Science*, 82, pp. 2750-2759, ISSN 1097-4628

Yang, C.P. & Su,Y.Y.(2003) properties of organosoluble aromatic polyimides from 3'-trifluoromethyl-3,4'-oxydianiline, *polymer*, 44, pp. 6311-6322, ISSN 0032-3861

Yang, S.C. & Kim, J.P. (2008) Flame retardant polyesters.III.Fibers, *Journal of Applied Polymer Science*, 108, pp. 2297-2300, ISSN 1097-4628

Zhu, Q. & Han, C.C. (2007) synthesis and crystallization behaviors of highly fluorinated aromatic polyesters, *polymer*, 48, pp. 3624-3631, ISSN 0032-3861

Chapter 3

Degradation and Recyclability of Poly (Ethylene Terephthalate)

S. Venkatachalam, Shilpa G. Nayak, Jayprakash V. Labde,
Prashant R. Gharal, Krishna Rao and Anil K. Kelkar

Additional information is available at the end of the chapter

1. Introduction

The physical and chemical properties of polymers depend on the nature, arrangement of chemical groups of their composition and the magnitude of intra or intermolecular forces i.e primary and secondary valence bonds present in the polymer. Degradation process occurs due to the influence of thermal, chemical, mechanical, radiative and biochemical factors occurring over a period of time resulting in deterioration of mechanical properties and colour of polymers. The degradation occurs due to changes accompanying with the main backbone or side groups of the polymer. Degradation is a chemical process which affects not only the chemical composition of the polymer but also the physical parameters such as colour of the polymer, chain conformation, molecular weight, molecular weight distribution, crystallinity, chain flexibility, cross-linking and branching. The nature of weak links and end groups in the polymers contribute to stability of polymers. The degradation process is initiated at the terminal units with subsequent depolymerization. For example paraformaldehyde with hydroxyl terminal starts to degrade at about 170°C whereas the same polymer with acetyl terminals decomposes at about 200°C [1]. Replacement of carbon main chain with hetero atoms like P, N, B increases the thermal stability e.g. PON polymers containing phosphorus, oxygen, nitrogen and silicon.

The exposure of polymeric materials to environmental factors over a period of time will lead to deterioration of physical, chemical, thermal and electrical properties. The degree or measurability of deterioration of these properties depends on the extent of degradation, nature of chemical processes involved during the degradation. Degradation can be considered as a type of modification of polymer chain that may involve the main chain back bone or the side chain or groups. The modification could involve rupture of primary valence bonds leading to lowering of molecular weight, crosslinking, cyclisation and thus this type of cleavage could be irreversible. There could be degradation involving secondary valence bonds within the polymer chain e.g. hydrogen bonds in proteins which are influenced by heat, pH changes , chemical agents etc. and such type of degradative process could be reversible. Since degradation is a chemical process it affects not only the chemical composition of the polymer but also various physical parameters such as chain conformation, molecular weight, molecular weight distribution, crystallinity, chain flexibility, crosslinking, branching, colour of the polymer, haziness etc. Different polyester compositions exhibit wide variation in their response to degradative agents depending on the nature of the repeating units, chemical composition and structure e.g branching, size, shape, crystallinity.

2. Poly (ethylene terephthalate)

Poly (ethylene terephthalate) known by the trade names Mylar, Decron , terylene, Recron, has high crystalline melting temperature (260°C), and the stiff polymer chains in the PET polymer imparts high mechanical strength, toughness and fatigue resistance up to 150-175°C as well as good chemical, hydrolytic and solvent resistance. Poly (ethylene terephthalate) fiber has a very outstanding crease resistance, good abrasion resistance and can be treated with cross-linking resin to impart permanent wash and wear properties [2-4]. The fiber can be blended with cotton and other cellulosic fibers to give better feel and moisture permeation. Thus the fiber is used for applications such as wearing apparel, curtain, upholstery, thread, tire cord filaments, industrial fibers and fabric for industrial filtration.

The polymer is also used for making blow molded bottles for soft drinks, beers, spirits, other food-products and pharmaceuticals. This is because of the outstanding barrier properties of poly(ethylene terephthalate) . The film applications include photographic, magnetic, X-ray films or tapes , metalized films and electrical insulation. PET also finds use as an engineering plastic where it replaces steel, aluminum and other metals in the manufacture of precision moldings for electrical and electronic devices, domestic and office appliances and automobile parts. In the engineering applications the polymer is reinforced with glass fiber or compounded with silicones, graphite or Teflon to improve strength and rigidity. The polymer reinforced with glass fiber are rated for continuous use at temperatures up to 145-155°C. The properties and usefulness of the final polymer depends on controlling its structure by control of the process parameters during the polymerization and subsequent processing of the product.

The polymer is generally obtained by melt –phase polymerization to get resins of inherent viscosity in the range of 0.5-0.7 dL/g. For getting polymer with higher molecular weight i.e. inherent viscosity greater than 0.7 dL/g, solid state polymerization (SSP) is carried out. The latter process involves heating of solid low molecular weight melt-phase polymer below its melting point but above its glass transition temperature (Tg). It is very difficult to polymerize higher IV polymer in the melt-phase because of the thermal degradation reactions occurring simultaneously and competing with the poly condensation reactions.

Polyethylene terephthalate (PET) is a semi crystalline polymer possessing excellent chemical resistance, melt mobility and spinnability [2-4]. The polymer is composed of repeating units as shown in the Figure 1. Each unit having a physical length of about 1.09 nm and a molecular weight of ~200. When it is produced from the reaction of terephthalic acid and ethylene glycol, it is capped on the left by H- and the right by –OH. Polymerization is thus accompanied by the production of water which is removed under elevated temperature and vacuum. Accordingly the presence of water in the molten state will rapidly depolymerize the structure so that thorough drying of the polymer prior to melt spinning of fibers is necessary.

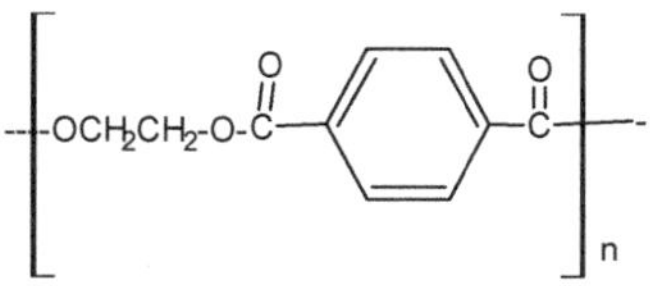

Figure 1. Structure of PET

The aromatic ring coupled with short aliphatic chain makes the polymer a stiff molecule as compared to other aliphatic polymers such as polyolefin or polyamide. The lack of segmental mobility in the polymer chains results in relatively high thermal stability. A textile grade polymer will have an average number of 100 repeat units per molecule so that the extended length of the typical polymer chain is about 100nm with a molecular weight about 20,000. Higher levels of polymerization produces higher strength fibers but the melt viscosity and stability of the melt to even tiny amounts of moisture causes hydrolytic degradation. The measurement of average degree of polymerization is done either by molten viscosity (by measuring the pressure drop through a calibrated orifice) or the viscosity of the diluted polymer in an appropriate solvent [3]. The latter is measure of polymer chain length known as Intrinsic viscosity or IV and the value for a typical fiber grade polymer is 0.6 dl/g in 60/40 w/w mixture of phenol and tetrachloroethane solvent [2-4]. The IV in the latter solvent is related to M_v (Viscosity average molecular weight) of the polymer by the Mark Howink equation (Equation 1) .

$$[\eta] = 7.44 * 10^{-4} (M_v)^{0.648} \quad (1)$$

It is very difficult to polymerize higher IV polymer in the melt-phase because of the thermal degradation reactions occurring simultaneously and competing with the poly condensation reactions. Thus during processing the polymer is subjected to temperatures in the range 280-300°C, which results in various types of degradations. The main degradations that can occur include thermal degradation, oxidative degradation and hydrolytic degradation. Radiation induced or photo degradation leading to free radical reactions and enzymatic catalysed reactions leading to logical degradation are also possible. In addition to these there can be chemical degradation reaction of polyester initiated by specific chemicals like glycol, ammonia or amines or other such reagents. Besides these there can be weathering ageing which could be the combined effect of exposure to temperature, moisture, chemical, UV and visible light and other conditions such as exposure to grease, oil. Polyester can also undergo stress induced degradation reactions when subjected to mechanical stress. The degradation of polyester can lead to several changes in the articles made out of the polymer. These changes include discoloration, chain scissions resulting in reduced molecular weight, formation of acetaldehyde and cross-links or gel formation and fish-eye formation in films. The thermal and thermo oxidative results in poor processibility and performance characteristics in the products. Discoloration is due to the formation of various chromophoric systems following prolonged thermal treatment at elevated temperatures. This becomes a problem when the optical requirements of the polymer are very high, such as in packaging applications.

The initial stage of thermal degradation is a random scission of the in-chain ester linkage resulting in formation of a vinyl ester and carboxyl end groups. Transesterification of the vinyl ester then occurs to give the vinyl alcohol, which is transformed immediately to acetaldehyde. The polyester chain is thus regenerated and an average degree of polymerization maintained. The net result of such a reaction is the replacement of hydroxyl end-groups by carboxylic acid end- groups, producing in the process an equivalent amount of acetaldehyde. Hydrogen atom abstraction may also occur to some extent when impurities in the polymer generate macro-radical sites. These will react with oxygen, producing peroxy radicals and subsequently hydroperoxides, which are themselves thermally and photochemically unstable and will induce further breakdown. The presence of moisture and acid/alkaline impurities will affect hydrolysis. Polyethylene terephthalate is an essentially hydrophobic polymer and hence rate of hydrolysis is thought to be determined by the nature of its chain ends. An increase in the carboxyl end-group concentration will increase the rate of hydrolysis of the polymer. Thermal degradation is degradation induced by elevated temperatures in the absence of oxygen. The chemistry of thermal degradation is different from hydrolytic degradation [6].

Thermal degradation results in different types of end-groups on the polymer chain. The initial step is the scission of the chain of the ester linkage resulting in a decrease in molecular weight either through random scission at the ester linkages or through chain ends and an increase in carboxyl end-groups. The methylene group which is located at the β position to the carbonyl group is the main point where the decomposition process is initiated. The main side reaction occurring is the β scission of the ester linkage results in the vinyl terminated carboxylate unit and carboxyl-terminated units as shown in Scheme 1.

The formation of acetaldehyde is explained generally through a Mc Lafferty rearrangement involving a six membered transition state occurring through inter or intra molecular hydrogen shifts in the methylene group which is located at the β position to the carbonyl group , as shown below [7] :

Apart from aldehyde CO, CO_2, ethylene , benzene, biphenyl are also identified as degradation products and the degradation products are analyzed by sub-ambient thermal volatilization analysis (SATVA) and pyrolysis gas chromatographic methods. Mechanism of ethylene formation is explained by the Scheme 3 [8-11].

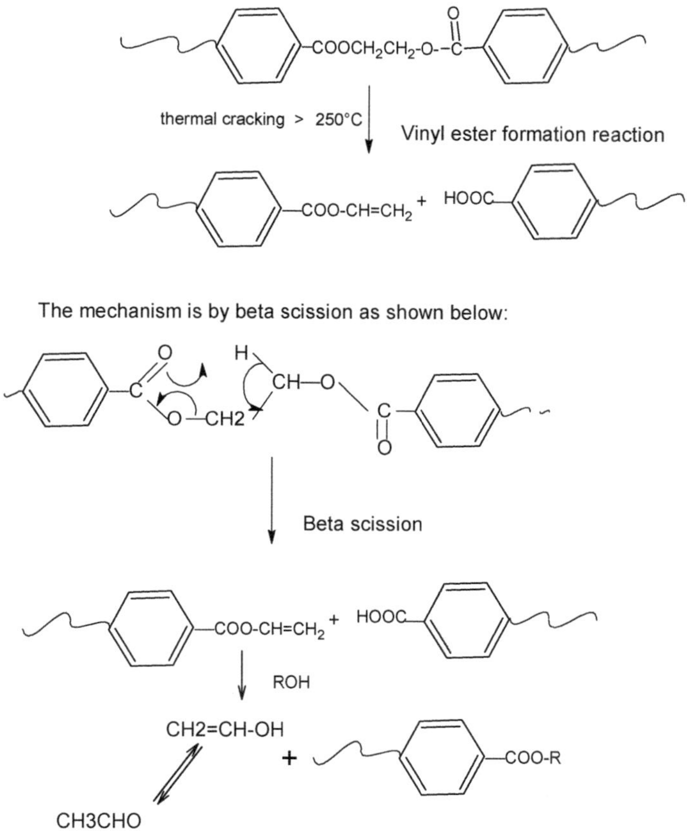

Scheme 1. Mechanism of β –scission followed by acetaldehyde formation

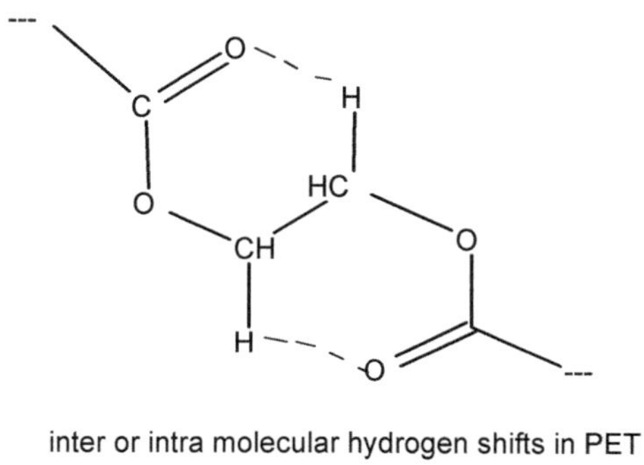

Scheme 2.

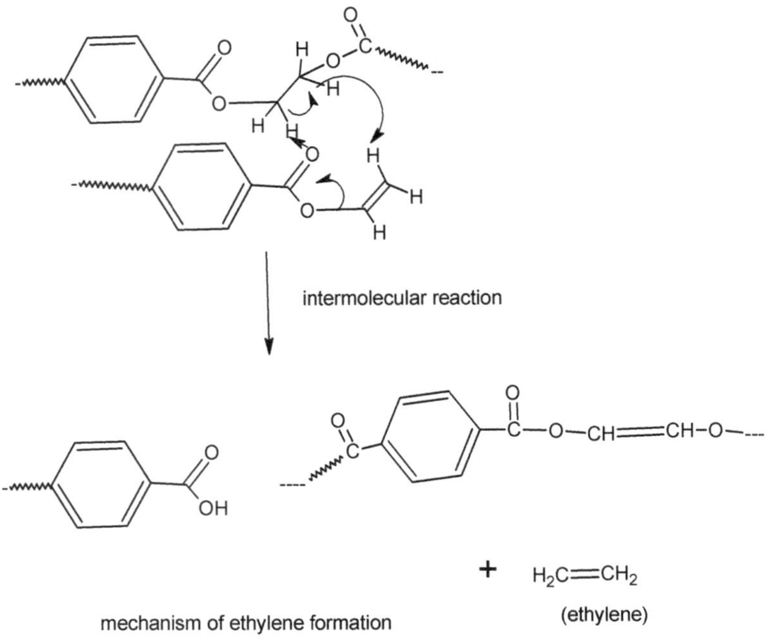

Scheme 3. Mechanism of ethylene formation

The overall degradation and cleavage can be represented by the following scheme 4 [12].

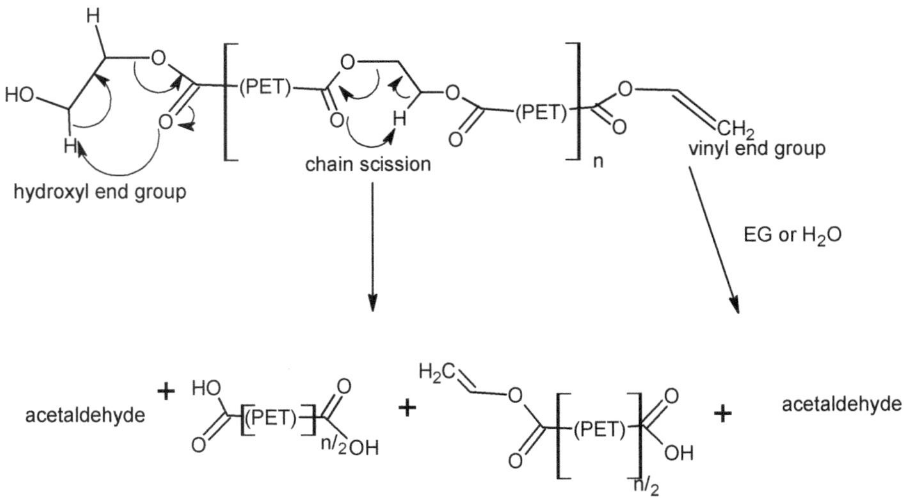

Scheme 4. Overall Mechanism of thermal degradation

The thermooxidative degradation involves the reaction of oxygen at elevated temperatures. This starts with formation of hydroperoxide at the methylene group in the diester linkage of polyester chain. It is not completely understood and is believed to follow free radical mechanism involving formation of hydroperoxides as shown Scheme 5 [13]:

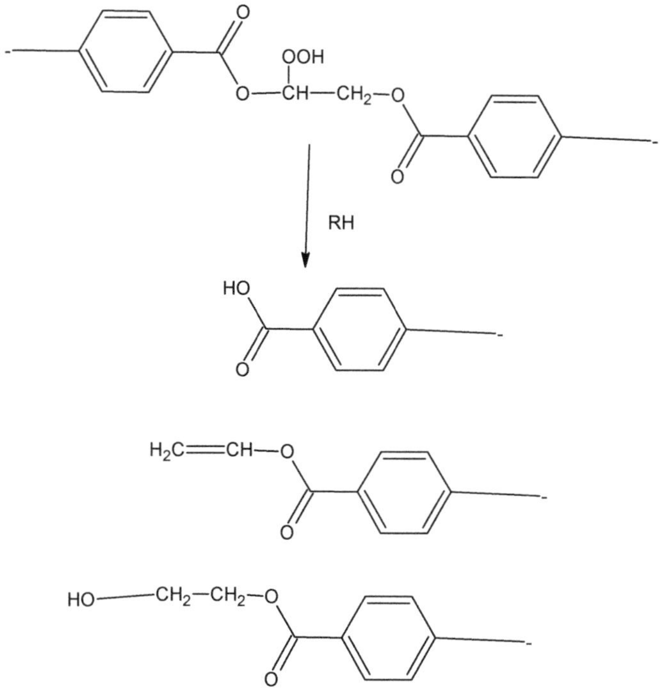

Scheme 5. Thermo oxidative degradation of polyester

Photodegradation of PET occurs on exposure to near ultraviolet light, resulting in both chain scission by analogues of Norrish-type I and II processes shown in the reaction scheme or a photo-Fries-type reaction. This leads to cross-linking there by making the polymer films and fibers, brittle, and discolored. The films also develop crazed surfaces. PET when exposed to UV light degrades rather rapidly leading to deterioration in physical and mechanical properties and develops intense yellow colour [14]. Infrared and gas chromatographic analysis, GPC and colour measurements have shown large changes in hydroxylic region and broadening of carbonyl peaks . These are attributed to perester or anhydride formation. Both hydroxyl and carboxylic end-groups along with carbon monoxide and carbon dioxide are the main products of photodegradation [14]

The rate of degradation under these conditions depends largely on the reaction temperature as well as of the kind and amount of metal compounds used for the tranesterification and polycondensation catalysts. Thermal degradation of ultra high molecular weight polyethylene terephthalate having IV exceeding 2 dl/g has a greater degradation rate than the conventional PET with IV 0.6 dL/g[5]. The high degradation rate of the higher molecular weight polymer is interpreted by the difference in the terminal group concentrations. Zimmerman et al [15] have shown that rate of thermal degradation in the closed system is about three times higher than that in open system.

Measurement of carboxyl end-groups is one of the fundamental methods for measuring number average molecular weight of polymers. Pohl has shown as early as in 1954 that hydrolytic degradation of polyester chains can be studied by end-group measurements, which consists of dissolving the sample in benzyl alcohol at high temperature followed by titration using standard alkali [16,17]. There are reports on investigation of degradation using measurements of viscosity, optical microscopy, FTIR and UV methods [18-22]. W. Chaouch et al [23] have investigated the effects of hydrolytic aging of a series of PET vascular prostheses and compared their chemical properties between the virgin prosthesis using nuclear magnetic resonance spectroscopy (1H-NMR) and have used the latter technique to determine the OH group concentrations and the number average molecular weight of the polymer subjected to *in-vitrio* conditions and validated the data using the classical titration and viscosity methods. They have observed that the polyester yarns undergo chain rupture by hydrolytic and oxidative degradation during texturizing.

If moisture is present before the polymer is melted hydrolytic degradation will occur. Each water molecule will break the chain so increasing the total number of chains by one. The effect on average molecular weight will be as shown in Equation (2) [3]:

$$M_n^{'} = \frac{M_n}{\left[1 + x * (\frac{M_n}{1800})\right]} \tag{2}$$

where Mn is the starting average molecular weight, $M_n^{'}$ s the average molecular weight after reaction with water and x is the water content (weight %). This relationship can be used to determine effect of moisture on IV and melt viscosity.

It can be seen from the above relationship that higher IV's produce increased sensitivity of the polymer to hydrolytic degradation. Hydrolytic degradation of polyethylene terephthalate is an autocatalytic reaction, being catalysed by the resulting carboxyl end-group. It is accompanied by an increase in hydroxyl end-groups and there is no discoloration of product and also there is no evolution of volatile products as shown in Scheme 6. This process is known to begin at temperatures ~100 °C. Hydrolytic degradation is reported to be 10,000 times faster than that of thermal degradation in the temperature range 100-120°C [13].

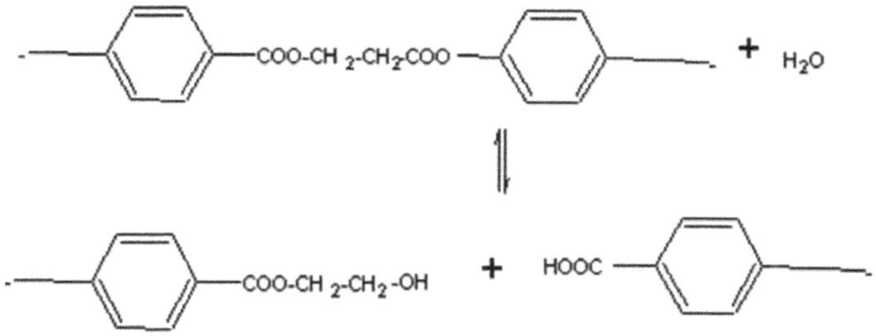

Scheme 6.

Suppose we consider the equilibrium reaction [16] shown in Scheme 7

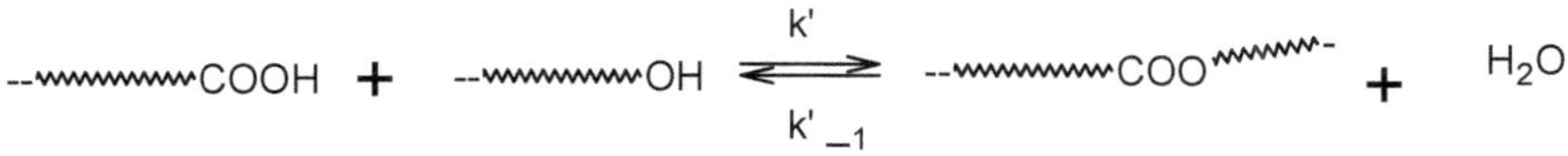

Scheme 7.

k′ is the rate constant for the forward reaction and k′ -1 is the rate constant of the reverse or depolymerization or hydrolysis reaction. The ratio of k′ to k -1′ is K, the equilibrium constant for the above reaction.

$K = \frac{k'}{k'_{-1}}$ it is given by equation (3) ,

$$K = \frac{[COO]_E \, [H_2O]_E}{[COOH]_E \, [OH]_E} \tag{3}$$

The values of K are in the range 0.1 to 1, for normal polyester formation. Hence the reversal of polymerization is a faster reaction and it will lead to degradation of the polyester. Under this conditions K is simplified as shown in Equation (4)

$$K = \frac{p\left[H_2O\right]}{M_0\left(1-p\right)^2} \tag{4}$$

where p is the extent of reaction[16] , since X_n=1/(1-p), (X_n is degree of polymerization which is ~90-100 for fiber grade polyester), it can be rewritten as shown in Equations (5) and (6).

$$K = \frac{p\left[H_2O\right]X_n^2}{M_0} \tag{5}$$

$$X_n\left(X_n - 1\right) = \frac{KM_0}{\left[H_2O\right]} \tag{6}$$

square of X_n is inversely dependant on $[H_2O]$ i.e higher the water concentration the degree of polymerization will decrease much faster.

2.1. Oligomer formation

The other side reaction (shown in Scheme (8)) that can take place is the formation of cyclic oligomers, which can be a major nuisance during melt spinning and dyeing.

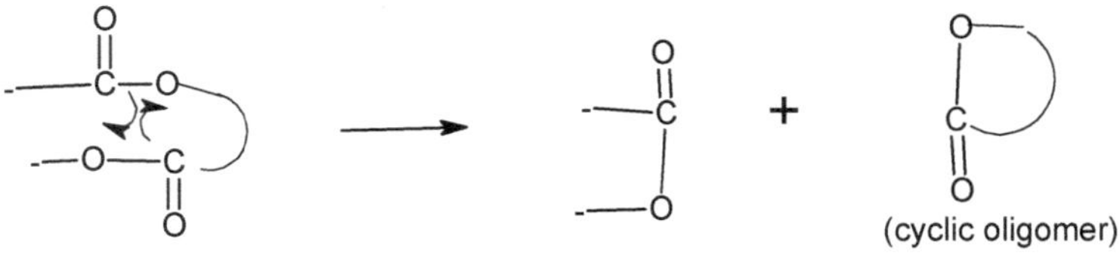

Scheme 8.

The cyclic trimer exudes from the hot polymer surface and coats the spinnerrette plates on the melt spinning units. It can even build up enough to cause breakdown of the thread line. The cyclic trimer exudes form the fibers (which have high surface /volume ratio) during dyeing. The exuded cyclic trimer can float in the bath liquor and contaminate the dyeing equipment [4] depending on the dye bath temperature.

2.2. Causes of discolouration in polyester

The vinyl ester ends also act as cross-linkers and gelling agents. They polymerize and the polymers thermally degrade to give yellow or brown polyenes that discolour the final polymer. The formation highly conjugated species is catalysed by carboxyl groups. The formation of coloured species is followed by increase in more carboxyl terminated species. Hence the product having higher carboxyl value gives more yellowing. The mechanism is given in Scheme 9 [24]:

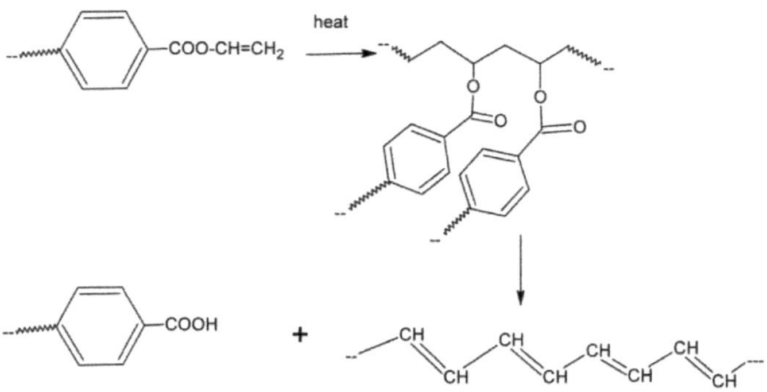

Scheme 9. Formation of conjugated structures leading to coloured species

The carboxyl end-groups are formed by the random thermal cleavage of the chains and a formation of double bonds.. The carboxyl groups catalyze the hydrolysis of ester groups i.e this can be auto catalytic. Intermolecular reaction leading to increase of acid value, ethylene formation and unsaturated species in the polymer backbone as explained in Scheme 9.

2.3. Experimental studies on degradation reactions

In our experimental studies on thermal degradation of PET having cationic dyeable comonomer units viz. 5-sulphoisophthalate moieties are tracked by measuring the mechanical properties of the textured yarns produced from the partially oriented yarn (POY) spun at different residence times [11]. The surface properties of the POY spun at different times are also examined by SEM studies. The degradation reactions of homopolymer, polyethylene terephthalate (Polyester A.) shown in Figure 2 is compared with that of cationic dyeable polyester containing 5-sulphoisophthalate comonomer units (SIPM) that shown in Figure 3 (Polyester B).

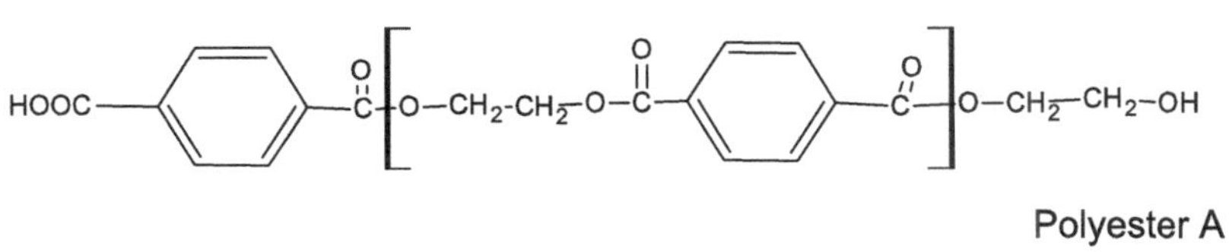

Figure 2. Structure of homopolymer (polyethylene terephthalate)

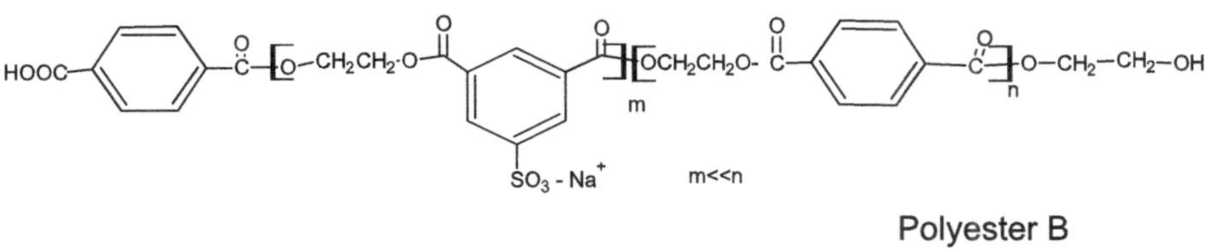

Figure 3. Structure of 5-sodiumsulphonato isophthalic acid copolymer of PEt

Sulphonate groups present in Polyester B contributes to ionomer interaction and steric hindrance thereby increasing the melt viscosity [25,26]. During the melt polymerization of polyethylene terephthalate (PET) there occurs an unavoidable side reaction due to coupling of the hydroxyl end-groups by dehydration forming diethylene glycol (DEG) units in the chain as shown in the following Scheme 10. [24] In the Scheme 10, the intermediate (A) gives acetaldehyde. The cyclic intermediate shown in Scheme 10 C, leads to both hydroxyl terminated intermediates (B) and polyester having diethylene glycol units in the backbone (D).

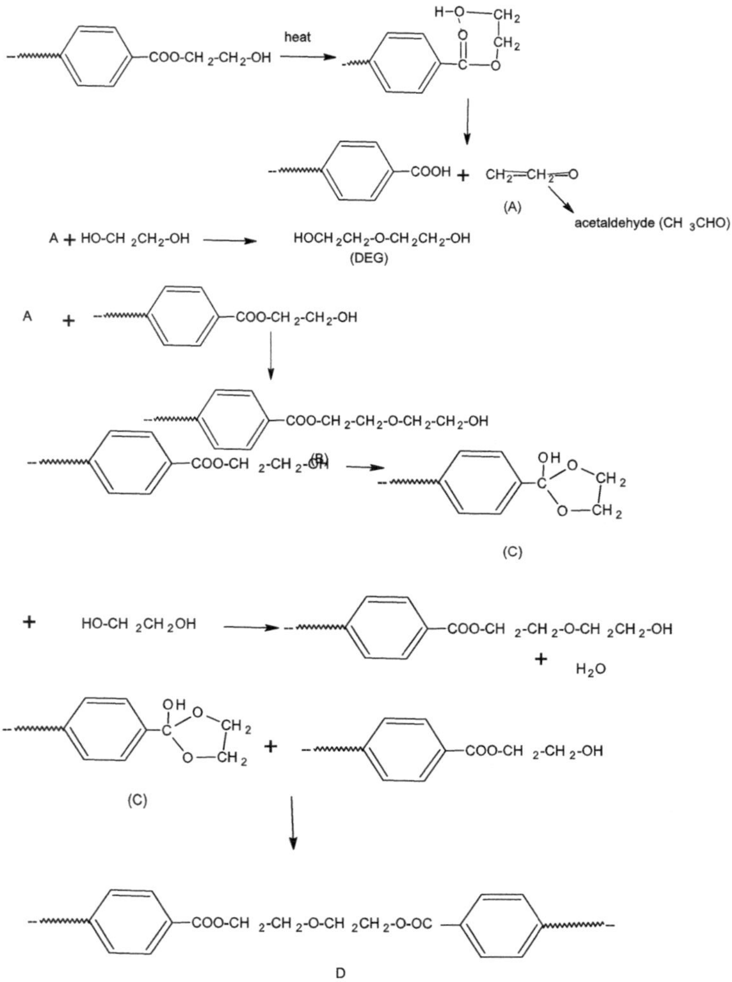

Scheme 10. Mechanism of DEG formation

The presence of DEG units depresses the melting point (ΔTm) which is given by the empirical rule based on Van't Hoff equation given by $\Delta Tm = -2.2\ m\ °C$ (where m is the molar percentage of DEG) . The Tm is measured accurately by differential scanning calorimetry (DSC). The presence of DEG units in the polymer reduces crystallinity, lowers softening point, thermal and hydrolytic stability. About 1 to 1.5 mole % of DEG is always present in PET homopolymer4. Hence DEG content of the polymer becomes an important parameter to be defined. Kinetic studies [11] of polyester containing 5-sulphoisophthalate units indicate that the copolymer degrades faster than homo PET. The rate of degradation depends on the residence time and temperature to which the molten polymer is subjected Rate of degradation of polymer melts is generally represented [27] by the equation (7)

$$\frac{1}{N}=\frac{1}{N_0}+k_0\,\mathrm{e}\ -\mathrm{E}/\mathrm{RT}t \tag{7}$$

where, N and N_0 are the final and initial number average degree of polymerization respectively. Where, k_0 is a constant, E is activation energy, R is the universal gas constant, T is temperature in Kelvin, t is time in minutes.

The degree of polymerization is not an easily measured variable. More commonly, polymer properties are measured in terms of intrinsic viscosity and the degradation equation will have to be written using this variable. The intrinsic viscosity, η can be related to degree of polymerization by the Mark Howink equation given by equation (8).

$$\eta = bN^{\beta} \tag{8}$$

where, η is intrinsic viscosity, b and β are constants depending on the type of polymer.

Substituting equation (7) in equation (8), the rate of degradation of polymer melts can be written as equation (9) , where η and η0 are final and initial intrinsic viscosities

$$\mathbf{1/\eta^{(1/\beta)} = 1/\eta_0{}^{(1/\beta)} + k_0\left(1/b\right)^{(1/\beta)} exp^{(-E/RT)}t} \tag{9}$$

The activation energies for the degradation are estimated from the kinetic data using the equation (9) are found to be 128.94 kJ $mole^{-1}$ for Polyester A and 59.22 kJ $mole^{-1}$ for Polyester B. These values indicate that the co-polymer is more prone to thermal degradation as compared to the homo-polymer. Holland and Hay [28] have shown by Thermal analysis and FTIR spectroscopy the average activation energy for the β C-H transfer process (loss of 1960,1730 and 1255 cm^{-1} bands in FTIR) was 230±10 kJ $mole^{-1}$ and 250±10 kJ $mole^{-1}$ for PET modified by DEG co polymerization and PET modified with both DEG and isophthalic acid IPA copolymerization respectively. They have also shown that the activation energy for the loss of $-O-CH_2-CH_2-OH$ i.e. ethylene glycol derived end groups (loss of 3440 cm^{-1} in FTIR) was ~ 160±10kJ $mole^{-1}$. Loss of ethylene glycol end-groups leads to the formation of a carboxyl end-groups, which promotes intra-molecular backbiting reactions. The favourable angle of 1,3 structure in the isophthalate unit of the 5- sulphoisophthalate co-monomer unit, facilitates the degradation process easily. The presence of bulky sulphonato group causes disorders in the fine structure of polyester fibre thereby lowering the hydrolytic and thermal stability [29].

Yarn produced with higher residence time has poor mechanical properties The SEM images of the yarn samples indicate that as the residence time in a particular temperature is increased more number of particles of degraded products are formed and they have the tendency to grow larger in size, as shown in figures 4a, 4b and 4c.

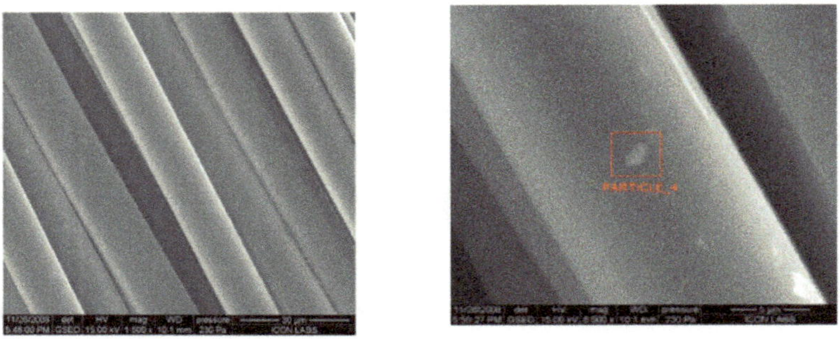

X1500 X6000

a: SEM picture of POY B Residence time 8min

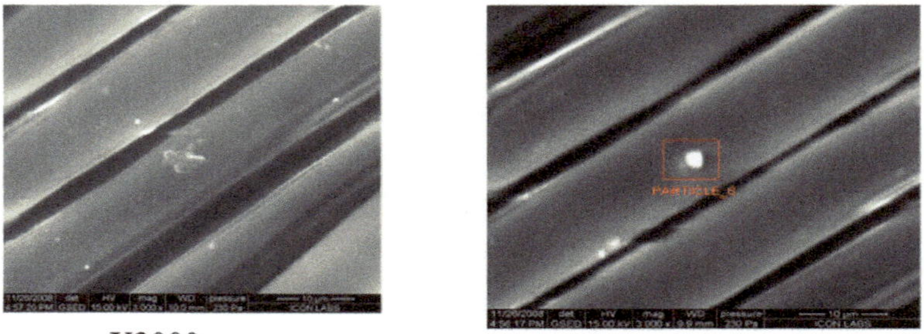

X3000

b: SEM picture of POY B Residence time 40min

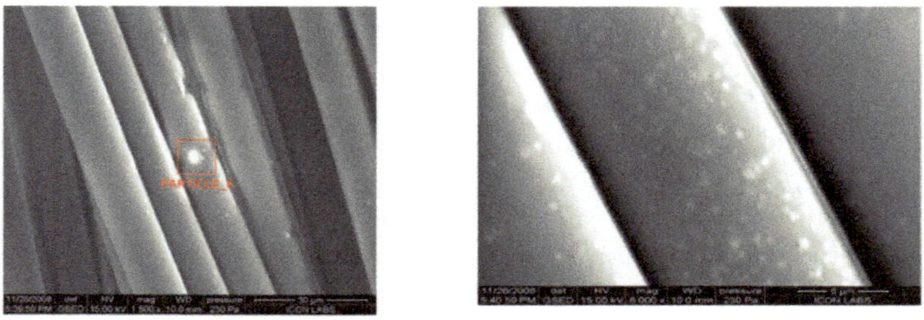

X1500 X6000

c: SEM picture of POY B Residence time 65min

Figure 4.

The degradation results in highly crystalline trimers and oligomers having carboxyl terminals, which occurs by three different routes involving the hydroxyl end-groups or vinyl end-groups or by the inexhaustible mid chain scission route [7,12]. Based on the mechanism of end-group scission of hydroxyl and/or vinyl end-groups and that of mid chain occurring simultaneously during thermal degradation, the by-product formed has two carboxyl terminals, indicating higher acid value in the resulting product. The proposed mechanism for degradation is given in Scheme 11.

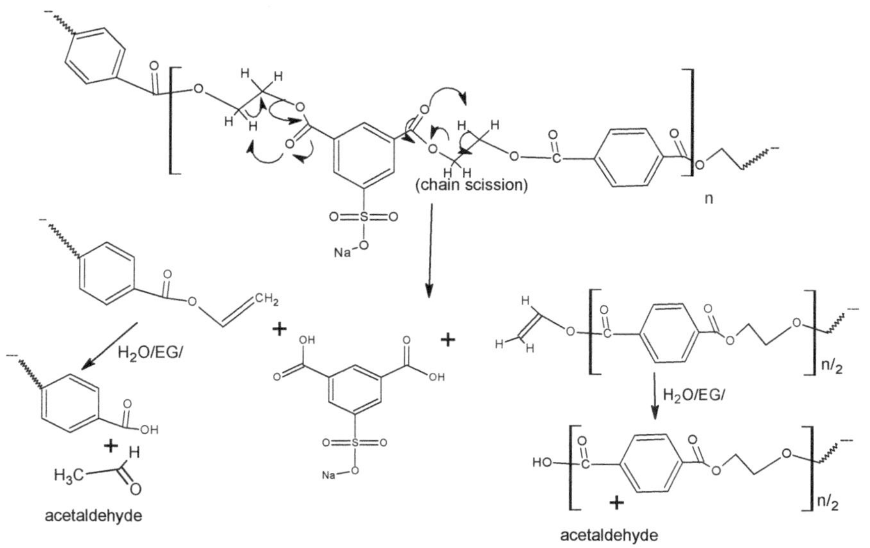

Scheme 11. Mechanism of degradation of polymer containing

The EDX analysis of these particles indicate presence of more organic material (i.e low molecular weight oligomers of larger particle size.) along with some external metallic impurities like Si, Fe, Ti etc. The particles as seen by EDX are organic and are due to more crystalline species. This is separately confirmed by measuring the surface cyclic trimers (shown in Table 1) in the yarn samples produced with different residence times.

S.N.	Residence time (min)	Cyclic trimer (parts per 106 parts (ppm)
1	8	195
2	40	221
3	65	366

Table 1. Amount of cyclic trimer in the polyester Yarn [11]

When the yarn was washed with methanol, some of the particles gets washed off as seen by the SEM of the washed material in Figures 5a and b indicating the oligomers and cyclic trimers get washed off, with methanol. This confirms that the particles seen in the surface are due to surface cyclic trimers.

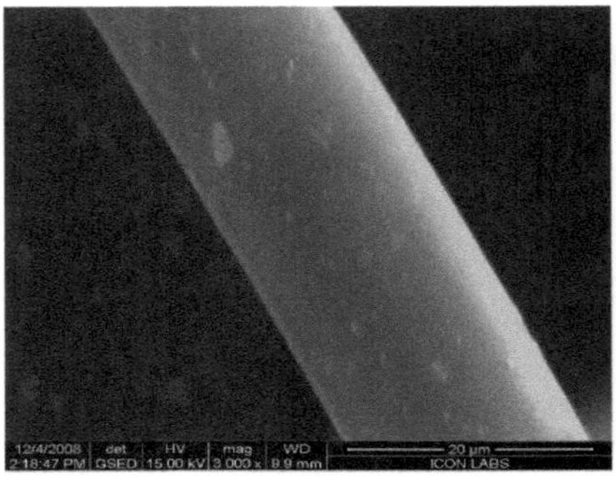

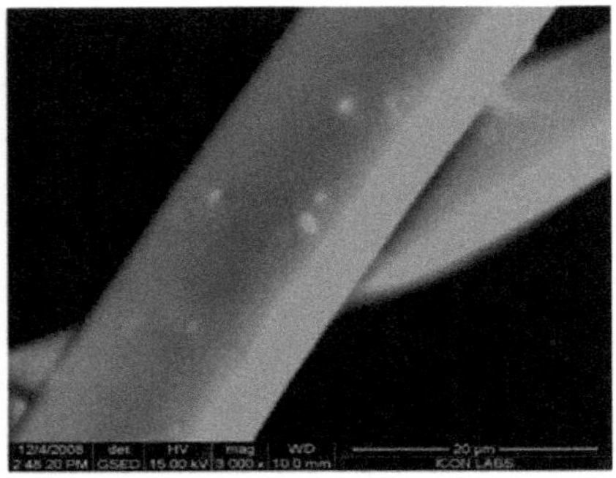

X3000 X3000

Figure 5. a and b : SEM picture of POY B (residence time 65 min) after washing with methanol.

2.4. Effect of thermal stabilizer on the thermal degradation

Phosphoric acid is added as thermal stabilizer to reduce the extent of degradation. It is reported that phosphoric acid interact with the polycondensation catalysts thereby modifying the catalyst [30-33]. Since the same catalyst catalyses the thermal degradation addition of phosphoric acid mitigates degradation reactions. The mechanical properties of the polyester textured yarn (PTY) produced with different amounts of phosphoric acid is given Table 2.

Quantity of Phosphoric acid added (parts per 10^6g)(ppm)	Tensile strength (g /d)	Elongation (%)
0	2.61	12.6
50	2.84	19.6
110	2.89	20.6
125	2.90	22.1
150	3.0	23.7

Table 2. Effect of phosphoric acid on the mechanical properties of the PTY

The data in Table 2 indicate that addition of phosphoric acid improves the mechanical properties of the yarn. But caution is needed on the amount of phosphoric acid used. Figure 6a and Figure.6 b show the SEM pictures of the polyester yarn prepared with different concentration of phosphoric acid in the polymer. The pictures indicated that when phosphoric acid is 125 ppm it causes more agglomerate formation, indicating that higher amounts of phosphoric acid results in more agglomerate formation which get oozed out to surface.

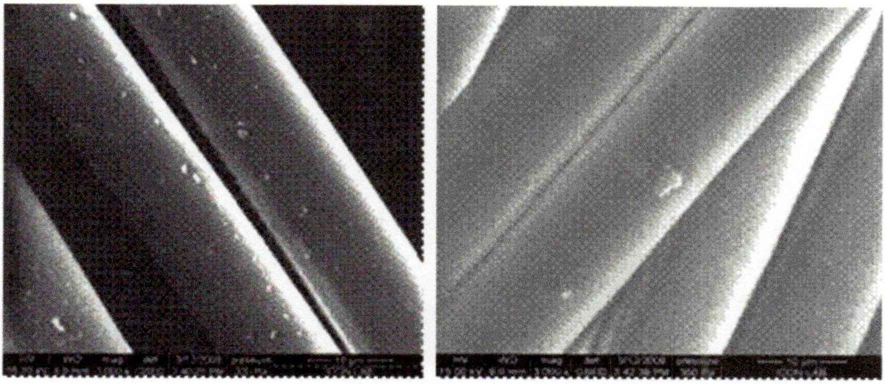

Figure 6. a SEM of polyester yarn with 110ppm H_3PO_4 Figure. 6 b SEM of polyester yarn 125 ppm H_3PO_4

2.5. Degradation studies on penta containing polyester

Degradation measurements were done on the polymer containing small amounts of pentaerithritol as a comonomer (to an extent of 0.15 % by weight). Figures 7 a-b, give the variation IV and carboxyl value for the polyester having penta erithritol units (Polymer C), when the dry polymer chips are melted and kept at temperatures 272°C, 278°C and at 285°C, in inert atmosphere at different times. The results indicate a drop in IV by 0.026 units at 272°C compared to 0.042 units at 285°C over the 15 min. & COOH end-group is increased more at 285°C(18 meq/Kg) compared 12 meq/kg at 272°C. Degradation involves the terminal hydroxyl groups having pentaerithritol moieties as shown Scheme 12. This process gets more easily facilitated in presence of alkali. In this process loss of two formaldehyde molecules and a ketene take place resulting in loss of pentaerithritol [34].

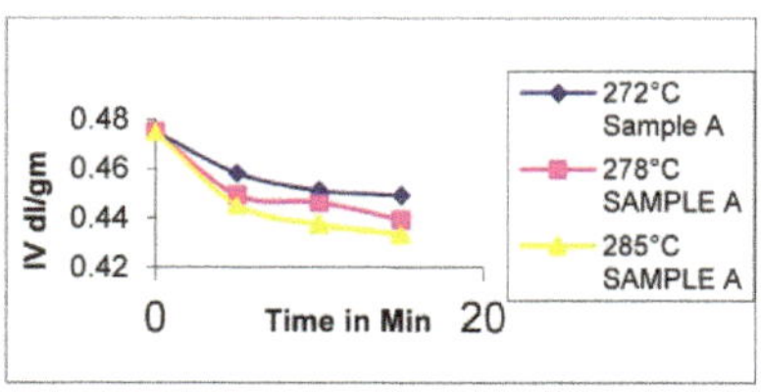

a variation of IV with time at 272,278 and at 285°C

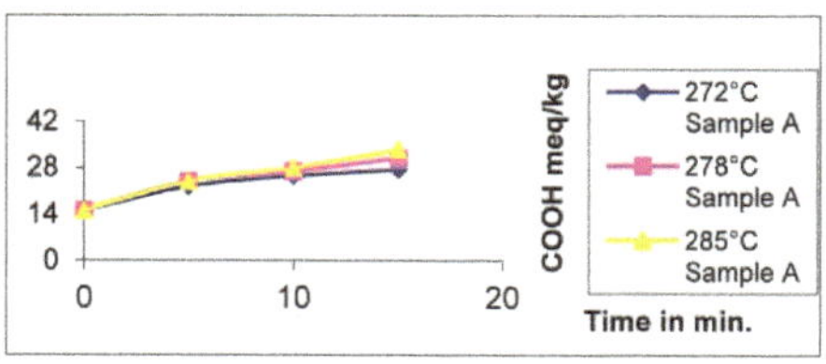

b variation of Carboxyl No. with time at 272,278 and at 285°C

Figure 7.

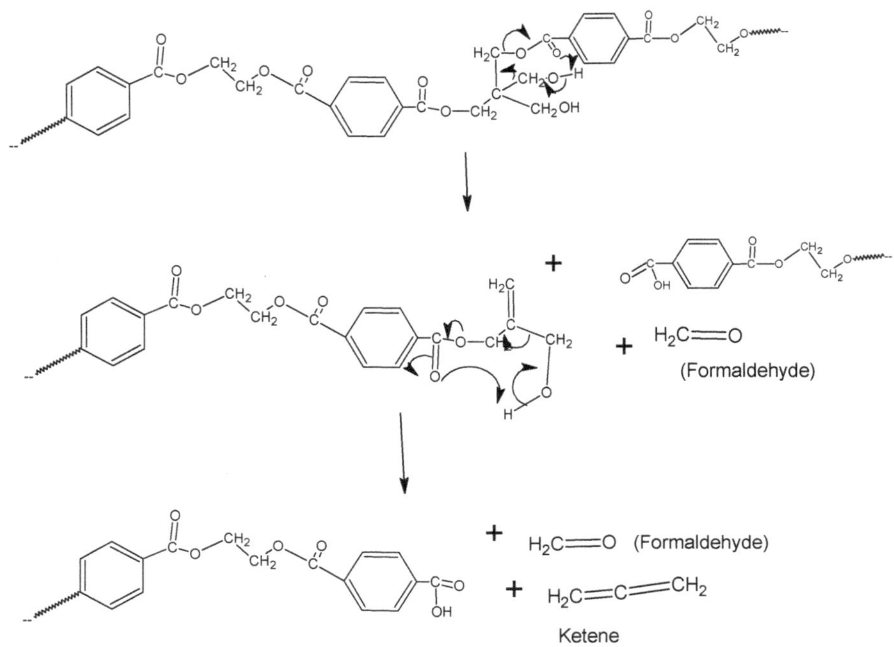

Scheme 12. Mechanism of Degradation in polyester containing penta erithritol moieties

The hydroxyl groups of the partly reacted penta units could become a chain scission source. This results in loss of molecular weight (decrease of IV) and increase in carboxyl terminated by products which is seen by increase in carboxyl value and decrease in IV. The loss of carbon in the samples were seen by EDX analysis in our earlier studies also confirms this mechanism. This also accounts for the loss of pentaerithritol [34]. IV decreases with time reaches an asymptotic behavior because of the result of equilibrium established in the molten polymer. The decrease of IV depends on how the equilibrium is established and how the hydroxyl terminals take part in the Mc Leferty type of fragmentation leading to formaldehyde and ketene loss as shown in Scheme 12. This depends on the residence time the molten polymer was kept at the particular temperature. This gets more aggravated in presence of alkali. Hence pentaerithritol containing polymer degrades more easily in presence of alkali.

From the degradation results it is clear that the higher residence time and temperature has caused degradation of the chips. The pentaeithritol units having unreached hydroxyl groups could become an easy source for initiating the thermal degradation. The hydroxyl terminals could have also formed due to hydrolysis by the moisture present in the system. It is reported that [16] hydrolytic degradation of polyester causes increase of elongation at break due to more chain scissions excepting samples degraded at higher temperature for more than 20 days in which elongation at break will decrease due to more weak points. Also, thermal degradation of samples will decrease elongation at break, because of the annealing process.

2.6. Effect of pentaerithritol units on alkali hydrolysis of polyester

Adding small amounts of multifunctional monomers results in cross-linking or branching of the main polymer chain [2,3,16]. However, randomly branched (hyper branched) polymers are formed [35] due to high reactivity of pentaerithritol, perfect branching is not achieved and additional linear units are present in the molecule. The extensive branching in the randomly branched polymer prevents crystallisation and results in amorphous material. They are generally brittle with low melt viscosity due to lack of long chains to form entanglements. Partial hydrolysis of branch sites in presence of alkali gives more conducive environment for getting randomly branched or hyper branched structures which are highly amorphous and makes the alkali penetrate the polymer structure easily and degrade it faster. It is known that hydrolysis of PET under room temperature conditions is an extremely a slow reaction. Incorporation of small amounts of pentaerithritol segments in the main chain of PET exerts a profound effect in the hydrolytic susceptibility of PET. Pentaerithritol containing units makes the polymer more susceptible to hydrolytic attack. In contrast, a higher degree of crystallinity hinders the reaction because the crystalline phase is inaccessible to water.

Hydrolytic attack on polyesters involves scission of an ester linkage in the main chain by water. Each chain scission uses up one water molecule and creates one carboxyl group and one hydroxyl end-group. In solid state the hydrolysis process depends on chain mobility and flexibility. A reduction in Tg of the polyester enhances the susceptibility of attack by water . The lower Tg also increases chain mobility and reduces energy required to achieve the transition state. Cagiao et al [36] have shown by wide angle X ray (WAX) studies that initial hydrolytic attack could be restricted to amorphous regions and crystallite edges, although lamellar stacks themselves are attacked. It is also known that chain scission of tie segments between crystallites results in further crystallization of amorphous phase giving rise to apparent increase in crystallinity, which could lead to decrease of tenacity. If the polymer is kept in the aqueous alkali bath for longer time, this becomes more degraded, as Scheme 13.

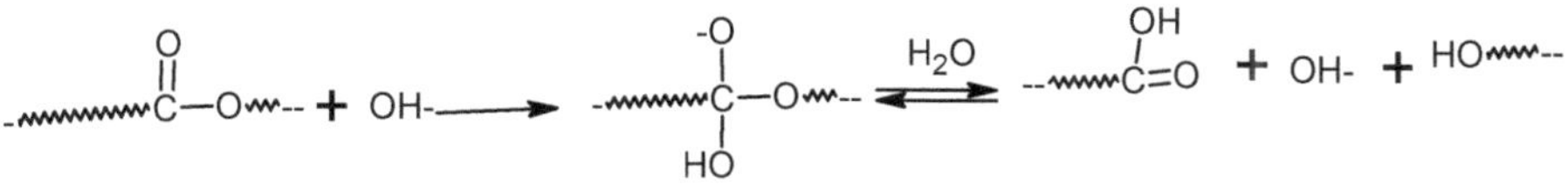

Scheme 13. Hydrolysis in alkaline medium

The depolymerization and thermo-oxidative degradation results in decrease of IV and a rise in the number of carboxyl end-groups [11]. Similar phenomena are reported for polyester copolymers at higher temperatures [7]. The degradation processes that occur are influenced by the moisture and oxygen present in the system. Polyethylene terephthalate is shown to give dual slopes of initial fast rate and later slow rate of degradation [37]. The initial rate is attributed to hydrolysis of residual water and the latter slow rate is attributed the thermal degradation initiated by thermal energy. Rate of degradation of polyethylene terephthalate polymer is studied by measurement of the evolved acetaldehyde by Khemani [12]. The degradation causes a decrease in molecular weight either through random scission at the ester linkages or through chain ends.

2.7. Recycling of polyester

The chemical nature of polyethylene terephalate permits easy recyclability by all known recycling methods. Recycling of polyester has become an important process from the environmental point of view and it has given commercial opportunity due to wide spread use and availability of PET bottles, packages and fibers [37-39]. While mechanical recycling is well established, chemical recycling is highly dependent of the manner in which the depolymerization is carried out.

The chemical recycling methods include processes namely methanolysis, glycolysis, hydrolysis, ammonolysis, aminolysis and other processes [40-41]. Chemical recycling of polyethylene terephthalate into yarn through a process where part of the virgin raw materials are replaced by washed post consumer polyester which is partially depolymerized before repolymerization [42]. Controlled hydrolysis of polyethylene terephthalate using dilute HNO_3 as catalyst is shown to produce cation exchange sites based on acidic groups and the acid sites are shown to adsorb heavy metal cations like Cd^{++}, large cationic dye molecules and the acid hydrolysis is shown to be much more efficient compared to neutral and alkaline hydrolysis [43].

Post-consumer PET bottles are generally sorted out at a material recovery facility, compressed into bales, washed and converted to clean dried flake [39]. The feed purity requirements of the intended use of the recycled PET makes recycling more challenging. Generally contaminants are not allowed in fiber or bottle applications due to breakage and aesthetic considerations. During recycling thermal and oxidative degradation products cause yellowing and diminishing of mechanical properties of the product. There are a number of other difficulties which are to be addressed during recycling PET. For example, the presence of ester group in the back bone of the polymer makes the polymer easily degradable with moisture hence recycling process requires special type of drying prior to processing and hence the conventional blow moulders or sheet extruders used for PE or PP couldn't be used in PET applications. PET drying temperature strongly affects the processing characteristics of the PET containing PVC. For example [39] drying at temperatures below decomposition temperature of PVC (120°C 24 hrs) results in clear PET without significant black speck formation upon extrusion but the rheological stability of PET is poor due to the contamination of HCl catalysed hydrolysis reactions. On the other hand, drying at very high temperatures (230°C, 4h) the majority of HCl is getting removed thus improving the rheological stability of the material, although the extrudate shows excessive black speck formation. While recycling PET traces of label adhesives (based on rosin acids and esters) cause PET to lose clarity and the contamination of impurities such as glue, dirt, paper cause severe deterioration of properties of the recycled polymer. The yet another factor to be considered is the difficulty in getting consistency in the batch-to batch quality of the polymer obtained [39]. It is also not easy to separate contaminants like PVC, PVDC, rosin adhesives, glues, EVA etc which generate acidic compounds which catalyze the hydrolysis of the back bone ester linkages of PET . While physical contaminants like dirt, glass fragments PE are removed the ingrained particulate materials embedded by mechanical abrasion and mechanical grinding during baling, transport and handling are difficult to dislodge. Such impurities pass through the mechanical recycling process and cause stress concentrations (eg.gels, blobs, black specks) that can create problems during fabrication such as excessive fiber breakage during spinning or blow-outs in the wall of blown bottles. Some contaminants such as degraded rubber and wood ash also pass through the extremely fine screens and melt filters and these lead to black specks to the recycled polyester.

State of the art recycling technology by M/s Teijin Limited, Osaka, Japan , has won the Honor award [44]. By this technology, valuable materials are recovered from PET bottles wastes which are crushed, washed and then dissolved in ethylene glycol at its boiling point under pressure of 1 bar to depolymerize to BHET which is later reacted with methanol in an efficient way to produce dimethyl terephthalate and ethylene glycol by ester exchange reaction at the boiling point of methanol.

The recycle value of polyester plastic is found to be second to aluminum and the conditions needed to effect degradation and the extractability of valuable products becomes necessary. New degradation test methods are needed to evaluate the same [37-39].

3. Conclusion

The characteristics needed to understand the wide range of susceptibility of the various agents that facilitate degradation e.g. oxidative, thermal, mechanical, chemical and their dependence on chemical composition and structure of polymers in general and polyethylene terephthalate in particular are discussed. The kinetic studies on polyester containing cationic dyeable comonomer units indicate that the copolymer degrades faster than homo PET . The rate of degradation depends on the residence time and temperature to which the molten polymer is subjected. The SEM images of the yarn samples indicate that as the residence time in a particular temperature is increased more number of particles of degraded products are formed and they have tendency to grow in large size. The degradation results in highly crystalline trimers or oligomers having carboxyl terminals. Addition of phosphoric acid is able to control degradation. The chemical nature of polyethylene terephalate permits easy recyclability by all known recycling methods. Recycling of polyester has become an important process from the environmental point of view and it has given commercial opportunity due to wide spread use and availability of PET bottles, packages and fibers. The effects of contaminants have deleterious effects on degradation and colour of the polymer while recycling of polyester.

Author details

S.Venkatachalam, Shilpa G. Nayak, Jayprakash V. Labde, Prashant R. Gharal,
Krishna Rao and Anil K. Kelkar
Reliance Technology Group, Reliance Industries Limited, B4, MIDC Industrial Area, Patalganga, Raigad District, Maharashtra State, India

Acknowledgement

The authors wish to thank Reliance Industries Limited for their financial support and permission to publish this work.

4. References

[1] SS Stivala and Leo Reich "Sructure vs Stabilityin polymer degradation" Polymer Engineering and Science 20, 6540660 (1980)
[2] Principles of polymerization IV edition, Gerge Odion, John Wiley Interscience 2004, pp 92-97
[3] Reese G "Polyester Fibers" in "Encyclopedia of Polymer science and Technology 3rd Ed.: Wiley :new York 2003, Vol 3 pp 652-678
[4] East A.J. In synthetic Fibers,Nylon, Polesters, acrylic,polyolefine: McIntyreJE Ed., The Textile Institute, Wood Head Publishing Limited, Cambridge,2005 pp 95-167
[5] S Tate and H Narusawa "Thermal degradation of UHMWPET Polymer 37, 1583-87 (1996)
[6] Ref 3 p 656
[7] Roslaniec Z, Pietkiewicz D "Synthesis and Characterization of polyester based thermoplastic elastomers: Chemical aspects" Hand Book of thermoplastic Polyesters, Ed Stoyko Fakirov, Wiley VCH, Vol 1, 2002, p 607-608; p 612-613
[8] Vijayakumar C.T, Ponnusamy E, Balakrishnan T and Kothanda Raman H , "thermal and pyrolysis studies of copolyesters" J Polym.Scienc, Polym Chem Ed., Vol 20, 2715-1725 (1982)
[9] Adams R E, "Pyrolysis mass spectrometry of terephthalate polyesters using negative ionization" J polym scien polym Chem ed., 20,119-129 (1982)
[10] Yoshihiro Sugimura and Shin Tsuge, " Studies on Thermal Degradation of Aromatic Polyesters by Pyrolysis-Gas Chromatography" J. Chromotagraphic Science, 1 , 269-271 (1979)
[11] Shilpa Nayak, Jayprakash Labde, Santosh Geedh, SanjivK Jaisingh, Krishna Rao, S.Venkatachalam, Anil K Kelkar, l Study on degradation reactions in polyethylene terephthlate containing 5-sulphoisophthalte moieties, J Applied Polym.science, Vol 118, issue 5, p2791-2800 (2010)
[12] Kishan C Khemani "A novel approach for studying the thermal degradation and for estimating the rate of aldehyde generation by chain scission mechanism in ethylene

glycol based polyesters and copolyetsers" Polymer degradation and stability Vol 67, 91-99 (2000).

[13] Culbert B and Christel A , "Continuous Solid state poly condensation of Polyesters" Chapter 4 , p 150, in Modern polyesters, chemistry and technology of polyesters and copolyester Ed John Schiers and Timothy E Long J Wiley 2003

[14] D.R.Fagerburg and H .Clauberg , "Photodegradation of Poly(ethylene terephthalate) and poly (ethylene/1,4 cyclhexylenedimethylene terephthalate)" Chapter 18 , p 609, in "Modern polyesters, chemistry and technology of polyesters and copolyester" Ed John Schiers and Timothy E Long J Wiley 2003

[15] Zimmerman, H., and Nguyen, T. K., Investigation on Thermal and Hydrolytic Degradation of Poly(ethylene Terephthalate), Polym. Eng. Sci. 2608,0 (1980)

[16] Ref 2, p 66-67

[17] Ronald Garmon, "End group Determinations" in Polymer Molecular weights part I , Philip E Slade Jr. (Ed.) Marcel Dekker,,Inc. New York ,

[18] Pohl, A. "Determination of Carboxyl end groups in Polyester Polyethylene terephthalate" Anal Chem 1954, 26, 1614.

[19] Berkowitz, S. Viscosity-Molecular Weight Relationships for Poly(ethylene Terephthalate) in Hexafluoroisopropanol-Pentafluorophenol Using SEC-LALLS, J Appl Polym Sci 29, 4353, (1984)

[20] Duh, B. Effects of the Carboxyl Concentration on the Solid-State Polymerization of Poly(ethylene terephthalate) J Appl Polym Sci 83, 1288-1304 (2002)

[21] Zhang, H.; Rankin, A.; Ward, I. M., "Determination of the end-groupconcentration and molecular weight of poly(ethylene naphthalene-2,6-dicarboxylate) using infra-red spectroscopy", Polymer 37, 1079. (1996),

[22] Postma, A.; Davis Thomas, P.; Donovan, A.; Richard, L.;Moad, G.; Mulder, R.; O'shea Michael, S, "A simple method for determining protic end-groups of synthetic polymers by 1H NMR spectroscopy", Polymer 2006, 47,1899.

[23] W. Chaouch, F. Dieval, D. Le Nouen, A. Defoin, N. Chakfe, B. Durand , "Nuclear Magnetic Resonance Spectroscopy Spectroscopic Investigation of the Aging Mechanism of Polyethylene Terephthalate Vascular Prostheses" J. Appl. Polym. Sci. , Vol. 113, 2813–2825 (2009)

[24] E.Marechal, "Polyester synthesis and Chemical aspects" "Synthesis and Characterization of polyester based thermoplastic elastomers: Chemical aspects" Hand Book of thermoplastic Polyesters, Ed Stoyko Fakirov, Wiley VCH, Vol 1, 2002, p 29

[25] Xiao Guang Ma; Xian-ying Guo; Lixia Gu. "Rheological behavior in blends of PET with ionomeric polyester" Europ.Polym.J. 2007, 43, 3613-3620

[26] Xian-ying Guo; Li-xia Gu ; Xiao-xia Feng, "The Glass Transition, Crystallization, and Melting

[27] Characteristics of a Class of Polyester Ionomers" J Appl Polym Sci. 2002, 86, 3660-6

[28] Mark H, Tobolsky AV "physical Chemistry of Polymeric systems" 2nd Ed.' Interscience;1950 Vol II

[29] Holland BJ,Hay JN, "The thermal degradation of PET and analogous polyesters measured by thermal analysis-Fourier transform infrared spectroscopy" Polymer, vol43, 1835-1847 (2002)

[30] Hsiao K J, KuoJL, Tang J W, Chen LT, "Physics and Kinetics of alkaline Hydrolysis of cationic dyeable poly (ethylene terephthalate) (CDPET) and polyethylene glycol(PEG) modified CDPET polymers: Effect of simethyl 5-sulphoisophthalate sodium salt/PEG content and the number average molecular weight of PEG" J Appl Polym Sci., 98, 550-556 (2005)

[31] Marechal, E. Poly ester -Synthesis and chemical Aspects In Hand Book of Thermoplastic polyester Stoyko Fakirov., Ed., Wiley VCH, 2002, Vol 1, Chapter 1, p.25.

[32] Cheoung , M.F.; Carduner, K.R.; Golovoy, M.; Van Oene , H.J Ind and Engg Chem Res 1989, 28, 476.

[33] Golovoy, M. ; Cheung M.F.; Carunduner K.R.; Rokosz, M. Polym.Eng.Sci., 1989, 29,12.

[34] Kamatani, H.; Konagaya, S.; Nakamura, Y. Polymer Journal 1980, 12, 125-130

[35] S. Venkatachalam etal Unpublished results.

[36] M.E.Rogers, T.E Long and SR Turner. "Introduction to synthesis methods in step growth polymerization" in "synthetic methods in Step growth Polymers" Martin E Rogers and Timothy E Long (Eds.)

[37] M.E. Cagiao and F.J. Balta Calleja, C.Vanderdonckt and H.G Zachmann, "Study on Morphology of Semicrystalline Poly(ethylene terephthalate) by hydrolysis etching" Polymer Vol 34 , 2024 -2029 (1993).

[38] Seo KS, Cloyd JD. Kinetics of hydrolysis and thermal degradation of polyester melts. J Appl Polym Sci 1991;42:845-850.

[39] F.G.CGallagher , "controlled Degradation of polyesters" Chapter 17 , p 591-608, in Modern polyesters, chemistry and technology of polyesters and copolyester Ed John Schiers and Timothy E Long J wiley 2003

[40] John Scheirs, "Recycling PET" in Polymer Recycling, Science ,technology and Applications, Chapter 4 p121-182,, John Wiley &Sons, New York, 1998.

[41] Daniel Paszun and Tadeuusz Spychaj , Chemical recycling of PET , Ind.Eng. Chem. Res. 36, 1373-1383 (1997)

[42] D.D Cornell, "Recycling polyesters by Chemical depolymerization" chapter 16 p 566-590, in Modern polyesters, chemistry and technology of polyesters and copolyester Ed John Schiers and Timothy E Long J wiley 2003

[43] Prasad C Upasani, Ashwin K Jain, Ninad Save, Uday S Agarwal and Anil K Kelkar "Chemical recycling of PET flakes into Yarn" J appl Polym.Scien, vol 123, Issue 1 520-525 (2011).

[44] Marcelo G. Rosmaninho, Erika Jardim, Flavia C. C. Moura, Gilmara L. Ferreira, Viviani Thom, Maria I. Yoshida, Maria H. Araujo, Rochel M. Lago, "Surface Hydrolysis of Postconsumer Polyethylene Terephthalate to Produce Adsorbents for Cationic Contaminants" J. of Appl. Polym. Sci., Vol. 102, 5284–5291 (2006)

[45] "Recovering Valuable Materials from Waste PET" Chemical Engineering www.che.com november 2003 , 45

Chapter 4

Bioactive Polyesters in Marine and Plant Endophytic Fungi

Sadia Sultan

Additional information is available at the end of the chapter

1. Introduction

Fungi are the second largest group of organisms in the world after the insects. It is estimated that there are one and a half million types of fungi in existence. However, with just 5% of this total having being described [1], a huge, still unknown and untapped microbial pool remains, which promise the discovery of novel, useful and economically profitable compounds. Organisms such as fungi, which generally living in highly competitive environments, are considered as major producers of secondary metabolites. Fungal secondary metabolites are characterized, not only by their structural diversity [2], but also by diversity of biological activity.

Endophytic microorganisms are to be found in virtually every plant on earth [3]. The most widely accepted definition of an endophyte is given by Bacon *et al*; "microbes that colonize living, internal tissues of plants without causing any immediate, overt negative effects" [4]. The most frequently encountered endophytes are representatives of the fungi [5]. In the past few decades however, it has been realized that plants may contain countless, previously undetected, numbers of these microorganisms known as endophytes. This has prompted a worldwide scientific effort to isolate endophytes and to study their natural products. Scientists have since discovered that endophytes may represent as potential sources of novel natural products for exploitation in medicine, agriculture, and industry. A few examples are presented in this chapter, with the focus on bioactive polyesters isolated from fungi coupled with the dereprilication of extract.

2. New methodology for rapid isolation and identification of known bioactive polyesters

Polyester is a widely used and useful material. Its usefulness extends into many fields including the medical field, industry, textiles and as bioactive natural and synthetic compounds.

2.1. Dereplication

Dereplication is a process for the rapid identification of already known natural products. This is strategically important for scientists when screening crude extracts from natural sources for novel bioactive compounds. The continued demand to get new drugs to the market more quickly and more cheaply, requires that the analytical technologies that support this work keep up with, for example, the rate at which new chemical entities (NCEs) are synthesized for high-throughput screening programmes [6]. There are numerous approaches to dereplication based on hyphenated techniques, and each has its own advantages, be they sensitivity, resolution, or scale (mg vs μg) [7]. The most common approaches are LC-UV, LC-MS, LC-MS/MS and LC-NMR, or combinations thereof, and the increasing use of capillary and cryo-NMR probes [8-16]. Of course, any technique involving mass spectrometry will always potentially suffer from problems associated with ionization (or lack thereof) of the compounds being studied. Despite this, in the pharmaceutical industry these hyphenated techniques are very powerful for the monitoring, characterization and identification of impurities [17]. Take LC-NMR as an example. There are three main coupling technologies, on flow, stopped-flow and loop-storage. However, they all have disadvantages. On flow results in poor signal-to-noise (S/N) ratio for the NMR spectra unless a reduced flow rate is used. However, reduced flow can then reduce the effectiveness of the chromatographic separation which makes this method only suitable for the more intense signals arising from the major constituents. Stopped-flow has the advantage that a number of chromatographic peaks can be studied, but the frequent stops then necessary for data acquisition can disturb the quality of separation, and concentrated samples from the major components can contaminate the NMR detection cell. Therefore, this approach is most suitable for mixtures having only a small number of constituents. In the loop-storage mode, the chromatographic run is not interrupted; instead each analyte is stored in a separate capillary loop in order for NMR acquisition to be carried out at a later stage. A prerequisite for this technique however, is that the analyte must be stable during the long NMR analysis time [18]. Also, it is very reliant on the sensitivity of the NMR instrument. For assisting with the sensitivity problem, the use of a cryoprobe has been a recent advancement in LC-NMR. In NMR cryoprobes, the electronic components are cryogenically cooled to ~20K while the sample remains at ambient temperature which reduces the electronic noise [19] thus gaining a better S/N. Cryoprobes provide quite significant sensitivity gains.

2.2. Advantages of dereplication with CapNMR

In the past, without CapNMR, the dereplication process was based mainly on data from the mass and UV spectra. However, both techniques have disadvantages and are not totally reliable. The mass spectrum can contain impurities which makes it difficult to assign the correct molecular mass. Furthermore, the spectrum may be dominated by the preferential ionization of a minor component, again making it difficult to assign a correct molecular mass to the major component. The UV spectrum only gives definitive structure information for those compounds with strong chromophores. The UV library of the Marine Group within the Dionex analytical HPLC system only contains the known compounds which the group has worked on, and to date this is not yet a comprehensive collection. Furthermore, many entries in the AntiMarin database do not have reported UV data for the compound listed. However, even typical UV profiles may also give wrong answers since similar chromophore may give the same UV spectra. It is for these reasons the NMR data is so important because it can deliver definitive structural information. The AntiMarin database includes information on, for example, the number and type of methyl groups, that can be recognized from ^{1}H NMR spectra. With the introduction of CapNMR ^{1}H NMR spectra can be obtained from a single HPLC-MT plate collection derived from only 200-500 μg of crude extract. The ^{1}H NMR spectrum can then provide specific structural information for searching the AntiMarin database. The most readily interpreted information available is that for methyl groups which can be described as singlets, doublets or triplets depending on their environment. Other features, such as the type of substituted benzene rings, are also easily recognized. From the HPLC separation of 200-500 μg of crude extract, a master MT plate can be prepared for the CapNMR experiments and daughter plates made for P388 bioactivity tests to locate the bioactive components and ESMS (Electrospray mass spectra) measurements. The UV profiles are obtained while collecting the MT plate. Therefore, it is now possible to determine from ~ 500 μg if a crude extract contains bioactive new compounds, or known compounds. This task can be easily completed in one or two days. Small volume NMR flow probes were first constructed in the laboratories of Sweedler and Albert in the 1990s [20-22]. These were developed for coupling to various chromatographic methods, resulting in so-called "hyphenated" techniques [23]. They were designed especially for coupling to capillary electrophoresis (CE) and capillary HPLC (CapLC) for the detection of the small volumes associated with these techniques.

A diagrammatic view and photograph of the Protasis capillary probe NMR (CapNMR) system is presented in **Figure 1**

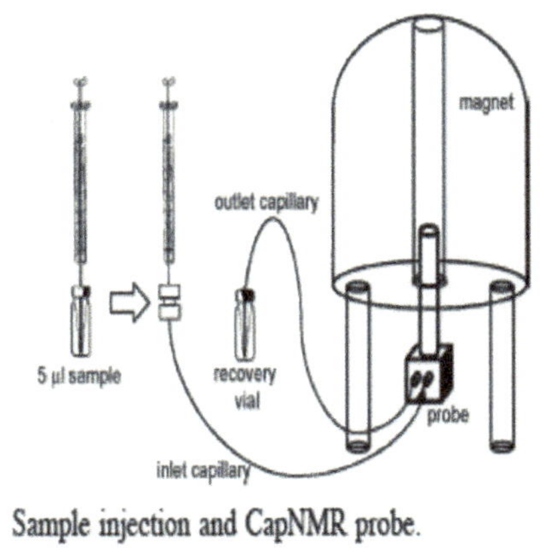

Figure 1. Capillary probe set-up and the Protasis capillary probe injection module.

2.3. Isolation of known bioactive polyesters using dereplication strategies

The natural products chemistry group at the University of Canterbury has been focused on natural products of fungal and marine origin. As well as searching for new bioactive natural products, the group has also focused on the development of new methodologies for dereplication.

Fungal extract 9PR2 which was isolated from the internal root tissue of *Callophyllum ferrugineum* (Guttiferae) [24] has provided a good example with which to demonstrate the new dereplication method, including the use of CapNMR and database searching.

Endophytic extract was diluted at 1 mgL^{-1} with methanol. 30 μL of the diluted extract was injected into HPLC. 10 % standard gradient programme was chosen with 80 % of 0.5 % formic acid plus distilled water and 20 % acetonitrile of HPLC grade.

The HPLC analysis revealed that the extract contained seven compounds. From the similarity of their UV profiles (Fig 2) the seven compounds were related. Based on the HPLC-UV profiles the assumption was made that the compounds contained a highly conjugated or aromatic system. These seven compounds were isolated from the appropriate wells in the MT plate (Fig 3) and each examined by CapNMR to obtain their ^{1}H NMR spectra. Compound **1** displayed three doublet methyl and two aromatic proton signals in the ^{1}H NMR spectrum. Because the compound had the same UV profile as the recorded for the lasiodiplodins, it was considered highly likely to also contain a 1,2,3,5-tetrasubstituted benzene system.

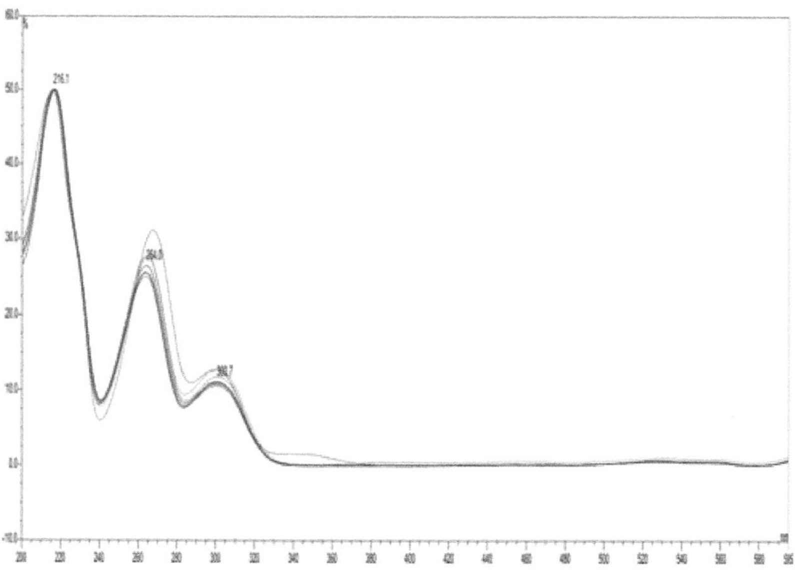

Figure 2. UV profile of the major peak for polyesters **1** to **7**

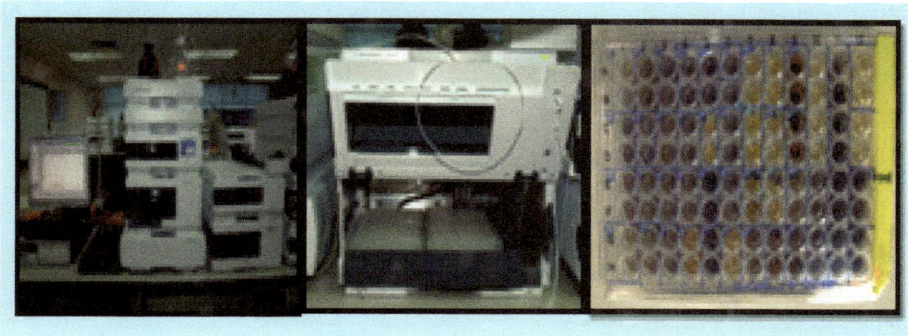

Figure 3. HPLC with DAD and ELCD detectors with fraction collector and microtitre plate.

These features were entered into the AntiMarin database, together with the supposed mass of 384 Da Two matches were found with the same structure, a polyester named 15G256V [25]. The literature NMR data for this compound also matched those obtained for compound **1**. Therefore, **1** was quickly identified as a known compound.

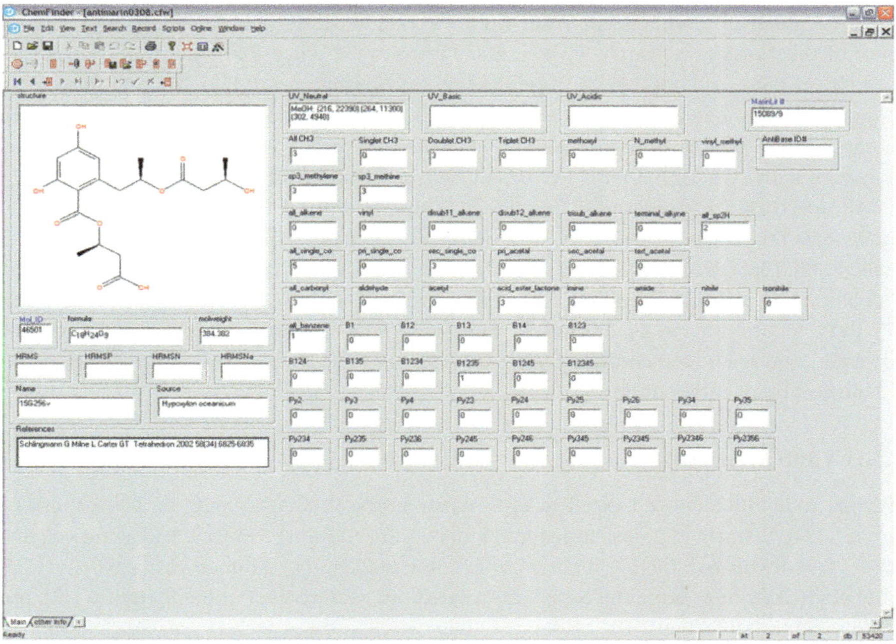

Figure 4. AntiMarin search result for **1**

The next compound examined, had signals for a doublet methyl group and for a 1,2,3,5-tetrasubstituted benzene ring in its ^{1}H NMR spectrum. These features, together with the supposed mass 194 Da, were used in an AntiMarin search. The NMR data for a (+)-6-hydroxymellein [25] matched with the data for Compound **2**. Therefore, **2** was also shown to be a known compound.

The ^{1}H spectrum of **3** was more complex, containing signals for four doublet methyl groups and four aromatic protons. These four aromatic protons were considered to come from two individual aromatic rings, and based on the UV profile, the two aromatic rings were both 1,2,3,5-tetrasubstituted. These features, together with the supposed mass of 680 Da, were used to initiate a search in AntiMarin. These searches found the polyester 15G256α [25] with matching data for compound **3** and have been previously reported. Therefore, compound **3** was identified as a known. Compound **4** had similar features to those for **3** in its ^{1}H NMR spectrum. It displayed signals for four doublet methyl groups as well as for two 1, 2, 3, 5-tetrasubstituted benzene rings. The mass, 578 Da, was different from **3**, and was put into an AntiMarin search. Again the polyester, 15G256π [25], matched the data for compound **4**. Therefore, compound **4** was also identified as known polyester.

The next compounds **5** and **6** had the same mass (662 Da), and both contained signals for four doublet methyl groups and four aromatic protons which could also be considered as 1,2,3,5-tetrasubstituted benzene rings from their ^{1}H NMR spectra. These features formed part of the AntiMarin search. Two similar macro cyclic polyesters were found from this search 15G256α and 15G256α-1, described by Schlingmann *et al* [25]. The literature data suggested that compound **5** was 15G256α and compound **6** was 15G256α-1. Both compounds (**5** and **6**) were thus readily identified as known Five doublet methyl groups together with two 1,2,3,5-tetrasubstituted benzene rings features were noted in the ^{1}H NMR spectrum of compound **7** (Figure-5). These structural features, together with the mass 646 Da, were included in an AntiMarin database search. Macrocyclic polyester called 15G256β [25] was found as a match, also from the previous literature by Schlingmann *et al*. The literature data were consistent with those observed for compound **7**, thus identifying it as a known compound [25]. This study demonstrates the dereplication method, including the use of CapNMR and database search for isolation and identification of compounds.

2.4. Cytotoxicity test

Fungal code 9PR2 showed excellent cytotoxicity in the P388 assay with an IC_{50} value 42.8 μg/mL [24]. Extract 9PR2 was tested for cytotoxic effect against a murine leukaemic cell line P388 and incubated for 72 h after which the MTT [3-(4, 5-dimethylthiazole-2-yl)-2,5-diphenyltetrazolium bromide] assay was carried out as described in the literature [12], but with minor modifications. The cytotoxic activity was expressed as the mean concentration of extract required to kill 50% of the cell population (IC_{50}).

2.5. Antifungal activity

Polyesters **5-7** were also isolated previously from marine fungus Hypoxylon aceanicum (LL-15G256) [25] and were assayed for antifungal activity. In vitro tests using *Neurospora crassa* OS-1 (nikkomycin (MIC=2 μg/mL) as positive control) have demonstrated that compound **7** to be most potent with MIC=0.5 μg/mL while compound **5** and **6** with MIC = 2μg/mL.

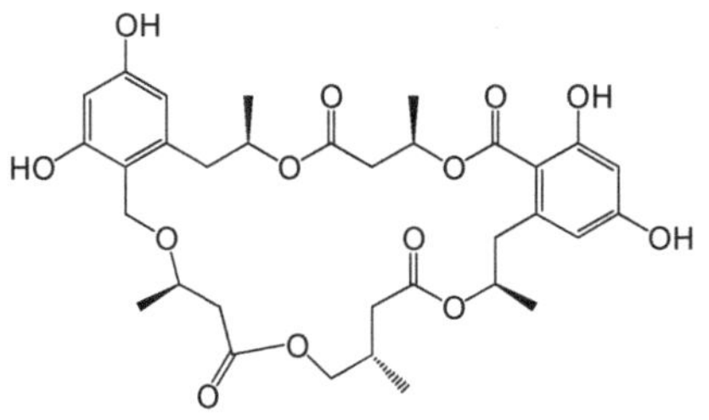

15G256α−1(**6**)

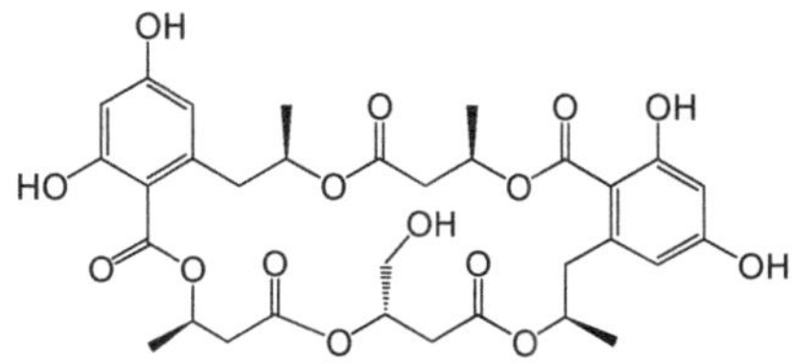

15G256β-1(**7**)

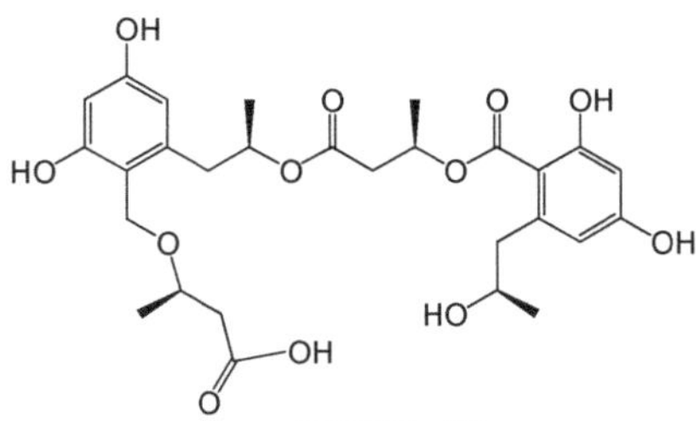

15G256π (**4**)

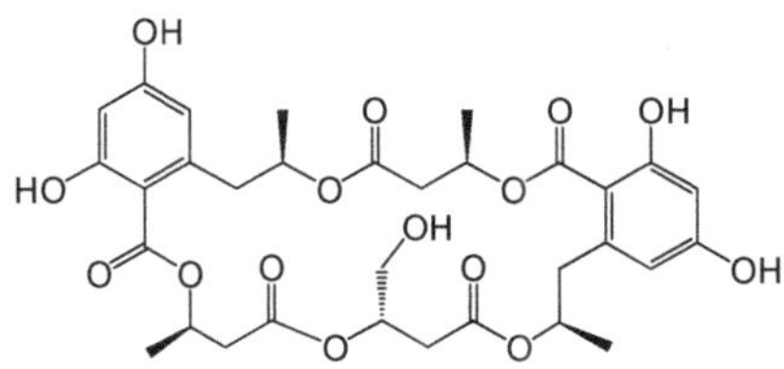

15G256α (**5**)

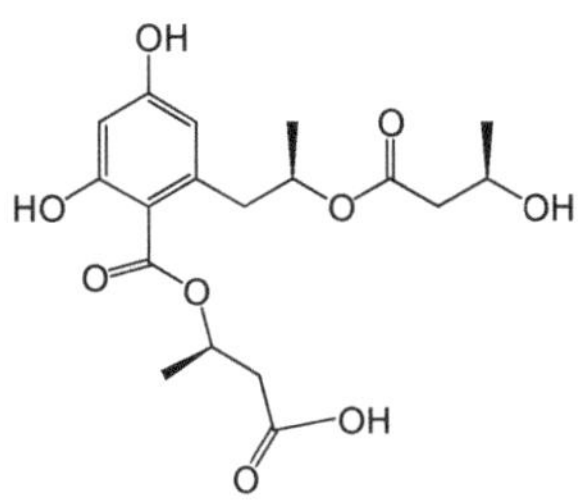

15G256 (**1**)

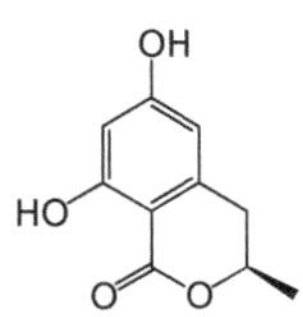

(*R*)-6-Hydroxymellein (**2**)

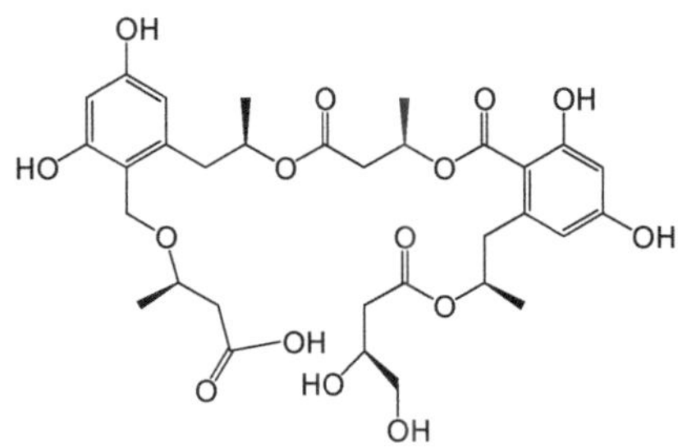

15G256α−2 (**3**)

Figure 5. Structures of Polyesters

3. Novel macrocyclic polyesters from fungus menisporopsis theobromae

In a study to uncover the chemical diversity of plant associated microorganisms, the influence of culture conditions on metabolite production of the fungal strain *Menisporopsis theobromae* BCC 4162 has investigated. (Figure-6). Studies on optimization of culture conditions led to the isolation of new linear polyesters menisporopsin B [26], along with the known macrocylic polyester, menisporopsin, from the seed fungus *Menisporopsis theobromae* BCC 4162.

The fungus was collected from Khao Yai National Park, Thailand and was fermented in peptone-yeast extract-glucose medium (PYGM) under static condition. Chemical investigation led to the isolation of a novel macrocyclic polylatone menisporopsin A (**8**) which possesses an unusual 2,4-dihydroxy-6-(2,4-dihydroxy-*n*-pentyl)benzoic acid residue [27].

A recent study showed that fructose instead of glucose in PYGM is more suitable as carbon source. Shaking not only enhance the production of menisporopsin but also reduced the incubation time [26]. On the basis of these two factors fermentation of BCC 4162 was further studied which led to the isolation of new analogue, menisporopsin B (**9**) in the time profile studies.

Compound **8** exhibited antimicrobial activity, with an IC_{50} value of 4.0μg mL^{-1}, and antimycobacterial activity (MIC value of 50μg mL^{-1}. While compound **9** exhibited antimalarial activity with an IC_{50} value of 1.0μg mL^{-1}

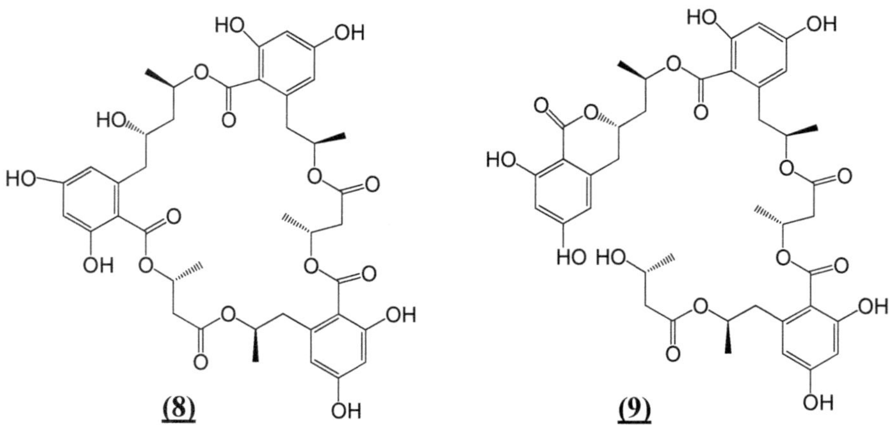

Figure 6. Structures of Menisporopsin A (**8**) and B (**9**)

4. Macrosphelides A and B, novel inhibitors of cell-cell adhesion molecule from microsphaeropsis sp. FO-5050

Cell adhesion molecules play important roles in various types of pathological conditions such as tumor, allergy and inflammatory diseases in the course of a screening program aimed at cell adhesion inhibitors obtained from microorganisms; macrosphelide was discovered in the fermentation broth of *Microsphaeropsis* sp. FO-5050 [28], which was isolated from a soil sample.

Macrosphelide is a newly discovered anti-cell adhesion substance, which is, a low molecular weight, unique 16-membered macrolide antibiotics possessing three ester bonds.

Macrosphelide **A** (**10**) and **B** (**11**) (Figure-7)were isolated by Satoshi et al in 1995 from culture broth of *Microsphaeropsis* sp. FO-5050 which had been isolated from a sample collected in Shizuoka prefecture and their structures were elucidated by spectroscopic methods and by chemical transformations [28]. Macrosphelides **A** and **B** with three esters in their molecules were classified as 6-membered macrocyclic compounds. Macrosphelide **B** was found to be a corresponding oxidative product of Macrosphelide A at C-14 position these Macrosphelides were classified as the first natural products bearing three lactones group in the molecule among the 16-membered macrocyclic antibiotics.

4.1. Biological activities

Anti-adherent activity: Mcrosphelides were assayed in an adhesion assay system using HL-60 cells and HUVEC cells both Macrosphelide **A** and **B** dose-dependently inhibited the adhesion of HL-60 cell to HUVEC stimulated with LPS. Macrosphelide **A** inhibited the adhesion of HL-60 cells to LPS-activated HUVEC monolayer (IC_{50}, 3.5 μM); Macrosphelide **B** also inhibited HL-60 adhesion but to a lesser extent (IC_{50}, 36 μM) in cell adhesion assay.

Antimicrobial activity: The antimicrobial activity of Macrosphelides was determined by the agar dilution method using paper disks. Macrosphelides **B** was active against bacteria, yeast and fungi, whereas Macrosphelide **A** showed no activities at a concentration of 1000 μg/ml against the microorganism tested [28].

5. Macrosphelides C and D, novel inhibitors of cell adhesion

For further research the fermentation of strain *Microsphaeropsis* sp. FO-5050 was carried out in the same way as reported [28] two new 16-membered macrocyclic compounds macrosphelids C (**12**) and D (**13**) were discovered. Compound **12** and **13** (Figure-7) were obtained in the yield of 1.8 mg and 8.0 mg, respectively, together with Macrosphelide A (580 mg) and B (16.1 mg) [29-30].

The molecular formula of **13** as determined as $C_{16}H_{22}O_7$ by HR-FAB-MS. The IR absorptions at 3462 cm^{-1} and 1732 cm^{-1} of **12** showed the presence of hydroxyl group and ester functions, respectively. In the ^1H NMR spectrum the signals at δ 2.36 (dddd, J = 13.9, 10.1, 9.5, 1.5 Hz,

H-8a) and δ 2.55 (m, H-8b) were newly observed compared with those of Macrosphelide A (**10**). The ^{13}C NMR spectrum of **12** showed methylene carbon signal at δ 38.8, for C-8 compared with that of **8** which was appeared at δ 74.7. These signals suggest the presence of methylene carbon instead of methane carbon bearing OH group. On the basis of this compound **12** was assumed to be 8-deoxy derivative of **10**. Further on the basis of HMBC correlations the structure of **12** was confirmed 8-deoxymacrosphelide A.

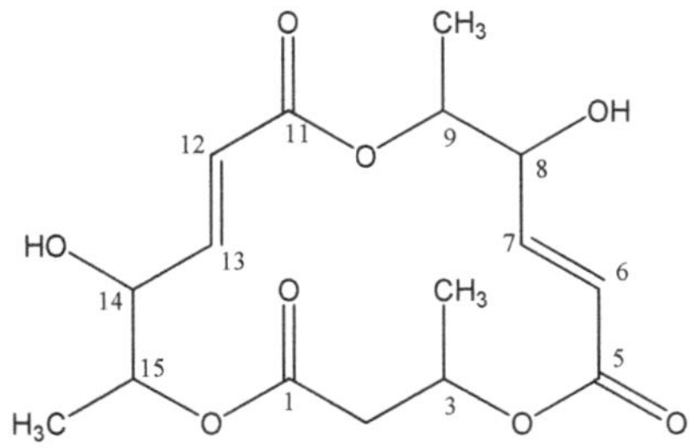

Macrosphelide A **(10)**

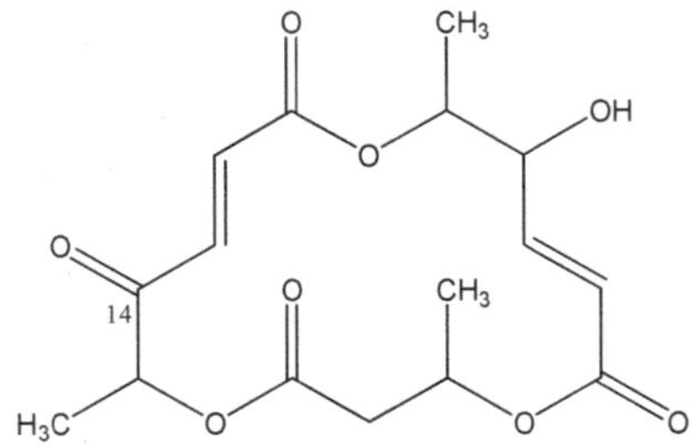

Macrosphelide B **(11)**

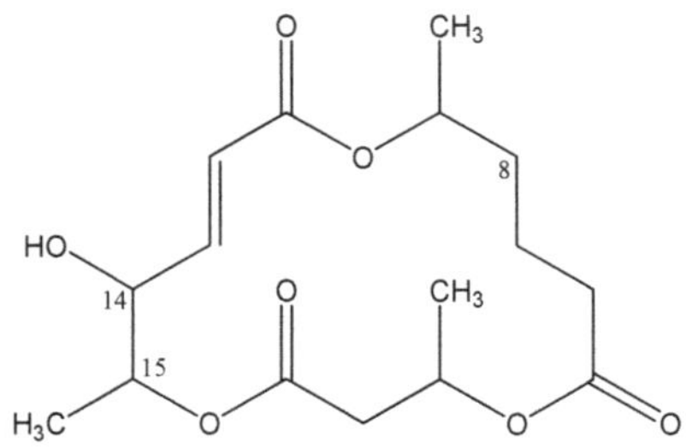

Macrosphelide C **(12)**

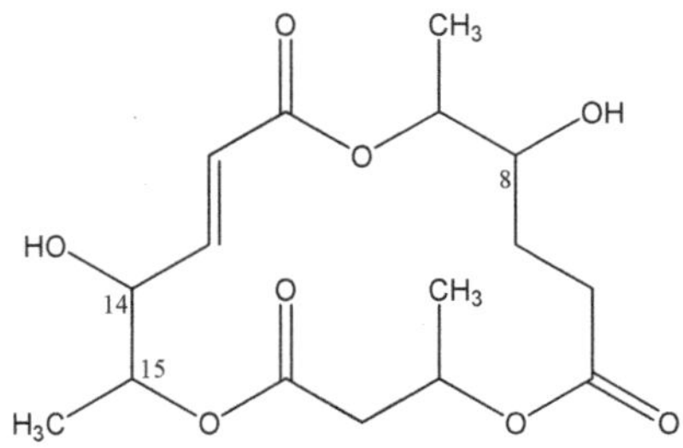

Macrosphelide D **(13)**

Figure 7. Structures of Macrosphelides **10-13**

Compound **13** gave the same molecular formula ($C_{16}H_{22}O_8$) as that of **8** based on the HR-FAB-MS but showed different on ^{1}H NMR spectra. The proton signals at H-14 (δ 5.05) shifted downfield of 0.92 ppm compared with that of **10**. On the other hand, the proton signals adjacent to the methyl carbon at H-15 (δ 4.06) shifted up field of 0.8 ppm. In the ^{13}C NMR spectrum of **13**, the chemical shifts of carbon signals showed similarly to those of **10**, except for signals of C-12, C-13, C-14 and C-15. In the ^{1}H NMR spectrum of compound **13** the coupling constants between δ 5.96 (d, J = 15.8, H-12) and δ 6.59, (d, J = 15.8, 8.6 Hz, H-13) showed the same *trans* configuration as that of **10**. On the basis of these Macrosphelide D (**13**) is presumed to be a stereoisomer of Macrosphelide A (**10**) at C-14 or C-15 positions.

5.1. Biological activity

Biological activities of Macrosphelides **C** and **D** were assayed according to the previous methods [28] in an adhesion assay system using HL-60 cells and HUVEC cells. The IC_{50} values of **12** and **13** were 67.5 μM and 25.0 μM, respectively.

6. Macrosphelides J (14) and K (15)

To clarify the structure-activity relationship of the Macrosphelides and to apply the results to the design of more potent inhibitors of cell-cell adhesion, more attention was devoted to the purification of other Macrosphelide derivatives from the broth of strain FO-5050 which resulted in the isolation of two new members of Macrosphelides J (**14**) and K (**15**) (Figure-8) [31].

The molecular formula of J (**14**)was determined to be $C_{16}H_{24}O_9$ by HR-FAB-MS. The IR absorbance at 3442 cm $^{-1}$ and 1730 cm $^{-1}$ of **14** indicated the presence of hydroxyl group and ester carbonyl group, respectively. From its ^{1}H NMR spectrum **14** was assumed to be a derivative of Macrosphelide B. The signals of δ 4.27 (dd, *J* = 18.9, 2.7 Hz, H-12) δ 2.66 ~ 2.81 (m, H-13) and δ 3.40 (s, 12-OMe) were newly observed when compared with those of Macrosphelide B. In addition, two olefinic protons of δ 6.73 and δ 7.03 which had been detected in Macrosphelide B were absent. The ^{13}C NMR spectrum of **15** showed one methine (δ 74.70, d, C-12), one methylene (δ 42.01, t, C-13) and one methyl (δ 58.79, q, 12-OMe) but the two olefinic carbons (C-12 and C-13 in Macrosphelide B) were absent. The low-field methyl signal showed the presence of a methoxy group. Thus, **14** was elucidated to be a 12, 13-hydro-12-methoxy-macrosphelide **B**.

The molecular formula ($C_{18}H_{26}O_9$) of **15** was assigned based on the HR-FAB-MS. The IR absorptions at 3442 cm^{-1} and 1730 cm^{-1} most likely correspond to a hydroxyl and an ester group. In the ^{1}H and ^{13}C NMR spectra, an additional methylene (δ_H 3.65, 3 47, δ_C 66.85, t, 12-CH_2) was observed compared with those of **15**. In addition the, low-field methyl proton of 5 at δ 3.40 (s, 12-Me) was shifted to high-field (δ 1.18) in **15**. From ^{1}H-^{1}H COSY, the direct connection between 12-CH_2 and 12-CH_3 was observed. By the PFG-HMBC experiment, the structure of **K (15)** was determined as shown below.

Biological Activities : These compounds were evaluated in an adhesion assay system using human leukemia cells (HL-60 cells) and HUVECS (human umbilical vein endothelial cells). The IC_{50} values of all compounds tested were greater than 100 μg/ml and did not indicate any effects on the cell growth of HL-60 and B16/BL6 melanoma when tested at 50 μg/ml.

SAR study of Macrosphelides: Macrosphelides J and K are compared to other neither showed inhibitory activity against HL-60 cells and HUVEC cells, the primary structural differences between the Macrosphelides J and K and Macrosphelides B was a double bond at C-12.

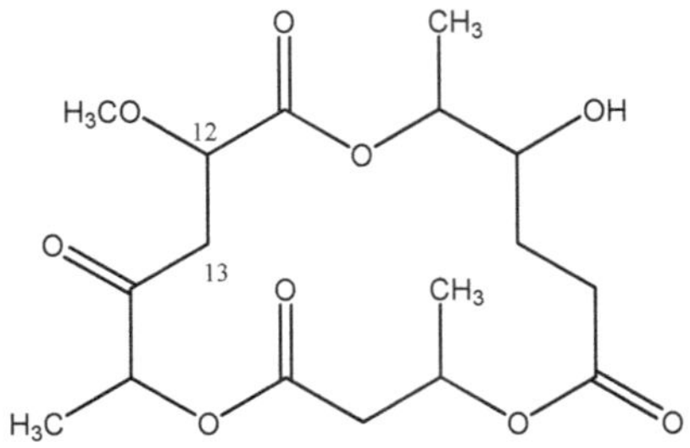

Macrosphelide J **(14)**

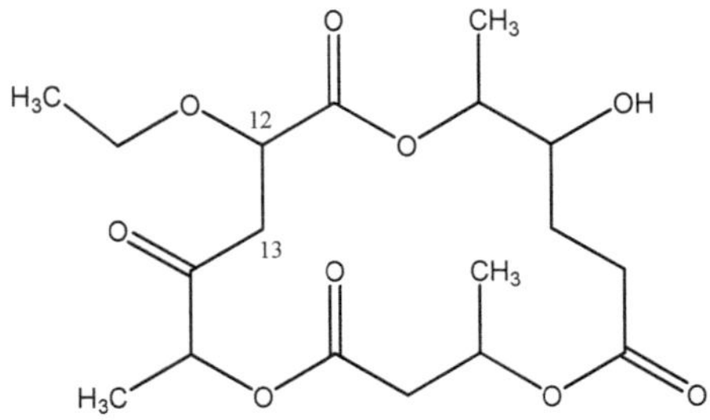

Macrosphelide K **(15)**

Figure 8. Structures of Macrosphelides **14-15**

On the other hand, Macrosphelides J and K were not artifacts of Macrosphelide B, because of their stability in MeOH or EtOH solution. These results show the double bond at C-12 to be important for the inhibitory activity and the analysis can help in the development of an inhibitor of the cell-adhesion molecule.

7. Conclusion

The Marine Group's dereplication method is a fast and reliable method for obtaining the information on compounds. An effort that would previously have taken possibly months of work, often with the unfortunate outcome of finding a known compound. This example using the 9PR2 extract illustrates just how quickly and efficiently the use of LC-UV-MS-NMR, through the intermediary of a microtitre plate and in combination with appropriate databases, can be used to identify the presence of known compounds starting from a very small amount of extract. The dereplication method developed by the Marine Group greatly saves the time taken to determine whether the compounds are known or unknown. Firstly, the sample collection amounts required are reduced from grams or kilograms down to milligrams only, which can provide a reduction in time from months to less than a week for working on the extracts. Furthermore, in the purification step, only several μg of compounds are required instead of the mg of compounds previously required, again reducing the time from weeks of effort to only one or two days using the MT plate HPLC method. The compounds are isolated from HPLC then can be analyzed by CapNMR and mass spectrometry to provide useful information for subsequent AntiMarin database searching to determine the novelty or otherwise of the compounds.

Author details

Sadia Sultan
Atta-ur-Rhaman Institute for Natural Product Discovery,
Faculty of Pharmacy, University Tecknologi MARA, Puncak Alam, Selangor, Malaysia

Acknowledgement

The author gratefully acknowledge her collaborators Lin SUN, Cole A.L.J., Blunt J.W., Munro M.H.G (Marine group) at University of Canterbury, Christchurch, New Zealand, for identification and characterization of known polyesters (**1-7**) using dereplication method and Aisha Adam, Nor Hadiani Ismail, J.F.F. Weber at UiTM Puncak Alam M.I. Choudhary at ICCBS (Karachi University) for their valuable support and guidance.

8. References

[1] Hawksworth, D. L. *Mycol. Res.* (1991), 95, 641-655.

[2] Turner, W. B.; Aldridge, D. C. *Fungal Metabolites II*; Academic Press INC. (London) LTD: London, (1983).

[3] Strobel, G.; Daisy, B.; Castillo, U.; Harper, J. *J. Nat. Prod.* (2004), 67, 257-268.

[4] Bacon, C. W.; James F. White, J. *Microbial Endophytes*; Marcel Dekker, Inc.: New York, 2000.

[5] Staniek, A.; Woerdenbag, H. J.; Kayser, O. *J. Plant Interact.* (2008), 3, 75-93.

[6] Bailey, N. J. C.; Marshall, I. R. *Anal. Chem.* (2005), 77, 3947-3953.

[7] Lang, G.; Mayhudin, N. A.; Mitova, M. I..; Sun, L.; van der Sar, S.; Blunt, J. W..; Cole, A. L. J.; Ellis, G.; Laatsch, H.; Munro, M. H. G.. *J. Nat. Prod.*(2008), 71, 1595-1599.

[8] Koehn, F. E. *In Progress in Drug Research; Petersen, F.,* Amstutz, R., Eds.; Birkhauser Verlag: Basel, Switzerland, (2008), Vol. 65, 177-210.

[9] Wolfender, J.-L.; Queiroz, E. F.; Hostettmann, K. *In Bioactive Natural Products: Detection, Isolation, and Structural Determination,* 2nd ed.; Colegate, S. M., Molyneux, R. J., Eds.; CRC Press: Boca Raton, (2007), 143-190.

[10] Nielsen, K. F.; Smedsgaard, J. *J. Chromatogr. A* 2003, 1002, 111–136.

[11] Wolf, D.; Siems, K. *Chimia* (2007), 61, 339–345.

[12] Fredenhagen, A.; Derrien, C.; Gassmann, E. *J. Nat. Prod.* (2005), 68, 385–391.

[13] Konishi, Y.; Kiyota, T.; Draghici, C.; Gao, J.-M.; Yeboah, F.; Acoca, S.; Jarussophon, S.; Purisima, E. *Anal. Chem.* (2007), 79, 1187–1197.

[14] Lambert, M.; Wolfender, J.-L.; Staerk, D.; Christensen, S. B.; Hostettmann, K.; Jaroszewski, J. W. *Anal. Chem.* (2007), 79, 727–735.

[15] Larsen, T. O.; Petersen, B. O.; Duus, J. O.; Sorensen, D.; Frisvad, J. C.; Hansen, M. E. *J. Nat. Prod.* (2005), 68, 871–874.

[16] Hu, J.-F.; Garo, E.; Yoo, H.-D.; Cremin, P. A.; Zeemg, I.; Goering, M. G.; O'Neil-Johnson, M.; Eldridge, G. R. *Phytochem. Anal.* (2005), 16, 127–133.

[17] Ermer, J.; Vogel, M. *Biomed. Chromatogr.* (2000), 14, 373-383.

[18] Exarchou, V.; Krucker, M.; van Beek, T. A.; Vervoort, J.; Gerothanassis, I. P.; Albert, K. *Magn. Reson. Chem.* (2005), 43, 681-687.

[19] Spraul, M.; Freund, A. S.; Nast, R. E.; Withers, R. S.; Maas, W. E.; Corcoran, O. *Anal. Chem.* (2003), 75, 1536-1541.

[20] Wu, N.; Peck, T. L.; Webb, A. G.; Magin, R. L.; Sweedler, J. V. *J. Am. Chem. Soc.* (1994), 116, 7929-7930.

[21] Wu, N.; Peck, T. L.; Webb, A. G.; Magin, R. L.; Sweedler, J. V. *Anal. Chem.* (1994), 66, 3849-3857.

[22] Behnke, B.; Schlotterbeck, G.; Tallarek, U.; Strohschein, S.; Tseng, L. –H.; Keller, T.; Albert, K.; Bayer, E. *Anal. Chem.* (1996), 68, 1110-1115.

[23] Jayawickrama, D. A.; Sweedler, J. V. *J. Chromatogr. A* (2003), 1000, 819-840.

[24] Sultan, S.; Shah, S.A.A.; Sun, L.; Ramasami, K.; Cole, A.; Blunt, John.; Munro, M.; Weber, J.F.F.; *Int. J. Pharm. Sci.* (2011), 3, Suppl 1, 6-9

[25] Schlingmann, G.; Milne, L.; Carter, G. T. *Tetrahedron*, (2002), 58, 6825-6835.

[26] Ruanglek V, Chokpaiboon S, Rattanaphan N, Madla S, Auncharoen P, Bunyapaiboonsri T, Isaka M. *J. Antibiot.* (2007), 60(12), 748-751

[27] Chinworrungsee M, Kittakoop P, Isaka M, Maithip P, Supthina S, Thebtaranonth Y. *J. Nat. Prod.* (2004), 67, 689-692.

[28] Hayashi M, Kim Y-P, Hiraoka H, Natori M, Takamatsu S, Kawakubo T, Masuma R, Komiyama K, Omura S. *J. Antibiot* (1995), 48, 1435-1439.

[29] Takamatsu S, Kim Y-P, Hayashi M, Hiraoka H, Natori M, Komiyama K, Omura S. *J. Antibiot* (1996), 49, 95-98.

[30] Takamatsu S, Hiraoka H, Kim Y-P, Hayashi M, Natori M, Komiyama K, Omura S. *J. Antibiot* (1997), 50, 878-880.

[31] Fukami A, Taniguchi Y, Nakamura T, Rho M-C, Kanaka K, Hayashi M, Komiyama K, Omura S. *J. Antibiot* (1999), 52, 501-504.

Chapter 5

Degradation of Polyesters in Medical Applications

Mashiur Rahman

Additional information is available at the end of the chapter

1. Introduction

The foundations of polyesters were laid by Carothers and Hill at the beginning of the 1930s [1]. Carothers and Hill synthesized polyesters with low melting temperatures (m.t.) and hydrolytically sensitive, such as the polyester produced from propylene glycol and hexadecanedicarboxylic acid, m.t. 75°C, *w*-hydroxydecanoic acid, m.t. 65°C, *w*-hydroxypentadecanoic acid, m.t. 95°C [2]. In the early 1940s, Whinfield and Dickson [3] in the United Kingdom, and Schlack [2] in Germany used terephthalic acid to develop polyesters with higher melting points. Poly(ethylene terephthalate) or PET, was first commercially produced by Imperial Chemical Industries (ICI) and DuPont and has become the most important man-made fibre in apparel and non-apparel applications.

The use of textiles in medical applications has a long history starting as early as 1952 with the first porous plastic arterial prosthesis Vinyon "N" fabric. It was made from 41 dtex yarn (yarn linear density) with a fabric count of 144×90 per square inch [4]. However, its use was discontinued due to the loss of more than 50% tensile strength within a year. Grafts from Ivalon, Orlon and Teflon textile fibers were also made, but discarded due to dilatation and loss in mechanical properties before anticipated lifetime [5]. Polyesters, because of their high strength, high modulus, low creep and biocompatibility, are being widely used for various medical applications as non-implantable and implantable materials, such as artificial tendon, artificial ligament, vascular grafts, artificial kidney, aortofemoral grafts, and many extra-anatomic bypass grafts (Table 1) [6]. However, the long term durability of polyester for implantable applications is in question because of slow degradation or aging of polyesters that occur in physiological conditions by physical and chemical processes that are synergistic in character. Both *in vivo* and *in vitro* studies showed that the cumulative effects of such slow degradation processes will ultimately lead implant failure via mechanical breakdown. The numerous occurrences of such failures of PET grafts were reported

Application	Polyester	Long term durability
Orthopaedic bandage	Polyester woven, polyester non-woven	Not applicable
Plasters	Polyester woven, non-woven, knitted	Not applicable
Non-biodegradable sutures	Polyester monofilament, braided	Applicable
Artificial tendon	Polyester woven, braided	Applicable
Artificial ligament	Polyester braided	Applicable
Vascular grafts	Polyester woven, knitted	Applicable
Heart valves	Polyester woven, knitted	Applicable
Artificial kidney	Polyester hollow fibre	Applicable
Surgical gowns	Polyester woven, non-woven	Not applicable
Surgical masks	Polyester non-woven	Not applicable
Surgical drapes, cloths	Polyester woven, non-woven	Not applicable
Surgical hosiery	Polyester knitted	Not applicable
Hospital blankets	Polyester woven, knitted	Not applicable
Uniforms	Polyester woven	Not applicable
Protective clothing	Polyester nonwoven	Not applicable

Table 1. Use of polyesters in different medical applications [6].

[7,8,9,10,11]. Table 2 shows the clinical data on patients who developed complications, some of whom died due to failure of PET Dacron grafts [7]; however, numerous Dacron graft failures were unpublished due to the fear of litigation [12,13]. The average time for graft failure was reported to range from 5.8 years and 7.4 years depending on the textile construction of the graft [7]. Of the five deaths reported (Table 2), two patients died just after 2 years following implantation due to the failure of a knitted PET graft and an unspecified PET graft. *In vivo* failure of the PET graft was mainly due to the breakdown of the PET fibers [5], which might have occurred because of manufacturing defects, residual stress and physiological environments [7]. Therefore, the objective of this chapter is to compare the *in vivo* polyester implant failures with the *in vitro* data in order to establish the failure mechanism.

2. Structure and properties of polyester for implant application

The polyester used for medical textile purposes, where long term durability is required, is different from the polyester used for apparel uses in respect of strength and fine structure. For implant applications, high strength, high modulus and low elongation are required (Table 3). Developments in polymer technology, spinning techniques, drawing techniques and machines were required to produce such high tenacity fibre [14].

Location	Failure time (y)	Type of graft	Defect/causes of failure	Outcome
Aortobifemoral	2	Knitted	Aortic tear, dilation & 3 cm rent	Hematuria, dead
Iliofemoral	2	Unspecified	Inguinal tear, developed a 2cm slit	Replaced, alive
Aortobifemoral	3.5	Weavenit	Bifurcation tear,1 cm tear along the guide line	Replaced, alive
Aortic tube	10	Knitted	Aortic rupture, 96.9% loss in tensile strength	Dead
Aortobifemoral	5	Knitted Dacron	Bifurcation tear Surface cracks	Dead
Aortobifemoral	8	Weavenit	Inguinal tear, Disintegration, holes, split etc.	Replaced, alive
Aortobifemoral	8	Tetoran woven	Aorti-ureter, cracks	Hematuria, dead
Aortoiliac/femoral	6	Cooley double velour	Right femoral aneurysm, broken graft fibres	Replaced, alive
Aortobifemoral	5	Double velour	Inguinal tear,	Replaced, alive
Aorto right femoral	7	Knitted double velour	Right femoral aneurysm, manufacturing faults	Replaced, alive
Aortobifemoral	7	Unspecified	Left femoral aneurysm, structural deficiencies	Replaced, alive
Aorta-femorals	5	Woven	Suture tear (polyester)	Dead
*Axillo-femoro-femoral	4	weavenit	Thrombosis	Dead
*Ilio-femoral (transversal)	10	weavenit	Thrombosis	Dead
*Aorot-femoral (bilateral)	25	weavenit	Thrombosis	Dead

Table 2. Clinical data on patients who developed primary Dacron graft failure [7]. *[7a]

The production of standard polyester involves the extrusion of the polymer melt by pumping it through narrow capillaries where it obtains fibrous geometry. The fibre modulus and strength increase and then decrease as the spinning speed is increased but elongation decreases continuously as the spinning speeds are increased. PET fibres melt spun at a take up velocity of around 1000m/min have an amorphous structure and are called low orientations yarns. A take up velocity of around 3000m/min gives the partially oriented yarns (POY) and above 4000m/min gives fully oriented yarns (FOY) [16]. The main structural changes occur during the drawing process as the molecules acquired a more ordered structure (Figure 1), conformational changes take place where disordered gauche conformations of the ethylene glycol unit are changed into the ordered trans conformations [17].

Polyester type	Breaking tenacity (g/d)	Tensile strength (psi)	Breaking elongation (%)	Elastic recovery (%)	Stiffness (g/d)	Toughness (g/d)	Intrinsic viscosity (dl/g)
Apparel grade Polyester	2.0-2.5	33,000-42,000	18-60	67-86 at 2%	7-31	0.28-1.5	0.6-0.7
Medical grade polyester	7.2-8.2	118,000-140,000	*7-10	99 at 1%	54-77	0.35-0.55	0.90-1.0

Table 3. Mechanical properties of apparel and medical grade PET [15]. *Trevira high tenacity, type 703.

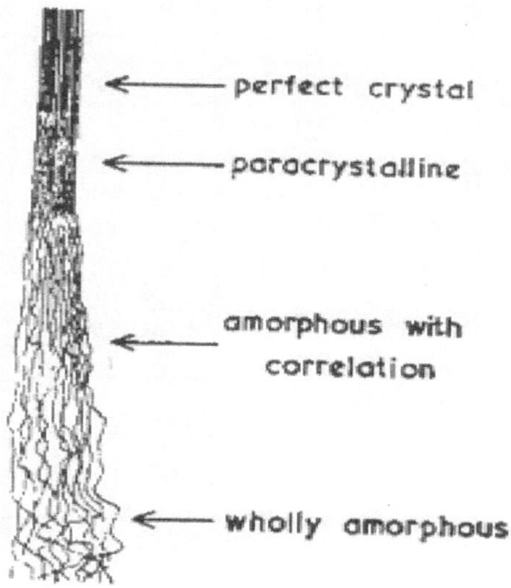

Figure 1. Range of molecular order in polymer packing [18].

For implant applications, higher molecular weight polymer is essential to obtain higher strength fibre. The molecular weight is usually determined from the intrinsic viscosity (IV) or the limiting viscosity number of the fibre in a given solvent at a stated temperature; then the relationship is used between limiting viscosity number and molecular weight based upon the Mark-Houwink relationship: $[\eta] = KM^a$, where 'K' and 'a' are constants which have been determined for the given solvent in relation to either number average or weight average molecular weights. The intrinsic viscosity of apparel grade polyester of tenacity: 2.0-2.5 g/d is about 0.60 to 0.70 g/dl in *o*-chlorophenol at 30°C [19] but to obtain high tenacity

(8.0-10.0 g/d) PET fibres, the intrinsic viscosity should be about 0.9dl/g and for a tenacity of up to 16.0-20.0 g/d this value should be 2.6dl/g [14,20].

Yarns for implant applications require more complex drawing and heat treatment in order to produce the required tenacity, elongation, thermal shrinkage, orientation of the molecules, and crystallinity of the fibres. In this process, machines with a larger number of godets than usual are used, where between the first pair of godets the yarns received some pre-orientation so that they could run smoothly in the later stages of processing [14]. The latest method to produce industrial yarn such as tyre cord consists of producing the yarns at speeds in excess of 3,000 m/mm to produce high modulus, low shrinkage yarn through subsequent drawing and heat treatment in several stages with heated godets and steam. This requires winding speeds of more than 6,000 m/min, depending on the viscosity of the melt and the yarn denier [21].

3. Degrading agents of polyesters

3.1. *In vivo* degradative agents

The degrading agents for implant applications can be physical (pulstaile force, high blood pressure), chemical (blood and physiological fluids), biological (thrombosis, infection), and thermal (body temperature: normal and elevated due to fever). Table 4 summarizes various degradative agents in physiological condition with their effects and possible unfortunate outcomes.

Agent	Source	Effect/outcome
Pulsatile force	Physiological stress, hypersensitiveness	Dilation, reduction in strength, death, cost of replacement
Residual stress	Textile processing, Physiological stress	Strength loss/rupture, death, cost of replacement
Physiological fluids	Patients body	Strength loss/rupture, death, cost of replacement
Thrombosis	Biological	Loss of implants' function (smooth blood flow), cost of repair, ongoing monitoring
Infection	Biological	Loss in physiological function, death, cost of replacement, ongoing monitoring
Temperature	Physiological	Loss in mechanical properties, premature rupture, death, cost of replacement
Defects	Textile processing	Rupture, strength loss, death, cost of replacement

Table 4. Various degrading sources of polyester in medical applications

3.2. The effect of stress on polyester durability

When polyester is used for implant applications, it would be expected to be under stress. Depending on the physiological location, two types of stresses that are encountered by polyesters *in vivo,* such as intermittent and continuous stress. However, in some location within the human body, the stresses encountered usually consist of several components acting simultaneously.

In general, oriented fibre forming polymers like polyesters when exposed to environmentally hostile conditions, show slow but negligible strength loss and slow loss in other properties over considerable lengths of time prior to sudden catastrophic failure [22], as shown in Figure 2.

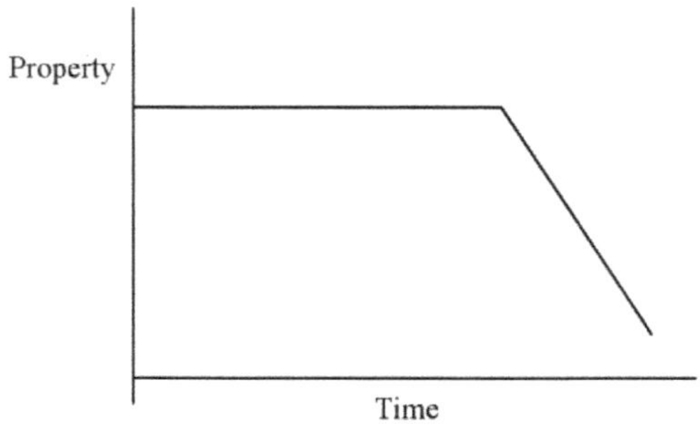

Figure 2. Changes of properties of polymers with time [22].

In case of polyester, it is reported that the yarn stressed to 70, 75, 80 and 85% of ultimate stress will have expected times-to-failure of $1x10^5$, $5x10^3$, 100 and 10 hours respectively [23]. The accelerated effect of stress and pigments on HDPE (High Density Polyethylene) for outdoor applications in the presence of various pigments and stabilizers, stress induced oxidation of polypropylene [24] and the accelerated effect of chemicals on HDPE geomembrane liners under load was reported [25]. Among the other industrially used polymers, the effect of stress during exposure to an NO_x environment for Nylon [26] and the effect of load on Kevlar fibres of poly(p-phenylene terephthalamide) was investigated [27]. For medical application, polyurethane fibres lost strength when treated under stress with water, ethanol and bovine serum after three months contact but under relaxed conditions there was no detectable strength loss [28].

Degradative agents	Short term resistance	Long term resistance
Dilute acid	Good resistance	Passable resistance
Concentrated acid	Moderate resistance	No resistance
Dilute alkali	Good resistance	Moderate resistance
Concentrated alkali	Moderate resistance	No resistance
Salt (brine)	Good resistance	Good resistance
Mineral oil	Good resistance	Good resistance
UV	Good resistance	Moderate resistance
UV (Stabilized)	Good resistance	Passable resistance
Heat (dry), 100°C	Good resistance	Good resistance
Steam, 100°C	Moderate resistance	No resistance
Moisture absorption	Good resistance	Good resistance
Creep tendency	Good resistance	Good resistance

Table 5. Resistance of polyester to common chemicals

3.3. Effect of chemicals on the degradation of PET

Despite the use of high tenacity PET in recent years, the main problems associated with polyesters are the hydrolysis in acid and alkaline media (Tables 5 and 6) and it was found that polyester is not suitable for use at even low alkaline pH (pH=9) [38].

Chemical compound	pH	Percentage strength retained after 1 year (20°C)
Sodium hydroxide, 0.1%	12.1	94
Sodium hydroxide, 2%	12.8	Severely degraded
Ammonium hydroxide	8-10	88
Ammonium hydroxide, 2%	11.4	Severely degraded
Calcium hydroxide, 15%	12.4	Severely degraded
Diethylamine	3.5	Severely degraded
Hydrazine, 2%	10.6	76
Hydrazine, 5%	10.8	Severely degraded
Potassium hydroxide, 0.1%	12.5	90
Potassium hydroxide, 2.0%	13.4	Severely degraded

Table 6. Alkali resistance of PET [31]

3.4. The hydrolysis of polyesters

Hydrolysis basically is the reverse reaction of the synthesis of PET. In hydrolysis the long chain linear molecule is split by a water molecule because of the scission of an ester bond. There are mainly three types of hydrolysis which may occur with polyesters during implant application:

a. Acid hydrolysis;
b. Hydrolysis in water;
c. Alkaline hydrolysis;

3.5. Acid hydrolysis of polyesters

In the acid hydrolysis, protonation of the in-chain oxygen atom of the ester group, followed by reaction with water produces one hydroxyl and one carboxyl end group, as shown in Figure 3.

It was found that 30% HCl halved the tenacity of the apparel polyester after 3 days of treatment at 70°C, whereas under these conditions 30% NaOH would destroy a yarn of any commercial diameter [33]. Further, in 50% nitric acid after 72 hours of treatment, the apparel grade fibres lost 14.5% strength at 40°C, but in 30% HCI at 40°C after 72 hours of treatment, no apparent loss of strength was found; at 60°C, for the same time, no significant strength loss was noticed in sulphuric acid of 70% concentration [32]. For industrial grade geotextile yarn, at 21°C after one year in 20% HCl (pH=0.1) the fibre is completely degraded whereas in 38% H_2SO_4 (pH=0.1) for the same conditions the strength of the fibres was almost

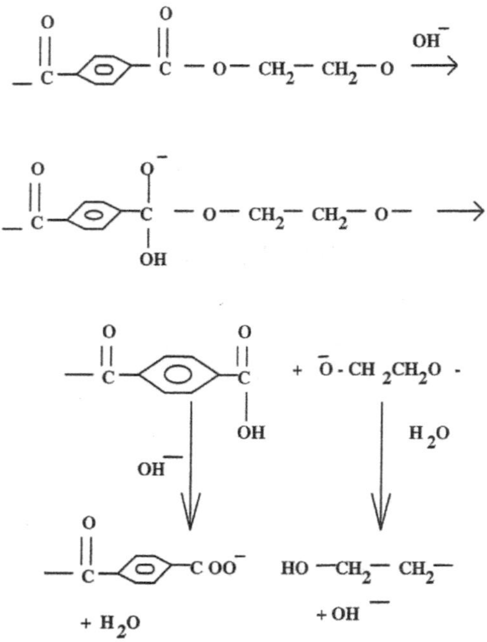

Figure 3. Reaction mechanism of the acid hydrolysis of polyesters [33].

unchanged [34]. Regardless of polyester types, usually, the reactivity of inorganic acids at equivalent concentration in aqueous media was in the order $HCl>HNO_3>H_2SO_4>H_3PO_4$ and attributed this order to the ability of the acid to diffuse into and dissolve in the PET [33].

Acid-catalyzed hydrolysis of esters is much slower at a given temperature than alkali-catalyzed hydrolysis if both are carried out homogeneously and in the heterogeneous hydrolysis that occurs in dilute solutions of strong acid and alkali; the surface attack on polyester fibres by acids is much slower than by alkali [33]. For implant application, generally the implants do not come into contact with the acid environment and therefore, further discussion on acid degradation will be very limited.

3.6. Hydrolysis of polyesters in water alone and other environments

PET is attacked by water or steam which causes hydrolysis of the chain. Reaction with water is very slow at lower temperatures, but the rate of attack increases with increasing temperature. For example, when treated with water at 70°C for four weeks there is no apparent effect on PET but 60% tenacity loss was observed after three weeks at 100°C and 64% after 10 hours of treatment in saturated steam at 150°C [39]. During the treatment with water the fibres do not lose any significant weight.

As well as other hydrolysis in the human body, bacterial attack may occur though after 17 years of micro-organisms environment *in vitro,* no effects was noticed [29]. Sato [35], however, reported an effect of a crude enzyme solution of Cladosporium cladosporioides FERM J-8, prepared from the fungus. The tensile strength, extension behavior, and relative viscosity were remarkably reduced at 30°C at the given pH values used (5.5, 7.0 and 8.0) after two weeks of treatment and these changes were believed to be due to the hydrolysis with esterase of Cladosporium cladosporioides. Such experimental conditions are likely to be obtained in the physiological environment, particularly the temperature and the alkaline pH, but it is not known if polyester implants will encounter such bacteria *in vivo.*

Generally, polyester implants encounter only alkaline condition *in vivo* which is much more severe than hydrolysis in acid and other hydrolysis conditions [33]. Considering the impact of alkaline chemicals (blood) and physiological stress, the chapter is devoted on the factors which impact alkaline hydrolysis and the synergistic effects of alkali and stress.

3.7. Alkaline hydrolysis

In 1952 Hall and Whinfield [36] reported that the effect of alkali on PET was to improve the handle of textiles; since then numerous attempts have been made to understand this subject. This alkaline hydrolysis has several advantages in the apparel industry, in that it produces light weight materials without using light weight yarns which reduces the cost of production and it also gives surface modification of the fibres. Such alkaline sensitivity of polyesters is also useful in countering the problem of cyclic trimer deposition during the high pressure dyeing of the fibres by using a small amount of alkali in the dye bath which hydrolyses this trimer [33].

The alkaline hydrolysis of PET can be expressed in the following way:

The hydroxide ions attack the electron-deficient carbonyl carbons of the polyesters to form an intermediate anion. Further reaction breaks the macromolecular chain and produces –COOH and O^--CH_2-CH_2- as shown in Figure 4. The final products of the hydrolysis are hydroxyl groups and carboxylate anion groups.

Although the alkaline degradation routes for polyesters have been studied for about 60 years for apparel polyesters, there are still some areas of disagreement in the results obtained by different groups of workers. For example, during alkaline hydrolysis, it is generally believed that the reaction occurs at the surface of the fibres so that the greater the surface area, the faster the reaction. Strong evidence for this view is the reduction in fibre diameter as the hydrolysis proceeds and the approximate constancy of the tenacity of the fibre. Where disagreement occurs, it may be that some variables in the polyesters, for example, types of polyesters, spin finish, fibre, yarn and fabric geometry, thermal history, comonomers, surface properties, delustring agents and solution type which lead to slower or faster hydrolysis. This variability in the rate is also applicable to medical polyester as recently as 2009, a lifetime between 156 months to 240 months for polyester implants (Aorto and Axillo bifemorale) was recorded [37].

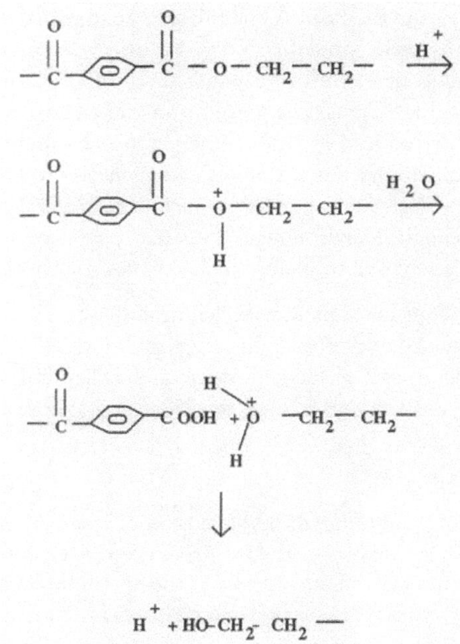

Figure 4. Reaction mechanism of alkaline hydrolysis [37].

3.7.1. Factors affecting alkaline hydrolysis

Most of the published literature on alkaline hydrolysis finds a linear relationship of weight loss with time [39,40,41,42,43,44,45,46,47,48,49,50]. However, other results have been obtained from time to time which show non-linear relationships of weight loss with time [40,51,52,53]. The non-linear relationship occurred due to the stoichiometric quantity of alkali, solution type (alcoholic or aqueous), and other factors (e.g. solvent, stress). If the hydrolysis is confined to the fibre surface and no other factors involved, then the tenacity, elongation, modulus, density and molecular weight of the hydrolyzed fibres should not be changed. But a tenacity loss is frequently reported in the literature [39-44,46,48,53-55), as well as a reduction in breaking elongation [44,4853-55], modulus [53] and molecular weight for explanted polyester implant [37] and apparel polyester [46,56] of the hydrolyzed fibres.

3.7.1.1. Effect of yarn geometry on alkaline hydrolysis

The hydrolysis of polyester fabric is dependent on the nature of the component yarns in the fabric and their construction details [42]. The authors ranked the fibres on the extent of weight reduction as follows: Bright/Textured>Rotoset bright/Textured > Rotoset/Crimped > Rototextured/Textured > Rotoset/Textured. Similarly, fabrics made from mixtures of spun yarn and filament yarn were degraded more rapidly than the wholly filament fabric [31]. Also, Regular textured yarn [54] and polyester implant made from textured yarn [37] degraded faster than the original flat yarn. For geotextile polyesters, PET needled non-woven (heavy weight) degraded more rapidly than the PET heat set non- woven fabric which was itself more rapidly degraded than the PET needled non-woven light weight fabric [57-58]. The authors attributed the differences to different molecular weight, crystallinity, and processing history of each type of polyester; other factors such as different fibre decitex were not considered.

3.7.1.2. Effect of heat setting temperature on alkaline hydrolysis

Medical grade polyester filament yarn requires more complex drawing and heat treatment. Generally the heat treatment is carried out after the drawing process to stabilize the yarn against thermal shrinkage during further processing and end use. The amount of shrinkage increases as the stretching in the yarn is increased. To remove the internal stresses, heat setting is carried out at about 140-220°C using heated draw rollers to give dimensional stability to the yarn which is most important for medical applications. However, to achieve the slowest rate of degradation *in vivo,* heat setting must be carried out at about 140-200°C under stressing condition [59-60] and heat setting temperatures above and below this temperature range increase the rate of degradation [39,42]. In contrast, heat setting of PET fibres at 120°C or higher prior to alkali treatment caused an increase in fibre density and increased the weight loss [61]. No explanation for these changes was offered and there is no clear unambiguous conclusion that follows from these studies. However, some of these discrepancies might be due to different heat setting procedures (free to relax or fixed dimensions) or different heat setting times.

3.7.1.3. Effect of filament cross-section

All so called first generation (up to 1970s) polyester implants were made from trilobal PET and later changed to circular cross-section as alkaline hydrolysis depends on the shape of the filament used [62]. For instance, fibres of bright round cross section are more resistant to hydrolysis than the delustured multi-lobal cross section fibres [43]; however, this does not distinguish between the two effects of cross-section shape and delustrant content. Investigation on the effect of different shapes of filament such as round, triobal, pentalobal and octalobal on the rate of alkaline hydrolysis showed that the weight loss (%) was highest for the octalobal fibres and lowest for round cross-section fibres [63]. But differences in yarn decitex and delustrant content made the interpretation of these results difficult. It was already known that delustred fibres are attacked more rapidly than bright fibres [60,64], so the effect of cross-section shape remains unclear. Given the greater surface area of the trilobal, pentalobal and octalobal fibres, a higher rate of hydrolysis would be expected and for implant application, in order to reduce the degradation rate, the cross-section of fibre should be circular.

3.7.1.4. Effect of additives

There are many types of materials which can be added to a yarn during and after the manufacturing period such as delustring agents, antistatic agents, spin finish, etc. The rate of alkaline hydrolysis of dull fibres which contains TiO_2 was greater than for equivalent bright fibres [60,64]. The surface pitting occurs due to the presence of TiO_2 [44,53,54] which eventually reduces the strength of the fibres.

The traditional definition of spin finish is a material which provides surface lubricity to the yarn so that smooth, high speed transfer over various metallic guides can take place with the minimum amount of fibre breakage [65]. A hydrophilic type spin finish can be used for apparel polyester processing and a hydrophobic type spin finish is used for industrial polyester. For example, to produce polyester tire cord yarns a mixture of polyoxyethylated-polyoxypropylated compounds of molecular weights of about 300 to 1,000 is very effective [66]. Another type of spin finish based on a polyalkylene glycol of molecular weight of at least about 300 is applied to tire yarns and subsequently the yarn is heated for 10 to 160 seconds at a temperature of 215°C to 230°C which improved the compression-extension durability of the fibres in rubber [67]. It may be that the manufacturers may use a special spin finish for the processing of medical textile yarns but no information on this is given in the literature.

3.7.1.6. Effect of comonomers in the polyester

The rate of alkaline degradation of polyesters is dependent on the chemical composition and structure of polymer. The cationic dyeable polyesters hydrolyze more rapidly than ordinary polyester, because such polyesters contain the strongly electron-attracting sulphonate group [68,69,70]. The most used chemical unit is the sodium salt of the 5-sulphoisophthaloyl group, incorporated as the corresponding dimethyl or bis(hydroxyethyl) ester into the polymer manufacturing process, and it is incorporated as the repeat unit at 2-2.5 mol% concentration. The electron density on the carbonyl groups is reduced and the electrostatic interaction between hydroxyl ion and carbonyl group is increased. So the hydrolysis of modified polyester containing this suphonate group is expected to be easier than for the ordinary polyester. Other copolymers such as adipate copolymers and the block copolymers containing polyether blocks also have higher hydrolysis rates, but on the other hand copolymerization with butane-1,4-diol units reduces the hydrolysis rates [33].

Hydrolysis of aromatic ester homologues made from β-hydroxy-ethyl aromatic acids have higher alkali resistance compared to that of PET [45] and aliphatic polyesters with substantial hydrocarbon:ester ratios have hydrolytic stability similar to PET but with higher ester group content and with esters derived from strong acids, very rapid hydrolysis can occur [33]. Such polyesters are polyglycolide, polydioxanone etc. can be used as absorbable sutures but some aliphatic polyester, such as polypivalolactone are less rapidly hydrolyzed than PET and can be used for implant application. Partially aromatic polyesters such as, poly(4-oxyethoxybenzoate), poly(ethylene diphenoxyethane-4,4′-dicarboxylate), poly(ethylene naphthalene-2,6-dicarboxylate) and poly(tetramethylene terephthalate) have higher hydrolytic stability than PET, whereas some liquid crystalline polymers such as that formed from 1,4-oxybenzoate, 1,4-dioxyphenylene and isophthaloyl units have a hydrolytic stability similar to that of PET [33].

4. *In vivo* rupture of polyesters

The failure of a polyester implant is defined as the inability of the implant fabrics such as woven, knitted and braided to retain its structural integrity after implantation for the anticipated lifetime. Scanning electron microscopic (SEM) examination of ruptured polyester implants revealed fibre thinning (reduction in diameter), transverse cracks, fractures and complete breakage of fibres, Figure 5 [5]. Further, scanning electron micrographs

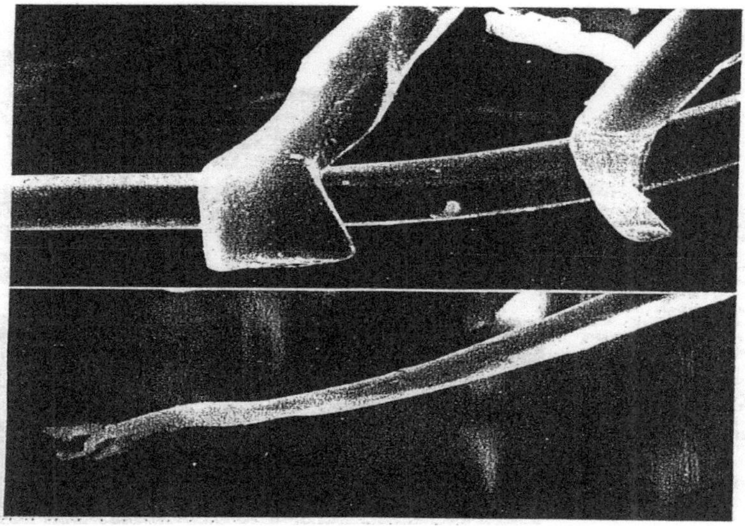

Figure 5. Broken, abruptly ruptured thinned out polyester filament [5].

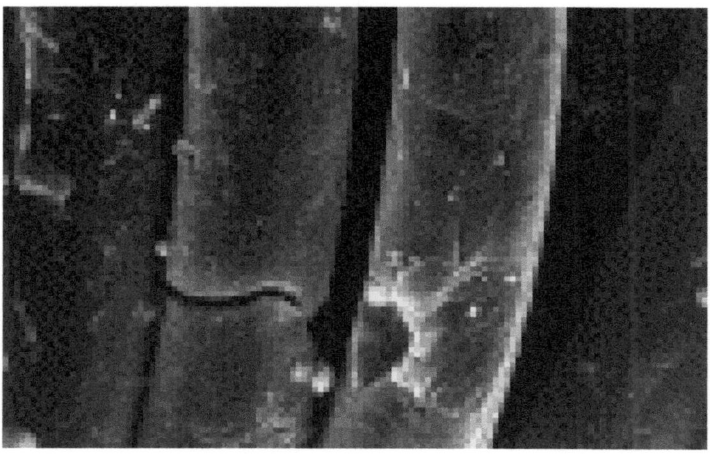

Figure 6. Cracks and holes in the explanted graft [62].

show fibre break down with resultant tears and full circumference transverse cracks as shown in Figure 6 [62], longitudinal splits, Figure 7 [71], and broken filaments, Figure 8 [8]. It is believed that the residual stress from manufacturing processes (for example, winding, warping, sizing, weaving/knitting and wet processing) and continuous hydrodynamic strain due to pulsatile blood flow, results in progressive stretching and thinning of the filament yarns, which causes cracking and gradual rupture of some filaments[5,72]. The entire load is then transferred to the remaining neighbouring filaments of the yarn, which were finally snapped (Figures 5 & 9).

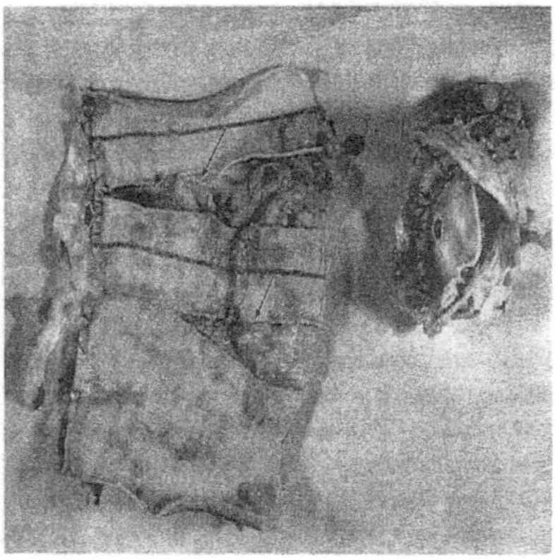

Figure 7. Ruptured PET graft, explanted after 15 years [71]

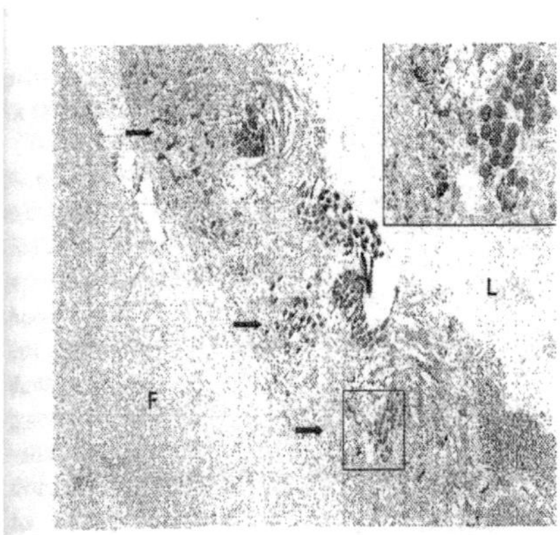

Figure 8. Broken filaments in the explanted polyester graft [8].

Other in vivo degradative phenomena reported were (Table 7) loss in bursting strength [73.74], yarn shifting [75], loss in cover factor [76], polymer degradation [12], dilation [13,62], loss in tensile strength [77], decreases in molecular weight and increases in end groups [37].

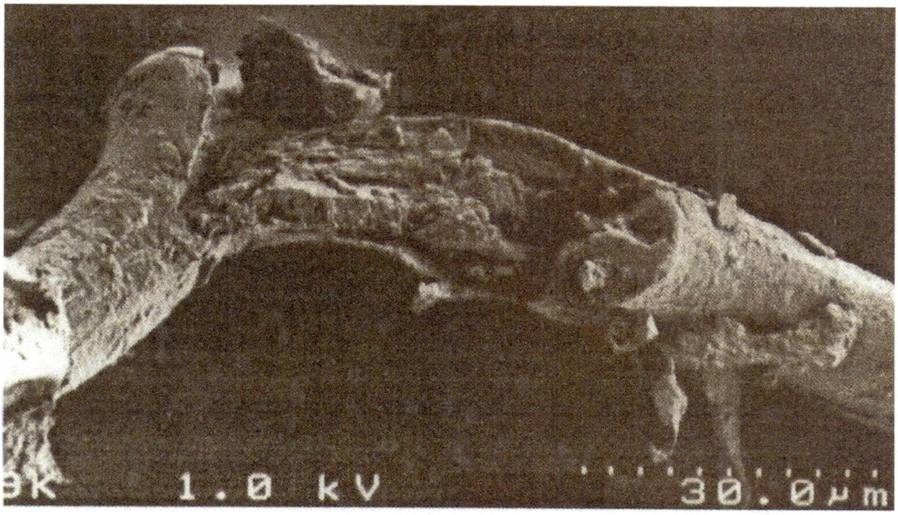

Figure 9. Filament of a 13-year-old vascular graft showing a fracture in the centre and the relatively smooth surface of the filament on the right side [73].

Changes in PET property	Duration (year)	Reference
Loss (25%) in Busting strength	13.5	[74]
Loss (55%) in Busting strength	5	[73]
Yarn shifting (2mm)	2.5	[75]
Cover factor (increase permeability)	5.5	[76]
Polymer degradation	16.0	[12]
Dilatation	9.0	[13]
Dilatation	--	[62]
Loss on tensile strength (40%)	14	[77]
Stiffness	14	[77]
Decrease in molecular weight (50%)	17	[37]
Increase in end groups, OH (100%) and COOH (170%)	17	[37]

Table 7. *In vivo* changes due to degradation and structural damage of polyester implant.

5. *In vitro* hydrolysis studies of implants:

In vivo ruptured (Section 4) PET showed numerous transverse cracks, reduction in diameter, loss in molecular weight and increase in end groups (Table 7). This is due to the internal degradation of PET in alkaline conditions (blood pH=7.4) under stressing environment (pulsatile force and force due to hypertension). During *in vitro* studies, we investigated the effect of applied stress and residual stress on PET implants at 40°C in the aqueous alkaline solution [78]. The temperature, chemical environment (alkaline pH), applied stress, and residual stress were chosen to mimic the physiological condition of the human body. To accelerate the process, the pH chosen was much higher (pH >13) than the pH of the human blood and extracellular fluid (pH = 7.4), [79,80]. The *in vitro* results produced with applied stress and residual stress in the alkaline chemical environment showed transverse cracks on the surface of the PET implants. This evidence also confirmed the findings of transverse cracks on PET prosthetic grafts removed from human bodies because of failures.

5.1. Effect of applied stress in alkaline environment

5.1.1. Weigh loss (%) and mechanical properties

Time (hour)	Applied load (MPa/kg)	Weight loss (%)		Loss in breaking load (%)		Loss in breaking strain (%)		% loss in work of rupture	
		[a]PET1	[b]PET2	[a]PET1	[b]PET2	[a]PET1	[b]PET2	[a]PET1	[b]PET2
24	0	3.4	5.6	6.4	4.0	5.0	4.3	30.6	10.4
	12.3/0.1kg	--	--	--	--	--	--	--	--
	61.5/0.5kg	3.5	5.4	14.9	8.1	19.9	12.9	36.2	26.2
	123/1 kg	3.4	4.8	18.2	10.2	21.6	18.2	39.2	30.7
	246/2 kg	4.5	5.4	18.4	22.8	25.5	30.5	43.6	51.5
48	0	8.0	11.0	16.0	10.9	7.5	4.5	27.9	18.0
	12.3/0.1kg	--	--	--	--	--	--	--	--
	61.5/0.5kg	8.2	10.8	19.0	17.9	23.2	19.1	42.2	37.6
	123/1kg	8.3	11.6	24.6	19.0	22.3	23.9	43.6	42.5
	246/2kg	8.5	10.5	32.1	44.4	36.6	44.3	62.7	72.5
96	0	15.3	18.5	23.0	22.7	12.9	12.0	35.8	32.3
	12.3/0.1kg	18.7	--	28.7	--	20.0	--	46.9	--
	61.5/0.5kg	16.5	17.6	28.9	26.2	25.6	19.6	53.8	45.8
	123/1kg	18.9	20.6	34.6	32.6	27.4	29.3	56.0	56.3
	246/2kg	20.2	25.2	58.6	73.1	51.5	65.6	82.0	86.7

Table 8. (%) loss in weight and mechanical properties of applied stressed polyester fibres in alkaline environment (10% NaOH), at 40°C [78].[a]PET1: Trevira (Hoechst) high tenacity PET; [b]PET2: Tergel Tech (Rhone Poulenc) high tenacity PET;

The weight loss (%) data of PET1 and PET2, are shown in Table 8. This table tracks the results of up to 96-h treatment period, in which relaxed and different stressing conditions were observed at 40°C. The weight loss (%) was approximately linear with time at a rate of 0.16% per hour for PET1 and 0.22% for PET2 for the relaxed ('0' load) and lower loaded PET samples (12.3–123 MPa). However, after 96 hour of treatment, weight loss was significantly higher for the highest loaded samples (246 MPa) than the relaxed and lower loaded samples.

The weight losses of PET2 samples were higher than the weight losses of the equivalent PET1 samples, even though PET1 is a slightly finer fiber (5.5 dtex) compared to PET2 (5.7 dtex). It must be realized that part of the weight loss for both samples is due to finish dissolving, although PET1 has the higher amount of finish (1.3%) than PET2 (1.0%).

The reason for the increased weight loss (%) due to highest applied stress is unknown. No *in vivo* weight loss data is available although thinning was reported in the ruptured polyester implant [5]. One suggestion is that the weight loss of the polyesters is related to the surface area of the fibre [45,69]. The increased surface area of the PET1 fibers which was found to be only ≈2.0%, by taking the extension of 5.0% due to the 2000 g (246 MPa) applied load from the load-elongation curve. Therefore, the slight increase in surface area for PET1 (≈2.0%) does not account for the greater discrepancy in the amount of weight loss for the relaxed and highest loaded samples of PET1 and PET2. Another factor that might have contributed to the higher weight losses was the fact that there were broken filaments similar to *in vivo* broken filaments as shown Figures 5 and 9 [5,73] in the hydrolysis bath of the highest loaded samples.

For both PET1 and PET2 samples, the breaking load decreased with an increase in treatment time as well as with increase in stressing on the fibers (Table 8). The breaking load loss of the stressed samples is higher than that of the relaxed samples and the most significant losses occurred at the highest loading conditions. For example, after 96 h of treatment, the PET1 samples which were under 246 MPa load, lost 58.6% of breaking load compared to only 23% for relaxed samples, under the same experimental conditions; PET2 samples lost 73 and 22.6%, respectively. It should also be noted that very similar loss in tenacity was observed for both PET samples.

Alkaline hydrolysis takes place only at the surface of the fibers resulting in thinner fiber, which was also reported in the *in vivo* ruptured implants; hence the percentage weight loss should be proportional to the percentage breaking load. However, a nonlinear relationship between weight loss and breaking load loss was reported due to the presence of solvents in the alkaline bath, titanium dioxide in the fiber, and crimp in the fiber [55,64,39]. The highest loaded samples (246 MPa) of both PET1 and PET2, in all treatment conditions, showed percentage loss in breaking load approximately three to four times than the percentage loss in weight. It is reported that in relaxed condition, the loss in breaking load for filament yarn can be between one and two times the percentage loss in weight [39,43,54], which the present *in vitro* study also confirms. For the highest loaded samples (246 MPa) the loss in breaking load and the loss in tenacity are very similar, therefore, the hydrolysis under applied loading conditions is no longer restricted to the surface. It is hypothesized that under higher loading conditions, the hydroxyl ions are able to penetrate deeper inside the PET fiber and causes internal hydrolysis. The scanning electron micrographs, shown later, display the formation of the deeper surface cracks under the heaviest loads.

The breaking strain (Table 8) and work of rupture (Table 8) decrease with the increasing treatment time and increasing loading condition, and the loss was most significant at the highest loading condition. For both PET1 and PET2, the modulus of the fibers decreased (~20%) significantly after 96 hour of treatment. However, the decrease in modulus was not significant between the samples of relaxed condition and lower loading conditions. For all PET fibers, loss of mechanical properties during alkaline hydrolysis, both in relaxed and loading conditions decreased in the following order: modulus<breaking strain<breaking load<work of rupture; with work of rupture being the highest losing property. The data

shown in the present study were obtained at 10% NaOH concentration which is unrealistic for medical applications; however, we have established that similar results could be found for more diluted concentrations of NaOH under stressing conditions [81].

Since the properties of the fibers decreased significantly when exposed to alkali and load, the PET samples were treated with water alone under loading conditions to understand the hydrolysis mechanism, and whether the loss of properties under such conditions was partly due to water hydrolysis under load. At room temperature, there were no significant changes in properties; however at 40°C, with the highest load of 246 MPa, PET1 samples shows a slight decrease (10%) in breaking load and tenacity, and a slight increase in modulus. A more significant decrease was observed in the breaking strain (20%) and work of rupture (30%). However, once again more significant decrease was observed for the breaking strain (20%) and work of rupture (30%) under highest loading conditions.

When treated with water and loading conditions, the physical appearances of the fibers did not change. The fibers retained their original luster after treatment in water for all loading conditions and at all temperatures. The handling properties of both relaxed and loaded samples in water were the same as the original fibers; whereas, alkaline-treated samples with higher loads tended to have a stiffer handle.

5.1.2. Effect of residual stress (pre-loaded samples) in alkaline environment

Weight loss (%) and Mechanical properties:

Treatment conditions	Time (hour)	Weight loss (%)	Breaking load (cN)	Breaking load (%)
PET1/NaOH/zero load	72	11.5	40.7±3.2	14.9
PET1/NaOH/zero load	72	13.6	44.9±1.8	14.3
PET1/NaOH/zero load	96	15.3	36.8±1.3	23.0
PET1/NaOH/zero load	96	18.5	40.5±1.6	22.7
PET1/NaOH/ 246MPa pre-load	48	7.9	33.8±2.3	29.3
PET1/NaOH/246 MPa pre-load	72	13.5	26.9±3.4	43.7
PET1/NaOH/246 MPa preload	96	19.4	22.2±2.1	53.6
PET2/NaOH/246 MPa pre-load	96	23.3	15.9±1.1	66.8

Table 9. Effect of residual stress (pre-loaded samples) in alkaline environment on PET weight loss (%), breaking load loss (%), and surface characteristics [82].

The weight loss (%), breaking load (cN) and breaking load (%) along with surface characteristics of PET1 and PET2 samples treated in aqueous NaOH in pre-loading conditions are given in Table 9. The weight loss (%) for both PET samples treated in aqueous NaOH in relaxed condition (no applied or pre-load) is increased with increasing treatment time. Similar trend in weight loss (%) increase with treatment time was observed for pre-loading (Table 9) conditions. However, like applied loaded samples in alkaline (NaOH) environment, the weight loss (%) of the preloaded samples was higher than the relaxed samples for the similar treatment conditions. For example, after 96 hours of treatment in aqueous NaOH, the relaxed PET1 samples (PET1/NaOH/zero load) lost 15.3% weight, whereas the pre-loaded samples (PET1/NaOH/246 MPa pre-load) lost 19.4% weight. The breaking load loss (%) for the relaxed samples treated in aqueous NaOH was similar to corresponding weight loss (%) except for 96 hours treated samples. The 96 hours treated samples in relaxed condition (PET1/NaOH/zero load) produced higher breaking load (%) than the corresponding weight loss (%). The breaking load loss (%) for the pre-loaded samples (PET1/NaOH/246 MPa pre-load; PET2/NaOH/246 MPa pre-load) was much higher than the corresponding weight loss (%).

5.1.3. *Effect of residual stress (pre-loaded samples) in acid environment and other alkaline (NH_4OH) environment*

The weight loss (%), breaking load (cN) and breaking load loss (%) for PET1 samples treated with H_2SO_4 and NH_4OH is given in Tables 10 and 11 respectively. The weight loss (%) was negligible for sulphuric acid treated samples in relaxed and applied loading conditions; however, the breaking load loss (%) was significantly higher and increased with increasing treatment time. For example, sample (PET-A/H_2SO_4/246 MPa applied-load) treated with 5% H_2SO_4 for 96 hours lost 70% of breaking load. Similar trend was observed with the NH_4OH treated samples (Table 11). The degree of attack on the mechanical properties of PET by the aqueous solution of NH_4OH is dependent on solution concentration and stressing conditions. While the weight loss (%) of NH_4OH treated samples was unchanged, the breaking load loss (%) was increased with the increasing applied loading conditions and concentration of NH_4OH. For example, at 2% NH_4OH and 246 MPa applied-load, the breaking load loss was 23.5% compare to 69.1% for 5% NH_4OH under similar treatment condition (Table 11). The weight loss (%) data agree with the data published in the literature as no weight loss (%) was obtained when PET was treated in aqueous ammonia at 30°C for 10 days [19] and n-butylamine at 21°C for 72 hours [39. However, a small amount of weight loss (3.3%) was reported when polyester was treated with 10% (w/w) aqueous solution of ethanolamine for 6 hour at 100°C [39].

Treatment conditions	Time (hour)	Weight loss (%)	Breaking load (cN)	Breaking load (%)
PET1/H_2SO_4/246 MPa applied-load	24	**NS	37.7±1.4	21.2
PET1/H_2SO_4/246 MPa applied-load	48	**NS	30.8±1.8	35.6
PET1/H_2SO_4/246 MPa applied-load	72	**NS	21.8±2.2	54.3
PET1/H_2SO_4/246 MPa applied-load	96	**NS	14.2±1.7	70.2

Table 10. Weight loss (%), breaking load loss (%), and surface characteristics of the H_2SO_4 treated samples in applied loading condition [82]. **NS: Not significant

Treatment conditions	Time (h)	Breaking load (cN)	Weight loss (%)	[c]Breaking load (%)
Control filament	0	45.9±2.8	0.0	0.0
[a]PET-A/NH_4OH /zero load)	120	43.5±4.2	0.5	5.2
[a]PET1/NH_4OH /123 MPa applied-load)	120	42.2±3.7	+2.6	8.1
[a]PET1/NH_4OH /246 MPa applied-load	120	35.1±3.1	+2.3	23.5
[b]PET1/NH_4OH /zero load)	120	31.3±2.5	0.7	31.8
[b]PET1/NH_4OH/123 MPa applied-load	120	24.2±2.8	+1.4	47.3
[b]PET1/NH_4OH/246 MPa applied-load	120	14.2±2.7	+1.8	69.1

Table 11. Weight loss (%), breaking load loss (%), and surface characteristics of the NH_4OH treated samples in relaxed and applied loading condition [82] [a]:concentration of NH_4OH:2%; [b]: concentration of NH_4OH:5%; treatment time:120 days. [c]:breaking load value in cN.

5.2. Scanning Electron Microscopy (SEM):

A remarkable phenomenon was observed when polyester fibers were hydrolyzed in alkali under stressing (applied and pre-stressing) conditions. After treatment with applied and pre-loads of 1.5 kg (184 MPa) and 2.0 kg (246 MPa), it was noticed that the fibers had lost their brightness, an effect obvious even to the naked eye. The delustred appearance was noticed at the loads of 184 MPa and 246 MPa and appeared to increase with increasing reaction time. Such a change in appearance was not observed when the samples were treated with alkalis alone, in either relaxed condition or lower level of loads (≤123 MPa), nor for the samples which were treated with the highest load of 2.0 kg in water alone. This delustring effect in the alkaline-treated samples is thought to be due to the presence of voids which are generated within the fibers, caused by exposure to both alkali and load.

The SEM micrographs of original PET1 (Figure 10) and PET2 (Figure 11) are cylindrical in shape and the surface of the fibers is smooth. Minor spots were seen on both PET samples which could be due to the particulate impurities during manufacturing of the fibers. When treated in relaxed condition with the alkali, the fiber surface remains quite smooth with a few defects detectable along the length of the fiber (Figure 12). These defects appear to be short, elongated pitting, parallel to the fiber axis. The presence of TiO_2 is responsible for this type of pitting on the surface as reported in the literature [60,64,83].

Figures 13-15 show the SEM microphotographs of PET1 treated with aqueous NaOH under 246 MPa stressing conditions at 40°C for different treatment times. The surface of these samples shows deeply penetrating and acutely elongated cracks which increased with treatment time. The fiber edge seen in the samples treated for 48 h (Figure 13) is smooth and unpitted. The same fiber at 72 h (Figure 14) shows more pitting. After 96 h of treatment, the sample had a weight loss of 20.2% and corresponding loss in work of rupture of 82.0%, and has also lost smooth linear edge of the fiber (Figure 15). The surface showed deep cracking and irregularities. Similar cracks and surface irregularities were observed in the sample treated with alkaline with a load of 184 MPa (Figure 16). By contrast, only shallow pitting on the surface (Figure 17) of the fibers was observed with the lower loaded samples (≤123

MPa). SEM microphotographs of PET2 samples which were treated for 48 h (Figure 18) and 96 h (Figure 19) respectively, with 246 MPa load also showed the same pattern of increased cracking and loss of smooth linear edge over time. The appearance of cracks on the room temperature treated samples is quite different than the appearance of cracks generated at 40°C. The cracks are rather 'dull' and 'unpronounced' at room temperature [81]; whereas, at 40°C, the heavily degraded samples of PET1 and PET2, the cracks are deeper and more acute, resembling a jagged 'corallite structures' (Figures 15 & 19).

Both PET1 and PET2 samples show the formation of large cracks of varying length and width across the surface of the fiber. Yet there are differences in the formation of cracks between PET1 and PET2 samples. In PET1 samples cracks are produced in two directions (Figure 15), longitudinal and transverse, with most of the cracks being transverse, at right angle to the direction of applied load. In the PET2 fibers (Figure 19) the cracks are transverse only, at right angles to the applied load. The cracks in the PET2 samples were also wider and longer. For example, the longest transverse crack in the PET1 sample was found to be about 6.7 μm (Figure 15), while in the PET2 it was about 7.5 μm (Figure 19). It is likely that the greater length of the cracks in the PET2 samples is responsible for the larger loss in mechanical properties. The differences between these two PET samples may reflect differences in processing history and in particular differences in the draw ratio and thermal history.

As the surface cracks formed only in the most heavily loaded samples, it was suspected that the load alone might have been responsible for the formation of cracks. However, SEM microphotographs of PET1 and PET2 samples which were treated with only water and 246 MPa load at 40°C temperature for 96 h show no cracks, nor any longitudinal marks such as are seen in alkaline-treated samples treated under relaxed conditions. PET sample treated with water and highest load are very similar to the original PET sample.

Since it is evident that the surface defects can be detected by treating PET with NaOH in relaxed and applied loading conditions, further investigations were carried out to find out whether residual stressing condition, which can be exist in the implant while in the human body, could be responsible for the formation of cracks. The residual stressing condition is created in the polyester by 'pre-loading' the PET samples [82] and then treating with alkaline solution in relaxed condition. This pre-loading condition creates residual stress in the samples. The residual stress in the textile structure is generated in two ways: (a) during textile transformation (spinning/winding/warping/weaving) and (b) during normal functioning in the human body due to intermittent pulsatile forces. It is reported that the residual stress in the textile structure is responsible for premature rupture of small caliber straight tubes (bifurcated grafts, femoro-popliteal grafts and axillo-femoral grafts) [84]. Figure 20 shows the scanning electron micrograph obtained from the pre-stressed sample of PET1 (PET1/NaOH/246 MPa pre-load) treated in 10% alkaline solution at 40°C. Both transverse and axial cracks can be seen in this micrograph. In contrast, only transverse cracks are present in the PET2 pre-stressed (PET2/NaOH/246 MPa pre-load) sample. The axial cracks in the pre-stressed sample (Figure 20) are much wider and deeper than the relaxed sample (Figure 12), however, both axial and transverse cracks are much less prominent than those of the 246 MPa applied-loaded sample (Figures 15 & 19). The axial cracks in the pre-stressed sample is much prominent than the transverse cracks. It seems that the axial cracks in all three alkaline treated PET1 samples (relaxed-Figure 12, applied-loaded-Figure 15 and pre-loaded-Figure 20) are parallel to each other and the angle between the axial cracks and filament axis is approximately about 2° to 5°. Similar axial pits and craters have been observed in the explanted Vanguard stent grafts [85].

No surface cracks were observed when the PET1 samples were treated with 5% sulphuric acid in relaxed condition and in stressing condition (Figures 21; PET1/H_2SO_4/246 MPa applied- load). Similar surface characteristics were observed when the PET1 samples treated with 2% and 5% NH_4OH in relaxed (PET1/NH_4OH/zero load; Figure 22), 123 MPa loading (PET1/NH_4OH/123 MPa applied-load; Figure 23) and 246 MPa loading (PET1/NH_4OH/246 MPa applied-load; Figure 24) conditions. The smooth appearance of the acid and ammonium hydroxide treated micrographs with 246 MPa loading conditions was surprising as both samples (PET1/H_2SO_4/246 MPa applied-load; PET1/NH_4OH/246 MPa applied-load) lost about 70% breaking load (Tables 10 & 11). The results suggest that the reaction of H_2SO_4 and NH_4OH in relaxed and stressed conditions (below and above yield loads) takes place throughout the actual substance of the fibres with a gradual chemical degradation of PET and splitting of linkages in the long chain polyester molecules. No cracks have been observed when polyester was treated with neat n-butylamine at 21°C for 72 hours [46]; however, formation of spiral, helical, transverse and axial cracks on various amine treated polyesters have been reported [46,86].

The implications of the *in vitro* study become significant and applicable when compared to PET prosthetic grafts removed from the deceased human bodies. In these cases, where PET failures resulted in deaths [5,7,37,72], similar transverse cracks and irregularities in fiber edge were observed. This *in vitro* study confirms that transverse cracks formed on the PET prosthetic grafts when the applied load exceeded the yield load in an alkaline environment. The yield load was measured using 'The Slope Threshold Method *[87]* which was found to be 1.22 ± 0.05 kg for PET1 and 1.24 ± 0.08 kg for PET2 which is lower than the applied load of 1.5 kg (184 MPa) and 2.0 kg (246 MPa). In certain parts of the human body, the PET prosthetic grafts remain under stressing conditions which is greater than critical level of load [72]. Also, due to the hypersensitiveness in some patients, PET implants encounter higher level of loads. Where the environment is similar to the alkaline condition (blood pH =7.4) and above yield loads, deep transverse cracks formed, which eventually cause failure in the PET prosthetic grafts, resulting in death. We speculate that the higher temperature (40°C) is responsible for an increased mobility of the OH^- ions, thus producing deeper penetration in PET fibers through the highly stressed ester bonds and resulting in the 'corallite structures' observed [78].

The surface characteristics of PET are different for different loading conditions and PET types. When the applied load was below yield load (12.3-123 MPa) of PET, only longitudinal cracks were observed in the PET1 samples and no visible cracks in either directions were observed in the PET2 samples except the 96 hours treated PET1 samples with 123 MPa applied loading condition (Figure 17) which shows the onset of crack formation . However, when the applied load was above the yield load which was 2000 g or 246 MPa applied-load and pre-load, severe longitudinal and axial cracks were generated in the PET1 samples (Figures 13-16, 20) and axial cracks were seen in the PET2 sample (Figures 18-19). The length and width of the cracks were increased with the severity of the hydrolysis. The longitudinal cracks were more prominent in the stressed and pre-stressed PET1 sample (Figures 13-16, 20) than the relaxed PET1 sample (Figure 12) for the similar hydrolysis conditions. The axial crack patterns (crack length, width and frequency) are different for PET1 and PET2 for the similar treatment condition. The difference in dimensions of crack size in PET1 and PET2 is due to the presence of different size crystals which depends on the processing and thermal history. The size and amount of crystals formation is dependent of the rate of nucleation rate. It is known that smaller and numerous crystals are formed with high nucleation rate while larger and few crystals are formed with slow nucleation rate [33]. Further, the differences in crack dimensions (width and length) in a particular PET are dependent on the fibrillar dimensions as well as stress in-homogeneity. For PET, the fibrillar transverse dimension was found to be between 150-1000 Å and the fibrillar length was 500-10,000 Å [88].

The transverse and longitudinal cracks in the stressed and pre-stressed PET samples are similar to those observed in the ruptured PET implants *in vivo* that resulted in numerous human deaths [5,72,85]. The longitudinal cracks in the PET1 sample might have been originated from the surface defects or due to the presence of TiO_2 as the sample is bright and contained about 0.1% TiO_2. Although TiO_2 should not be used in medical grade PET as it damages cell membranes [89] and increases the rate of degradation [60,64]. The scanning electron micrographs of titanium dioxide induced longitudinal cracks formation have been published for medical (Figure 25) and apparel grade polyesters [64,90]. Alternatively, the PET1 sample might have manufacturing defects, possibly from the heat setting treatments.

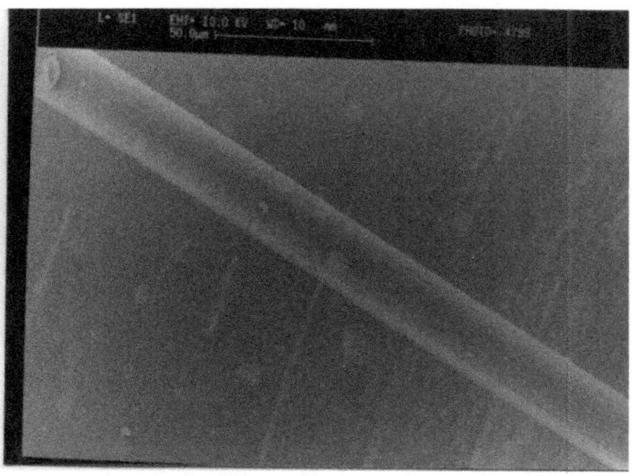

Figure 10. SEM micrograph of virgin PET1 [91].

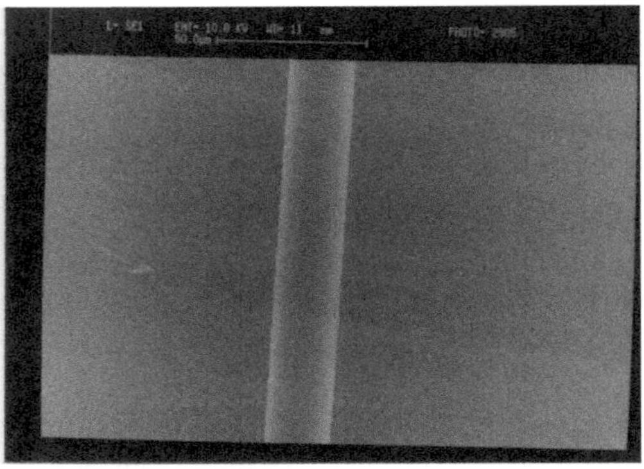

Figure 11. SEM micrograph of virgin PET2 [91].

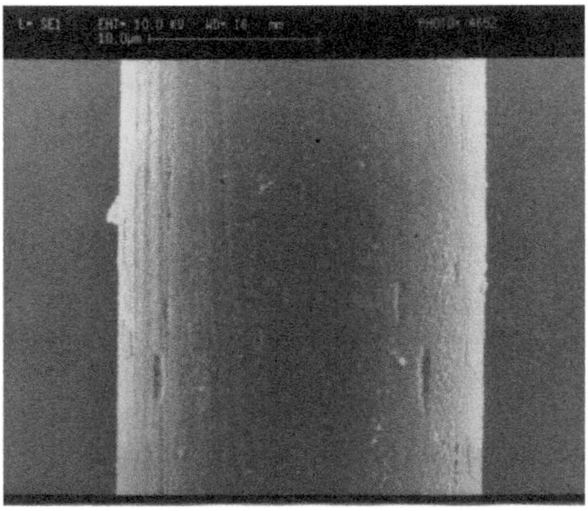

Figure 12. Scanning electron micrograph of PET1 treated in relaxed state with 10% NaOH, 96 hours/40°C [91].

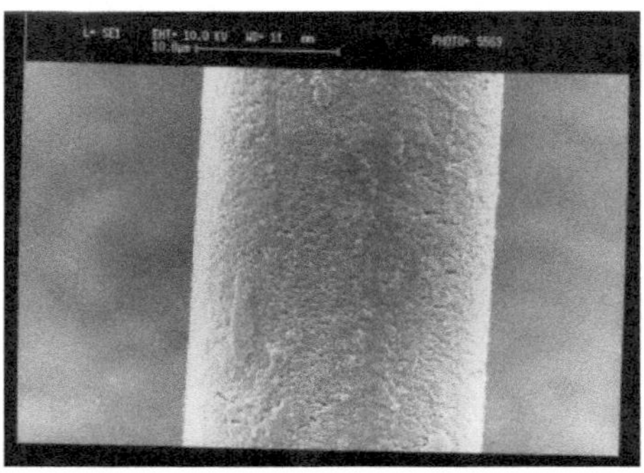

Figure 13. SEM micrograph of PET1 sample treated with 9.81% NaOH/48 h/40 ° C temperature/2 kg loads [78]

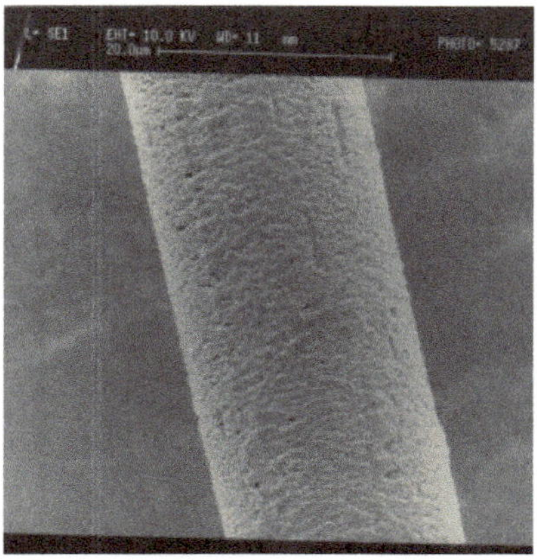

Figure 14. SEM micrograph of PET1 sample treated with 9.81% NaOH/72 h/40°C temperature/2 kg load [78].

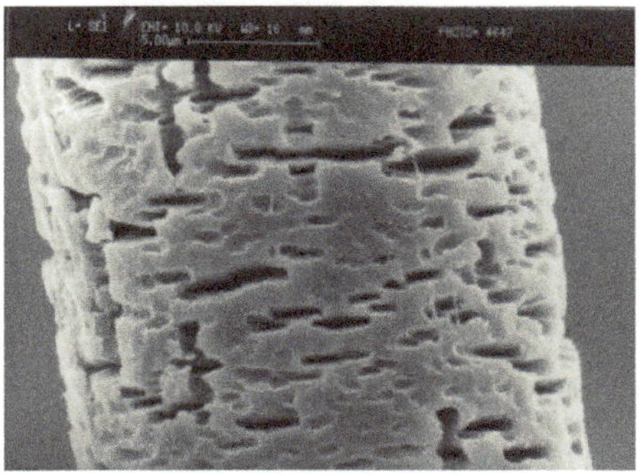

Figure 15. SEM micrograph of PET1 sample, treated with 9.81% NaOH/96 h/40°C temperature/2 kg load (246 MPa) [78].

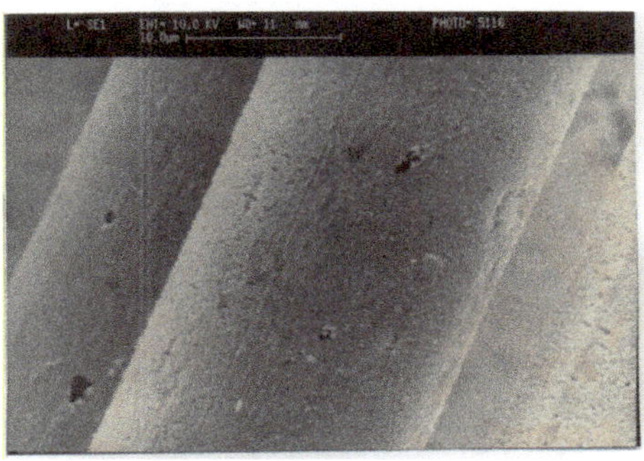

Figure 16. Scanning electron micrograph of PET1 treated at room temperature for 20days under 1500g (184 MPa) stressing condition in 10% NaOH [91].

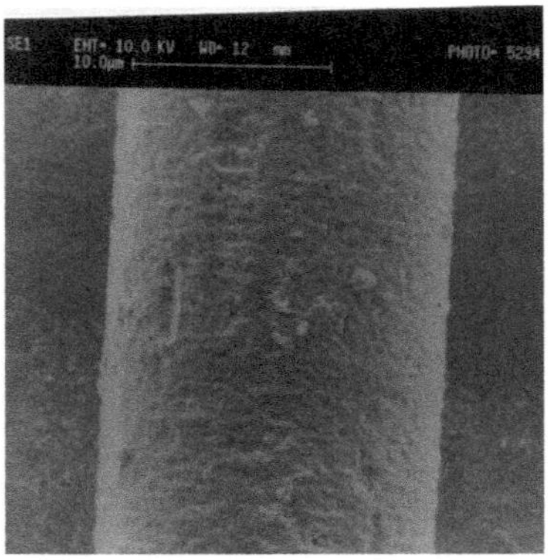

Figure 17. Scanning electron micrograph of PET1 treated at room temperature for 20days under 1000g (123 MPa) stressing condition in 10% NaOH [81].

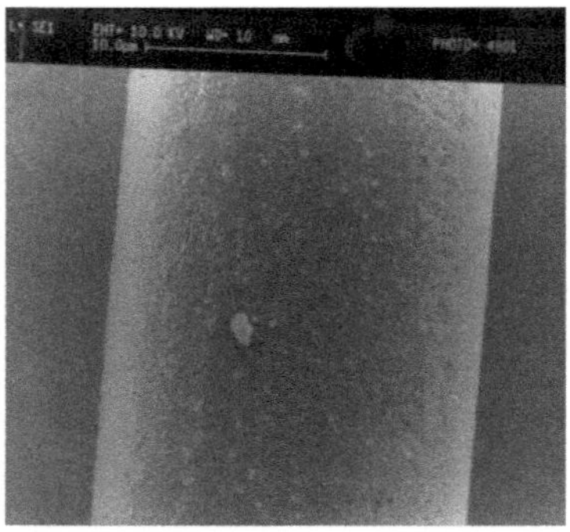

Figure 18. SEM micrograph of PET2 sample, treated with 10.06% NaOH/48 h/40°C temperature/246 MPa [78].

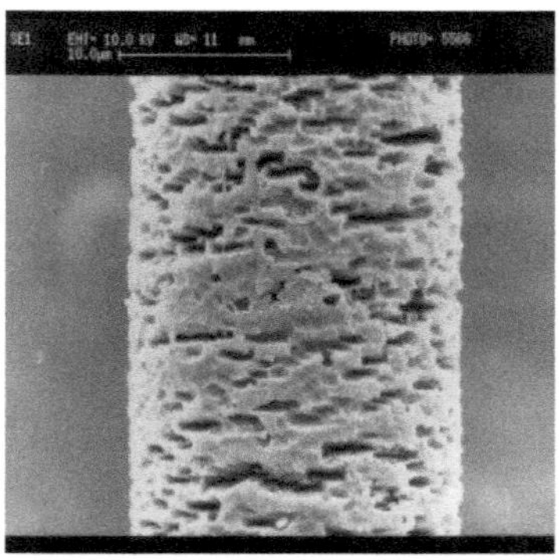

Figure 19. SEM micrograph of PET2 sample, treated with 10.06% NaOH/96 h/40°C temperature/246 MPa [78].

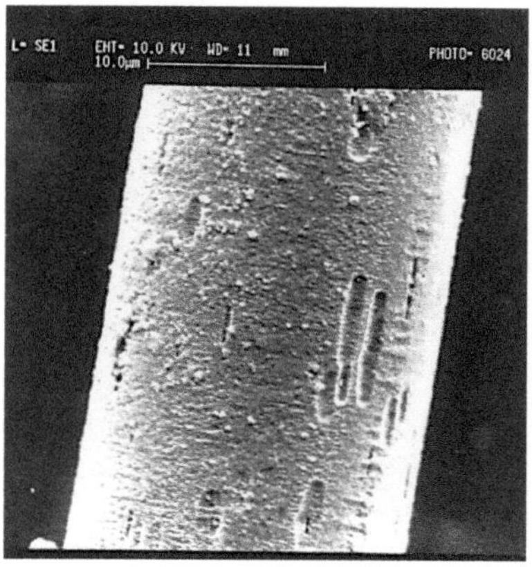

Figure 20. Micrograph of pre-stressed sample (pre-stressed for 48 hours) treated with 10% NaOH in relaxed condition, 40°C for 96 hours [82].

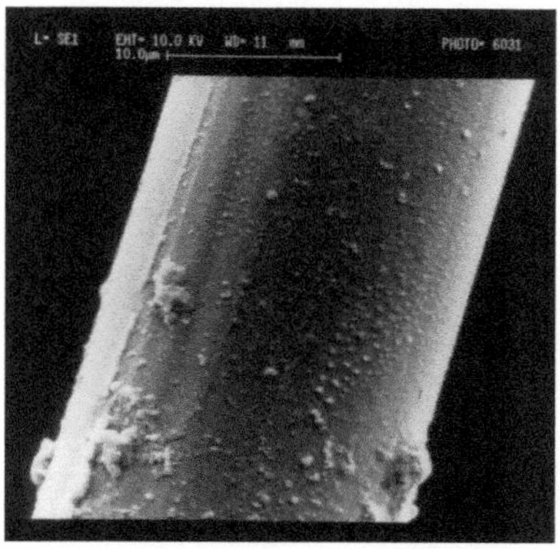

Figure 21. Micrograph of 2000g stressed sample treated with 5% H_2SO_4, 40°C/48 hours [82].

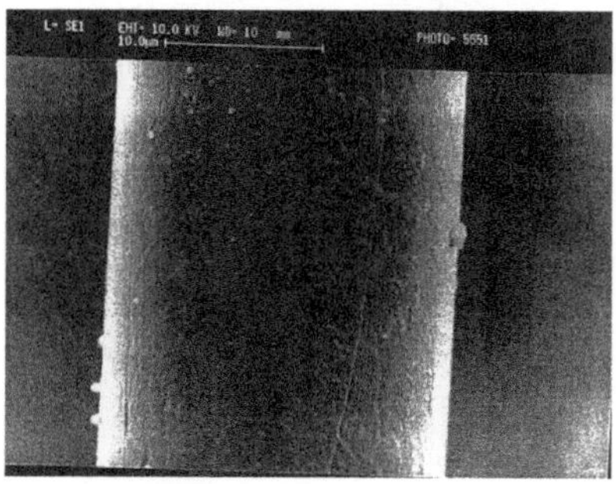

Figure 22. SEM micrograph of PET sample treated with 5% ammonium hydroxide, at room temperature, 120 days, under relax condition [91].

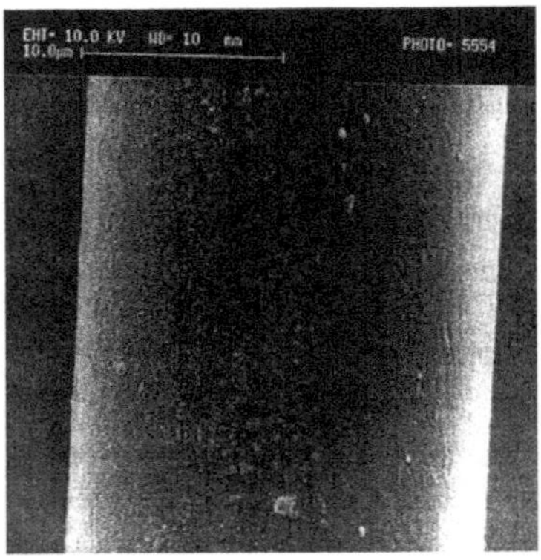

Figure 23. SEM micrograph of PET1 treated with 5% ammonium hydroxide, at room temperature, 120 days, under 1kg load [91].

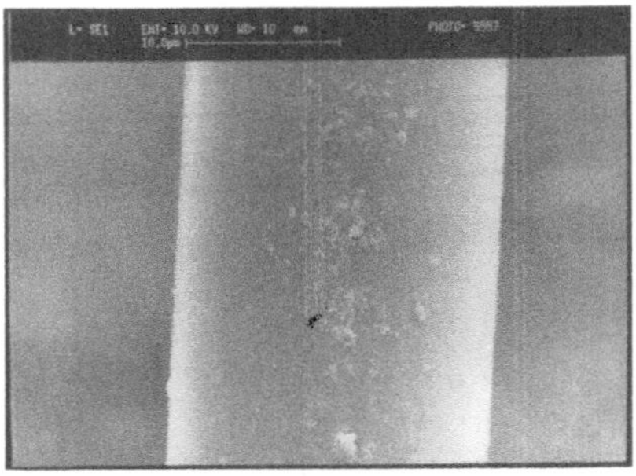

Figure 24. Micrograph of 2000g stressed sample treated with 5% NH_4OH, room temp./120 days [91].

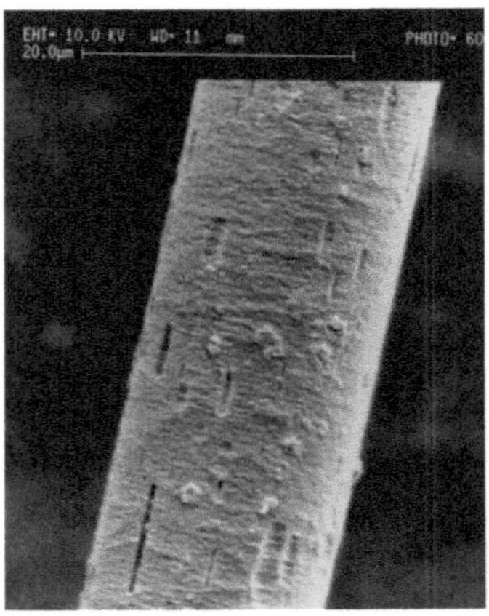

Figure 25. SEM micrograph of PET1 sample, treated with 4.0% NaOH, 4h at 100°C. (Titanium dioxide content: 0.1%) [90]

6. Accelerated degradation tests and lifetime prediction:

Due to the physical and chemical variability within the PET, it is very difficult to estimate how long PET implants will last in the physiological conditions. One suggestion is that when using PET medical devices for human spare parts, the surgeons must be committed to a follow-up of such a patient for his/her remaining life. Only by this means will the safety and effectiveness of the spare part become known. Others plea for reporting all graft failures, preferably with identifying details of material and construction and an attempt to evaluate factors responsible for failure.

Previously it was reported that the weight loss (%) of medical grade PET roughly fits the ICI formula [91] given in Equation (1). An empirical equation was also reported relating to loss in breaking load with treatment time and applied load for a specific NaOH concentration and treatment temperature [78]. The generic empirical equation is given in Equation (2) and the Equations (3) and (4) were computed by Statistical Package for Social Science (SPSS) 'Enter Method' for PET1 and PET2 respectively at 40°C and 10% NaOH concentration using breaking load (cN) data. The breaking load data for PET1 and PET2 is given in Table 8.

$$Weight\ loss(\%) = \frac{0.47t(x^2+6x)}{1.082^{100-T}\sqrt{d}} \qquad (1);$$

Where t = time in hour, x = concentration of NaOH (%); T = temperature in °C; d = decitex of fibre.

$$Y = \beta_0 + \beta_1\, T + \beta_2\, L^2 \qquad (2)$$

where Y = predicted breaking load (cN); T = Treatment time (day); L = applied load (g);

$$\begin{gathered} Y = 47.1 - 3.44\,T - 2.26 * 10^{-6}\,L^2 \\ (R^2 = 0.85, S = 2.88, 95\%\ confidence\ \lim = \pm 6.27) \end{gathered} \qquad (3)$$

$$\begin{gathered} Y = 56.4 - 4.79\,T - 4.36 * 10^{-6}\,L^2 \\ (R^2 = 0.90, S = 3.56, 95\%\ confidence\ \lim = \pm 7.77) \end{gathered} \qquad (4)$$

In the Equations (3) and (4), R^2 is the coefficient of determination, S is the standard deviation about regression and 95% confidence limits for the predicted breaking load Y are given by $\pm t_{0.025}$ S, where $t_{0.025}$ is obtained from probability points of the t distribution table using n-2(=12) degrees of freedom.

Using the Equation (3) for PET1, the lifetime for 50% loss in breaking load for 'zero applied loading' condition would be only ≈7 days in 10% NaOH at 40°C. Since the dependence of NaOH concentration with the weight loss (%) is (x^2+6x) according to Equation (1), assuming the rate of weight loss is proportional to breaking load loss, we can calculate the lifetime using lower concentration of alkali which is likely to encounter by PET implants. Hence, for 0.01% or 0.0025M NaOH (pH<11), the lifetime would be about 50 years in zero loading condition.

Similarly for PET2 sample, using the Equation (4) and taking the dependence of NaOH ($x^2 + 6x$), the life time would be about 45 years in 0.0025 M NaOH at 40°C. The calculated lifetime for PET varies between 16.5 years to 214 years depending on the exposure conditions (Table 12). Since the pH of blood is only 7.4, physiological fluids alone cannot be responsible for pre-mature rupture of PET implants as reported in the numerous medical literatures (Table 2). However, where the pre-mature rupture occurred, perhaps the main contributing factor was the site specific physiological loads that exceed the polyester yield load. Another reason for pre-mature rupture of PET implants was the use of apparel grade PET materials [10] as it was found that the retrieved PET implants in Hospital Saint-Joseph, Marseille, France, was manufactured from semi-dull, round cross-section, textured multifilament, Type 56T Dacron. This specific type of polyester was apparel grade and not to be used for implants. The textured and semi-dull polyesters degrade faster than the clear (bright) and flat filament polyester yarns [46].

Predicted life-time (yeas)	Types of PET	Conditioned considered	Reference
27	Bottled PET	Wet soil	[92]
214	High tenacity filament, 1100 detex, 200f	pH = 10 (Na^+), 25°C	[24]
16.5	High tenacity filament, 1100 detex, 200f	Ca^{++}, 25°C	[24]
100	High tenacity filament	Saturated soil, pH=2-10	[24]

Table 12. Life-time predictions for PET.

7. Conclusions

In order to predict a realistic lifetime of a PET implant, a far greater studies are required than the present one and any previous studies. The polyesters must be carefully selected before their implantation considering the exposure conditions and fibre physical and fine structures. Use of apparel grade polyesters to manufacture implants must be forbidden as changes in polymerization techniques, additives, and processing treatments all enhance the long term durability of polyesters. The textile properties of implants must be recorded before their implantation as it would be easier to compare with the ruptured explanted samples. There is a clear lack of communications between surgeons and textile scientists as most of the ruptured explanted samples analyzed without using any textile standards.

Thus far, the durability of polyester implants is still in questionable because of pre-mature ruptures. In an attempt to address the longevity issue, 'lifetime' of medical grade polyester has been calculated from empirical equations. However, to avoid short-term accelerated tests for long term physiological durability, it is recommended to test polyester in animal models.

Author details

Mashiur Rahman

Department of Textile Sciences, Faculty of Human Ecology, University of Manitoba, Winnipeg, Manitoba, Canada

8. References

[1] Carothers W H, Hill J W [1952] Studies of polymerization and ring formation XV: Artificial fibres from synthetic linear condensation super polymers. J. Amer. Chem. Soc. 54:1579-1587.

[2] Ludwig H (1971). Polyester Fibres Chemistry and Technology, Wiley Interscience, London, 1p.

[3] Whinfield J R, Dickson J T (1947). Brit. Pat. 578,079.

[4] Voorhees A B Jr, Laretzki A, Blakemore A H (1952) The Use of Tubes Constructed from Vinyon "N" Cloth in Bridging Arterial Defects (a Preliminary Report). Ann. Surg. 135:332– 336.

[5] Berger K, Sauvage L (1981) Late Fiber Deterioration in Dacron Arterial Grafts. Anna. Surg. 193:477-491.

[6] Rajendran S, Anand S C (2002) Development in Medical Textiles, Textil. Progr. 32:1–37.

[7] Wilson S E, Krug R, Mueller G, Wilson L (1997) Late Disruption of Dacron Aortic Grafts. Ann.Vasc. Surg. 11:383-386. 7a. Maarek JM, Guidoin R, Aubin M, Prud'homme RE, (1984) Molecular weight characterization of virgin and explanted polyester arterial prostheses. J. Biomed. Res. 18:881-894.

[8] Van Damme H, Depre M, Creemers E, Limet R (2005) Intrinsic Structural Failure of Polyester (Dacron) Vascular Grafts. A General Review. Acta. Chir. Beig. 105:249-255.

[9] Tomizawa Y (2003) Vascular Grafts: Basic Research and Clinical Applications. In: Tura A, editor. Vascular Grafts Experimenting and Modelling, Advances in Fluid Mechanisms, Vol: 34, WIT Press, Southampton, pp. 1–39.

[10] King M W, Zhang Z, Guidoin R (2001) Microsctructural Changes in Polyester Biotextiles During Implantation in Humans. J. Textil. Apparel. Technol. Manag. 1:1–8.

[11] Ottinger L W, Darling R C, Wirthlin L S, Linton R R (1976). Failure of Ultra light weight Knitted Dacron Grafts in Arterial Reconstruction. Arch. Surg. 111:146–149.

[12] Chakfe N, Riepe G, Dieval F et al. (2001) Longitudinal ruptures of polyester knitted vascular prostheses. J. Vasc. Surg. 33:1015-1022.

[13] Trippestad A (1985). Late Rupture of knitted Dacron Double Velour arterial Prostheses, Acta. Chir. Scand. 151:391-395.

[14] Falkai B V(1996)), Developments in the production of industrial PET filaments. Asian Text. Res. 5:26-38.

[15] Adanur S (1995) Fibre Properties and Technology. In: Adanur S. editor. Wellington Sears Handbook of Industrial Textiles, Lancaster, USA: Technomic Publishing Co. Inc. pp 555-607.

[16] Shimizu J, Kikutani T (1993) Polyesters 50 Years of Achievement, The Textile Institute, Manchester, UK, 166p.

[17] Ward I, Cansfield D O, Carr P (1993). Polyesters 50 Years of Achievement, The Textile Institute, Manchester, UK. pp 192-195.

[18] Hearle J W S, Greer R (1970) Fibre Structure.Text. Prog, 2:1-18.

[19] Zeronian S H, Collins M J (1988). Text. Chem. Colour. 20:25-28.

[20] Ziabicki A (1996). Text. Res. J. 66:705-712.

[21] Baur K H (1993). Polyesters 50 Years of Achievement, The Textile Institute, Manchester, UK.
[22] Tamblyn J W, Newland G C (1965). Induction period in the aging of polypropylene. J. Appl. Polym. Sci. 9:2261-2260.
[23] Voskamp W, Risseeuw P (1987) Method to establish the maximum allowable load under working conditions of polyester reinforcing fabrics. Geotext. Geomembr. 6:173-184.
[24] Horrocks A R, D'Souza J A (1992). Degradation of Polymers in Geotextiles and Geomembranes. In: Hamid S, Amin M, Maadhah A, editors. Handbook of Polymer Degradation, (Eds). New York: Marcel Dekker Inc, PP433-505.
[25] Stessel R I, Hodge J S (1995) Chemical resistance testing of geomembrane liners. J. Hazard. Mater. 42:265-267
[26] Smith L V, Devires K L (1993). Mechanical properties of polymeric fibres exposed to stress in a NO_x environment. Polymer, 34:546-550.
[27] Morgan R J, Pruenda C O, Buller N et al. (1984). 29th SAMPE Symposium, April 3-5, 891.
[28] Ely J L (1984) Polyurethane in Biomedical Engineering, Planck H, Egbers G, Syre L. editors. Elsevier Science, Amsterdam, 1-15.
[29] Lecrecq B, Schaeffner M, Delmas ph et al. (1990). Proc. 4th Int. Conf. on Geotextiles, Geommebranes and Related Products, Hoedt, G. D. (Edt), The Hague, Balkema, Rotterdam, pp 679-684.
[30] Van Zantem R V(1986) Geotextiles and Geomembranes in Civil Engineering, Balkema, Rotterdam, pp 213-241.
[31] Davis G W (1989) Aging and durability of polyester geotextiles. In: Koerner R. M. editor. Durability and Aging of Geosyntheticd. New York: Elsevier Applied Science. p65.
[32] Korshak V V, Vinogradova S V (1965) Polyesters, Pergamon Press, Oxford, England.
[33] McIntyre J E (1998 Polyester Fibres, In: Lewin M, Pearce E M, editor. Handbook of Fibre Science and Technology: Fibre Chemistry, Volume IV. NY: M. Dekker, NY, pp 1-73.
[34] Sprague C J, Davis G W (1991) High-Tech Fibrous Materials, Chapter-20. ACS-Series-457, pp 304-319.
[35] Sato M (1983). Sen-i Gakkaishi, 39, T209-219.
[36] Hall J, Whinfield J R (1952). U.S. Pat., 2,590,402.
[37] Chaouch W, Dieval F, Le Nouen, D, Defoin A, Chakfe N, Durand B(2009) ^{1}H-NMR spectroscopic study of the effect of aging vascular prostheses made of poly(ethylene terephthalate) on the macromolecular weight. J. Biomed. Mater. Res.: Part A, 91A, 939-952.
[38] Goedt G D (1990). Discussuin of polyester durability, Proc. 4th Int. Conference on Geotextiles, Geomembranes and Related Products, The Hague, Balkema, Rotterdam, 1171.
[39] ICI (UK), Chemical Properties of Terylene, The Effects of Alkalis, Industrial Fibre Manual, 1978; Section A1/3.
[40] Gawish S M, Ambroise G(1986) Amer Dyestuff Reporter. 5:30-32.
[41] Dave J, Kumar R, Srivastava H C (1987) Studies on modification of polyester fabrics I: Alkaline hydrolysis. J. Appl. Polym. Sci. 33:455-477.
[42] Teli M, Purkaystha A (1993) Amer. Dyestuff. Reporter. 82:34-40.
[43] Houser K D (1983) Text. Chemists Colourists. 15:70/37-72/39.
[44] Ellison M S, Fisher L D, Zeronian S H (1982) J. Appl. Polym. Sci. Physical properties of polyester fibers degraded by aminolysis and by alkalin hydrolysis. 27:247–257.

[45] Yamamoto Y, Sangen O, Nakano H (1984). Sen-i Gakkaishi, 40:T122-124.
[46] Zeronian S H, Collins M J (1989) Surface Modification of Polyester by Alkaline Treatments. Text. Prog. 20:1-26
[47] Namboori C G G, Haith M S (1968) Steric effects in the basic hydrolysis of poly(ethylene terephthalate). J. Appl. Polym. Sci. 12:1999–2005.
[48] Namboori C G G (1969). Text. Chem. Colourists, 1:50/24-51/25.
[49] Olson L M, Wentz M (1984). Text Chem Colourists, 16:48/35-54/41.
[50] Cho H, Huh M W, Furuhata K, Sakamoto M (1992). Sen-i Gakkaishi, 48:T610-618.
[51] Grancaric A M, Kallay N (1993) Kinetics of polyester fiber alkaline hydrolysis: Effect of temperature and cationic surfactants. J App. Polym. Sci. 49:175–181.
[52] Shenai V A, Lokre D B (1978). Text. Dyer Printer, 11:27-28.
[53] Sanders E M, Zeronian S H (1982) An analysis of the moisture-related properties of hydrolyzed polyester. J. Appl. Polym. Sci. 27:4477–4491.
[54] Maillo J, Gacen J, Naik A (1993). Melliand Textilber. 74: E291-293.
[55] Jan S S, Uen J L, Shyu G A, Wuu C L, Chen C C (1992). Amer. Dyest. Rep. 81:32-37&40.
[56] Mathur A, Netravali A N, O'Rourke T D (1994). Geotext. Geomemb. 13:591-626.
[57] Halse Y, Koerner R M, Lord A E, (1987a) Effect of high levels of alkalinity on geotextiles. Part I: $Ca(OH)_2$ solutions. Geotext. Geomemb 5:261-282.
[58] Halse Y, Koerner R M, Lord A E (1987b) Effect of high levels of alkalinity on geotextiles. Part II: NaOH solutions. Geotext. Geomemb, 6: 295-305.
[59] Niu S, Wakida T (1993) Effect of Heat-Setting Temperature on Alkali Hydrolysis of Poly (Ethylene Terephthalate) Fibers. Text. Res. J. 63:346-350.
[60] Collins M J, Zeronian S H, Meyer M S (1991) The use of aqueous alkaline hydrolysis to reveal the fine structure of poly(ethylene terephthalate) fibers. J. Appl. Polym. Sci. 42:2149–2162.
[61] Hashimoto T (1959). Sen-i Gakkaishi, 15:794.
[62] Nunn D B (1999) Structural Failure of Dacron Grafts. Semin. Vasc. Surg. 12: 83-91.
[63] Tsuji W, Nakao T, Doyama Y et al. (1984). Sen-i Gakkaishi, 40:T350-358.
[64] Solbrig C M, Obendorf S K (1991) Alkaline Hydrolysis of Titanium Dioxide Delustered Poly(ethylene Terephthalate) Yarns. Text. Res. J. 61:177-181.
[65] Postman W (1980). Text. Res. J. 50:444-453.
[66] Marshal R M, Dardoufas K C(1977). US Patent, 4,019,990.
[67] Marshal R M, Dardoufas K C (1976). US Patent, 3,962,516.
[68] Latta B (1984) Improved Tactile and Sorption Properties of Polyester Fabrics Through Caustic Treatment, Text. Res. J. 54:766-775.
[69] Guang L, Lixia G, Tong S (1993) *J. China Text. Univ.* 10:34.
[70] McIntyre J E (1985) *In:* Handbook of Fibre Science and Technology, *Vol. 4: Fibre Chemistry.*
[71] Berman S S, Hunter G C, Smyth S H et al. (1995) Application of computed tomography for surveillance of aortic grafts. Surgery, 118: 8-15.
[72] Yashar J J, Richman M H, Dyckman J, Witoszka M, Burnard R J, Weyman A K, Yashar J (1978) *Surgery,* 84: 659.

[73] Riepe G, Loos J, Imig H, Schroder A, Schneider E, Petermann J, Rogge A, Ludwig M, Schenke A, Nassutt R, Chakfe N, Morlock M (1997) Long-term in vivo Alterations of polyester Vascular Grafts in Humans. Eur. J. Vasc. Endovasc. Surg. 13:540-548.

[74] Rao T J, Pan C, Guidon R, Marceau D et al. (1991) Soft Filamentous Woven Polyester Arterial Prostheses from China. Biomaterials,12:335-344.

[75] Guidoin R, Marois Y, Douville Y, et al. (2000) First-Generation Aortic Endografts: Analysis of Explanted Stentor Devices from the EUROSTAR Registry. J. Endovasc. Ther. 7:105–122.

[76] Riepe G, Heintz C, Kaiser E et al. (2002) What can we Learn from Explanted Endovascular Devices. Eur. J. Vasc. Endovasc. Surg, 24:117-122.

[77] Ambrad-Chalela E, Shi Q, Berman A et al. (2001) Favorable Histologic Findings and Tensile Strength at 14 Years in Knitted Polyester Aortofemoral and Femoropopliteal Grafts in the Same Patient. Ann. Vasc. Surg. 15:578-581.

[78] *Rahman M, East G (2006)* Effect of applied stress on the alkaline hydrolysis of poly(ethylene terephthalate) at 40°C: Relevance to medical textiles. J. Appl. Polym. Sci. 102:4814-4822.

[79] Marieb E P (2000) *Essentials of Human Anatomy and Physiology,* 6th ed. San Francisco, USA:*Addison Wesley Longman.* 80p.

[80] Chan J C (1983) *Adv. Pediatr.* 30:401.

[81] East G, Rahman M (1999) Effect of applied stress on the alkaline hydrolysis of geotextile poly(ethylene terephthalate). Part 1: room temperature, Polymer, 40:2281-2288.

[82] Rahman, M, East G (2012). *In Vitro* Study of Poly(ethylene terephthalate) Implants for Long-term Durability. Text. Res. J. accepted for publication.

[83] Fujimoto K, Iohara K, Ohwaki S, Murase Y *J (1991) J. Appl. Polym. Sci.* 42:1509.

[84] Damme H, Depre M, Creemers E, Limet R (2005) Intrinsic Structural Failure of Polyester (Dacron) Vascular Grafts. A General Review. Acta. Chir. Beig. 105:249-255.

[85] Chuter T A M (2009) Durability of Endovascular Infrarenal Aneurysm Repair: When Does Late Failure Occur and Why? Semin. Vasc. Surg. 22:102-110.

[86] Holmes S A (1996). Surface Defect Geometry/Tensile Failure Relationships of Aminolyzed Poly(ethylene terephthalate) Text. Res. J. 66:214-218.

[87] *INSTRON Series IX (1992), Automated Materials Testing System, Version-5, D12–14.*

[88] Tucker P, George W (1972) Microfibres Within Fibres: A Review, Pol. Eng. Sci. 12: 364-377.

[89] Valant J, Drobne D, Novak S. (2011) Effect of ingested titanium dioxide nanoparticles on the digestive gland cell membrane of terrestrial isopods. Chemosphere, 87:19-25.

[90] Rahman M, East G (2009) Titanium Dioxide Particle-induced Alkaline Degradation of Poly(ethylene Terephthalate): Application to Medical Textiles. Text. Res. J. 79:728-736.

[91] Rahman M (1997). The Alkaline Degradation of Poly(ethylene terephthalate) Geotextiles, University of Leeds, UK.

[92] Allen N S, Mohammad M, Edge M, Jones K. (1991) Environmental Degradation of Poly(ethylene Terephthalate). Text. Res. J. 61:690-696.

Section 2

Polyester Composites

Chapter 6

Fibre Reinforced Polyester Composites

Salar Bagherpour

Additional information is available at the end of the chapter

1. Introduction

A composite material is a non uniform solid consisting of two or more different materials that are mechanically bonded together. Each of the various components retains its identity in the composite and maintains its characteristic structure and properties. Generally, the structure of a composite consists of two phases, matrix and reinforcement. The matrix is a continuous phase and the reinforcement is a discontinuous one. The duty of reinforcements is attaining strength of the composite and the matrix has the responsibility of bonding of the reinforcements. There are recognizable interface between the materials of matrix and reinforcements. The composite materials, however, generally possess combination of properties such as stiffness, strength, weight, high temperature performance, corrosion resistance, hardness and conductivity which are not possible with the individual components. Indeed, composites are produced when two or more materials or phases are used together to give a combination of properties that cannot be achieved otherwise. Composite materials especially the fiber reinforced polyester (FRP) kind highlight how different materials can work in synergy. Analysis of these properties shows that they depend on (1) the properties of the individual components; (2) the relative amount of different phases; (3) the orientation of various components; the degree of bonding between the matrix and the reinforcements and (4) the size, shape and distribution of the discontinuous phase. The material involves can be organics, metals or ceramics. Therefore, a wide range of freedom exists, and composite materials can often be designed to meet a desired set of engineering properties and characteristics [1].

There are many types of composite materials and several methods of classifying them. One method is based on the matrix materials which include polymers, metals and ceramics. The other method is based on the reinforcement phase which has the shape of fiber, particulate and whisker. Whiskers are like fibers but their length is shorter. The bonding between the particles, fibers or whiskers and the matrix is also very important. In structural composites, polymeric molecules known as coupling agent are used. These molecules form bonds with the dispersed phase and become integrated into the continuous matrix phase as well. The most popular type of composite material is the fiber-reinforced polyester composites, in which continuous thin fibers of one material such as glass, carbon or natural fibers are embedded in a polyester matrix. They are also called glass fiber reinforced polyester (GFRP), carbon fiber reinforced polyester (CFRP) and natural fiber reinforced polyester (NFRP). The objective is usually to enhance strength, stiffness, fatigue, resistance, or strength to weight ratio by incorporating strong and stiff fibers in a softer, more ductile matrix. The microstructure of a selected GFRP composite is shown in figure 1.

The usages of fiber reinforced polyesters are in airplanes, electronics components, automotives, rail ways and wagon systems and sporting equipments. Beside their desired mechanical properties, their resistance to corrosion is also a tempting factor to use these composite in different areas. Although they are sensitive to UV light, heat and moisture environments, good maintenance could increase their life time. In this chapter different phases of FRPs, the mechanical relationships between different components of FRPs, the mechanism of degradation and aging of FRPs and application of them is discussing [2].

Figure 1. Microstructure of glass fibre reinforced polyester composite.

2. Different phases of FRP composites

2.1. Fiber reinforcements

The fiber is an important constituent in FRP composites. A great deal of research and development has been done with the fibers on the effects in the types, volume fraction, architecture, and orientations. The fiber generally occupies 30% - 70% of the matrix volume in the composites. The fibers can be chopped, woven, stitched, and braided. They are usually treated with sizing such as starch, gelatin, oil or wax to improve the bond as well as binders to improve the handling. The most common types of fibers used in advanced composites for structural applications are the fiberglass, aramid, and carbon. The fiberglass is the least expensive and carbon being the most expensive. The cost of aramid fibers is about the same as the lower grades of the carbon fiber. Other high-strength high-modulus fibers such as boron are at the present time considered to be economically prohibitive.

2.2. Glass fibers

The glass fibers are divided into three main classes E-glass, S-glass and C-glass. The E-glass is designated for electrical use and the S-glass for high strength. The C-glass is for high corrosion resistance, and it is uncommon for civil engineering application. Of the three fibers, the E-glass is the most common reinforcement material used in civil and industrial structures. It is produced from lime-alumina-borosilicate which can be easily obtained from abundance of raw materials like sand. The fibers are drawn into very fine filaments with diameters ranging from 2 to 13 X 10^{-6} m. The glass fiber strength and modulus can degrade with increasing temperature. Although the glass material creeps under a sustained load, it can be designed to perform satisfactorily. The fiber itself is regarded as an isotropic material and has a lower thermal expansion coefficient than that of steel [3].

There are also the other fiber glasses which are used for FRP reinforcement as well as;

- A-glass, soda lime silicate glasses used where the strength, durability, and good electrical resistivity of E-glass are not required.
- D-glass, borosilicate glasses with a low dielectric constant for electrical applications.
- ECR-glass, calcium alumino silicate glasses with a maximum alkali content of 2 wt.% used where strength, electrical resistivity, and acid corrosion resistance are desired.
- AR-glass, alkali resistant glasses composed of alkali zirconium silicates used in cement substrates and concrete.
- R-glass, calcium alumino silicate glasses used for reinforcement where added strength and acid corrosion resistance are required.
- S-2-glass, magnesium alumino silicate glasses used for textile substrates or reinforcement in composite structural applications which require high strength, modulus, and stability under extreme temperature and corrosive environments.

Table 1 and 2 show the chemical and mechanical properties of different glass fibers respectively.

2.1.2. Aramid fibers

These are synthetic organic fibers consisting of aromatic polyamides. The aramid fibers have excellent fatigue and creep resistance. Although there are several commercial grades of aramid fibers available, the three most common ones used in structural applications are Kevlar 29, Kevlar 49 and Kevlar 149. The Young's Modulus curve for Kevlar 29 is linear to a value of 83 GPa but then becomes slightly concave upward to a value of 100 GPa at rupture; whereas, for Kevlar 49 the curve is linear to a value of 124 GPa at rupture (see Table 3). As an anisotropic material it's transverse and shear modulus are an order of magnitude less than those in the longitudinal direction. The fibers can have difficulty achieving a chemical or mechanical bond with the resin [4].

	A-glass	C-glass	D-glass	E-glass	ECR-glass	AR-Glass	R-glass	S-2-glass
SiO2%	63-72	64-68	72-75	52-56	54-62	55-75	55-60	64-66
Al2O3%	0-6	3-5	0-1	12-16	9-15	0-5	23-28	24-25
B2O3%	0-6	4-6	21-24	5-10		0-8	0-0.35	
CaO%	6-10	11-15	0-1	16-25	17-25	1-10	8-15	0-0.2
MgO%	0-4	2-4		0-5	0-4		4-7	9.5-10
ZnO%					2-5			
BaO%		0-1						
Li2O%						0-1.5		
Na2O+K2O%	14-16	7-10	0-4	0-2	0-2	11-21	0-1	0-0.2
TiO2%	0-0.6			0-4	0-4	0-12		
ZrO2%						1-18		
Fe2O3%	0-0.5	0-0.8	0-0.3	0-0.8	0-0.8	0-5	0-0.5	0-0.1
F2%	0-0.4					0-5	0-0.3	

Table 1. Chemical composition of different glass fibers.

	A-glass	C-glass	D-glass	E-glass	ECR-glass	AR-Glass	R-glass	S-2-glass
Density (gr/cm3)	2.44	2.52	2.11-2.14	2.58	2.72	2.70	2.54	2.46
Tensile Strength (MPa) at -196°C		5380		5310	5310			8275
Tensile Strength at 23 °C	3310	3310	2415	3445	3445	3241	4135	4890
Tensile Strength at 371 °C				2620	2165		2930	4445
Tensile Strength at 538 °C				1725	1725		2140	2415
Modulus of Elasticity (GPa) at 23 °C	68.9	68.9	51.7	72.3	80.3	73.1	85.5	86.9
Modulus of Elasticity at 538 °C				81.3	81.3			88.9
Elongation%	4.8	4.8	4.6	4.8	4.8	4.4	4.8	5.7

Table 2. Mechanical properties of different glass fibers.

Fibers	Tensile Strength (MPa)	Modulus of Elasticity (GPa)	Density (gr/cm3)
Kevlar 29	2920	83-100	1.43
Kevlar 49	3000	124	1.44
Kevlar 149	3450	143	1.47

Table 3. Mechanical properties of aramid fibers.

2.1.3. *Carbon fibers*

Carbon fibers refer to fibers which are at least 92 wt.% carbon in composition. They can be short or continuous; their structure can be crystalline, amorphous, or partly crystalline. The crystalline form has the crystal structure of graphite (Figure 2), which consists of sp2 hybridized carbon atoms arranged two-dimensionally in a honeycomb structure in the x-y plane. Carbon atoms within a layer are bonded by (1) covalent bonds provided by the overlap of the sp2 hybridized orbitals, and (2) metallic bonding provided by the delocalization of the p_z orbitals, i.e., the π electrons. This delocalization makes graphite a good electrical conductor and a good thermal conductor in the x-y plane. The bonding between the layers is Vander Waals bonding, so the carbon layers can easily slide with respect to one another; graphite is an electrical insulator and a thermal insulator perpendicular to the layers. Due to the difference between the in-plane and out-of-plane bonding, graphite has a high modulus of elasticity parallel to the plane and a low modulus perpendicular to the plane. Thus, graphite is highly anisotropic.

The high modulus of a carbon fiber stems from the fact that the carbon layers, though not necessarily flat, tend to be parallel to the fiber axis. This crystallographic preferred orientation is known as a fiber texture. As a result, a carbon fiber has a higher modulus parallel to the fiber axis than perpendicular to the fiber axis. Similarly, the electrical and thermal conductivities are higher along the fiber axis, and the coefficient of thermal expansion is lower along the fiber axis.

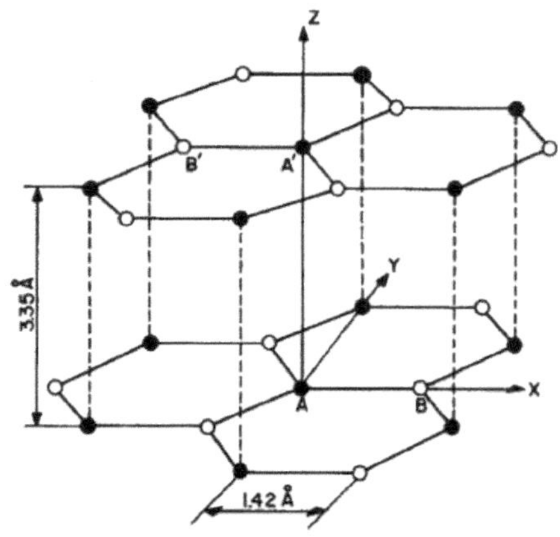

Figure 2. The crystal structure of graphite.

The proportion of graphite in a carbon fiber can range from 0 to 100%. When the proportion is high, the fiber is said to be graphitic, and it is called a graphite fiber. However, a graphite fiber is polycrystalline, whereas a graphite whisker is a single crystal with the carbon layer rolled up like a scroll. Because of their single crystal nature, graphite whiskers are virtually flaw-free and have exceptionally high strength. However, the production yield of graphite whiskers is too low for them to be commercially significant.

Commercial carbon fibers are fabricated by using pitch or polyacrylonitrile (PAN) as the precursor. Precursor fibers are fabricated by conventional spinning techniques, such as wet spinning for PAN and melt spinning for pitch. They must be converted to a form which is flameproof and stable at the high temperatures (>700°C) involved in carbonization. Therefore, before carbonization (pyrolysis), they are stabilized for the case of the PAN precursor, or infusiblized for the case of the pitch precursor. Both stabilization and infusiblization are carried out in an oxidizing atmosphere. After that, general-purpose and high-performance fibers are obtained by carbonization in an inert atmosphere, followed by graphitization at >2500°C in an inert atmosphere if a high modulus is desired, whereas activated carbon fibers are obtained by activating in a reactive atmosphere, such as steam at elevated temperatures. To enhance the preferred orientation in the high-performance carbon fibers, graphitization can be performed while the fibers are under tension. The higher the graphitization temperature, the greater the preferred orientation.

For the case of pitch as the precursor, isotropic pitch gives an isotropic carbon fiber, which belongs to the category of general-purpose carbon fibers, whereas anisotropic pitch (such as mesophase pitch) gives high-performance carbon fibers which have the carbon layers preferentially parallel to the fiber axis.

Surface treatments of carbon fibers are essential for improving the bonding between the fibers and the polymer matrix. They involve oxidation treatments and the use of coupling agents, wetting agents, and/or sizing (coatings). Carbon fibers need treatment both for thermosets and thermoplasts. As the processing temperature is usually higher for thermoplasts than thermosets, the treatment must be stable to a higher temperature (300-400"C) when a thermoplast is used.

Although there are many carbon fibers available on the open market, they can be arbitrarily divided into three grades as shown in Table 4. They have lower thermal expansion coefficients than both the glass and aramid fibers. The carbon fiber is an anisotropic material, and its transverse modulus is an order of magnitude less than its longitudinal modulus. The material has a very high fatigue and creep resistance.

Fibers	Tensile Strength (MPa)	Modulus of Elasticity (GPa)	Density (gr/cm3)
Carbon, High strength	2480	230	1.8
Carbon, High Modulus	1790	370	1.9
Carbon, ultra High Modulus	1030-1310	520-620	2.0-2.1

Table 4. Mechanical properties of carbon fibers.

Since its tensile strength decreases with increasing modulus, its strain at rupture will also be much lower. Because of the material brittleness at higher modulus, it becomes critical in joint and connection details, which can have high stress concentrations. As a result of this phenomenon, carbon composite laminates are more effective with adhesive bonding that eliminates mechanical fasteners [5].

2.1.4. *Polyethylene fibers*

A high strength, high modulus polyethylene fiber called Spectra™ was developed at Allied Signal Technologies during the 1980s. The Spectra is based on ultra-high-molecular-weight polyethylene (UHMWPE). It has a specific gravity of 0.97, meaning that it is the only reinforcing fiber available that is lighter than water. The Spectra is available in three classification (spectra 900, 1000 and 2000). Nominal properties are listed in table 5. The high specific strength of the fibers makes it appropriate for tensile application. The glass transition temperature of UHMWPE is in the range of -20°C to 0°C, and hence the fiber is in the rubbery state at room temperature and exhibits time-dependent (viscoelastic) behavior. This feature imparts outstanding impact resistance and toughness, but lead to undesirable creep effects under long-term sustained loading. The melting temperature of the fiber is 147 °C, and hence use of polyethylene fibers is limited to relatively modest temperatures.

Fibers	Tensile Strength (MPa)	Modulus of Elasticity (GPa)	Density (gr/cm^3)
Spectra 900	2600	70	0.93-1.45
Spectra 1000	3200	105	0.93-1.45
Spectra 2000	3400	115	0.93-1.45

Table 5. Mechanical properties of Carbon fibers.

2.2. Resin systems of the matrix

The resin is another important constituent in composites. The two classes of resins are the thermoplastics and thermosets. A thermoplastic resin remains as solid at room temperature. It melts when heated and solidifies when cooled. The long-chain polymers do not chemically cross link. Because they do not cure permanently, they are undesirable for structural application. Conversely, a thermosetting resin will cure permanently by irreversible cross linking at elevated temperatures. This characteristic makes the thermoset resin composites very desirable for structural applications. The most common resins used in composites are the unsaturated polyesters, epoxies, and vinyl esters; the least common ones are the polyurethanes and phenolics.

2.2.1. *Polyester resins*

Generally polyester resins can be made by a dibasic organic acid and a dihydric alcohol. They can be classified as saturated polyester, such as polyethylene terephthalate, and unsaturated polyester. To form the network of the composite matrix, the unsaturated group or double bond needs to exist in a portion of 8 the dibasic acid. By varying the acid and alcohol, a range of polyester resins can be made. Orthophthalic polyesters are made by phthalic anhydride with either maleic anhydride or fumaric acid. Isophthalic polyesters, however, are made from isophthalic acid or terephthalic acid. The polyester resin is usually dissolved in monomer (styrene is the most widely used), which will copolymerize with it and contribute to the final properties of the cured resin. The addition of catalyst will cause the resin to cure. The most frequently used catalyst is methyl ethyl ketone peroxide (MEKP) or benzoyl peroxide (BPO) and the amount varies from 1-2%. The catalyst will decompose in the presence of the polyester resin to form free radicals, which will attack the unsaturated groups (like C=C) to initiate the polymerization. The processing temperature and the amount of the catalyst can control the rate of polymerization, the higher temperature or the more the catalyst, the faster the reaction. After the resin turned from liquid to brittle solid, post cure at higher temperature may need to be done. The purpose of the post cure is to increase Tg of the resin by complete cross-linking. The properties of the polyester resin are affected by the type and amount of reactant, catalyst and monomers as well as the curing temperature. The higher the molecular weight of polyester and the more points of unsaturation in molecules, the higher is the strength of the cured resins. Orthophthalic polyesters are environmentally sensitive and have limited mechanical properties. They have been replaced in some applications by isophthalic polyesters due to the excellent environment resistance and improved mechanical properties of the latter.

The unsaturated polyester amounts to about 75% of all polyester resins used in the world. It is produced by the condensation polymerization of dicarboxylic acids and dihydric alcohols. The formulation contains an unsaturated material such as maleic anhydride or fumaric acid which is a part of the dicarboxylic acid component. The formulation affects the viscosity, reactivity, resiliency and heat deflection temperature (HDT). The viscosity controls the speed and degree of wet-out (saturation) of the fibers. The reactivity affects cure time and peak exotherm (heat generation) temperatures. High exotherm is needed for a thin section curing at room temperature and low exotherm for a thick section. Resiliency or flexible grade composites have a higher elongation, lower modulus and HDT. The HDT is a short term thermal property which measures the thermal sensitivity and stability of the resins. The advantages cited in the unsaturated polyester are its dimensional stability and affordable cost. Other advantages include ease in handling, processing, and fabricating. Some of the special formulations are high corrosion resistant and fire retardants. This resin is probably the best value for a balance between performance and structural capabilities [3,4].

2.2.2. Epoxies

The epoxies used in composites are mainly the glycidyl ethers and amines. The material properties and cure rates can be formulated to meet the required performance. Epoxies are generally found in marine, automotive, electrical and appliance applications. The high viscosity in epoxy resins limits it use to certain processes such as molding, filament winding, and hand lay-up. The right curing agent should be carefully selected because it will affect the type of chemical reaction, pot life and final material properties. Although epoxies can be expensive, it may be worth the cost when high performance is required. Table 6 shows mechanical properties of polyester and epoxy resins.

2.2.3. Vinyl esters

The vinyl ester resins were developed to take advantage of both the workability of the epoxy resins and the fast curing of the polyesters. The vinyl ester has higher physical properties than polyesters but costs less than epoxies. The acrylic esters are dissolved in a styrene monomer to produce vinyl ester resins which are cured with organic peroxides. A composite product containing a vinyl ester resin can withstand high toughness demand and offer excellent corrosion resistance [3].

	Composition (Amorphous)	Use	Tensile Strength (MPa)	Modulus of Elasticity (GPa)	Density (gr/cm^3)
Polyesters	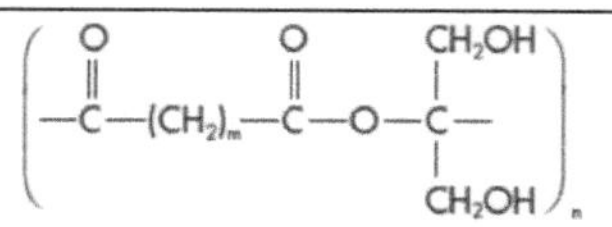	Fiber glass, laminate	40-85	1.3-4.5	1.1-1.4
Epoxies	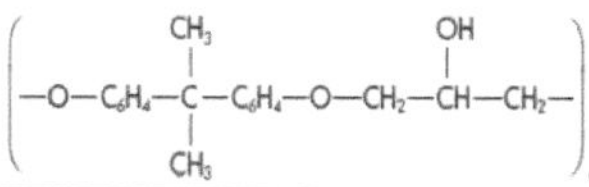	Fiber glass, adhesives	40-85	2.1-5.5	1.2-1.4
	Flexural Strength (MPa)	Compressive Strength (MPa)	Fracture Toughness K_{Ic} (MPa m $^{1/2}$)	Thermal Expansion (10^{-6} K^{-1})	
Polyesters	205- 690	140- 410	0.5	100-200	
Epoxies	1000-1500	150- 825	0.6- 1.0	55-110	

Table 6. Mechanical properties of polyester and epoxy resins.

3. Mechanical relationships in FRP composites

Improved strength, fatigue resistance, Young`s modulus, and strength to weight ratio are provided in most fiber reinforced composites by incorporating strong, stiff, but brittle fibers into a softer, more ductile matrix. The matrix material transmits the force to the fibers, which carry most of the applied force. The matrix also provides protection for the fiber surface and minimizes diffusion of species such as oxygen or moisture that can degrade the mechanical properties of fibers. The strength of the composite may be high at both room temperature and elevated temperatures.

3.1. Tensile properties

The mechanical properties of FRPs are related to the mechanical properties and fractional volume of each phases, matrix and reinforcement. This is called the rule of mixture. By using the rule of mixture the properties as well as strength, modulus of elasticity, density, electrical and thermal conductivity are predictable. For example if ρ_c is density of a composite and $\rho_c, \varrho_2, \dots, \varrho_n$ and $f_1, f_2, \dots f_n$ are the densities and volume fraction of each phases respectively in the composite, the ρ_c is according to relation 1.

$$\rho_c = \sum(f_j.\rho_j) = f_1\rho_1 + f_2\rho_2 + \cdots + f_n\rho_n \quad (1)$$

So according to the rule of mixture the density of a FRP composite is :

$$\rho_c = f_m\,\rho_m + f_f\,\rho_f \quad (2)$$

Where the subscripts *m* and *f* refer to the matrix and fiber. Note that f_m= 1- f_f .

In addition the rule of mixture is used to predict the modulus of elasticity when fibers are continuous and unidirectional. Parallel to the fibers, the modulus of elasticity may be as high as:

$$E_c \parallel = f_m\,E_m + f_f\,E_f \quad (3)$$

However, when the applied stress is very large, the matrix beings to deform and the stress-strain curve is no longer linear (figure 3). Since the matrix now contributes little to the stiffness of the composite, the modulus can be approximated by:

$$E_c \parallel = f_f\,E_f \quad (4)$$

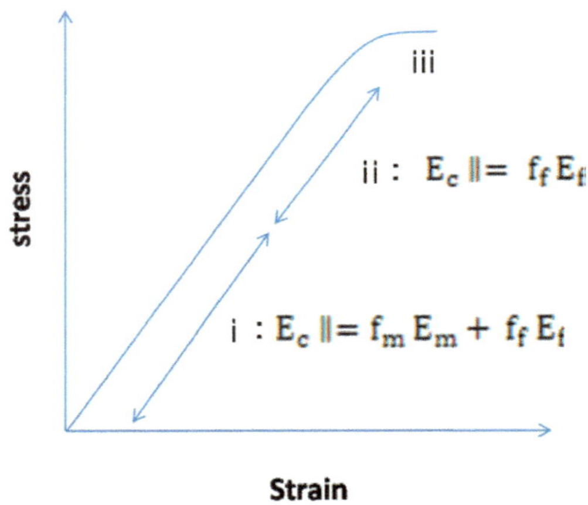

Figure 3. The stress strain curve for fiber- Reinforced composite. At low stresses (region i), the modulus of elasticity is given by the rule of mixtures. At higher stresses (region ii), the matrix deforms and the rule of mixture no linger obeyed.

When the load is applied perpendicular to the fibers, each component of the composite acts independently of the other. The strain of composite is according to the rule of mixture.

$$\varepsilon_c \parallel = f_m\,\varepsilon_m + f_f\,\varepsilon_f \quad (5)$$

And it could be change to relation 6.

$$\frac{\sigma_c}{E_c} = f_m\left(\frac{\sigma_m}{E_m}\right) + f_f\left(\frac{\sigma_f}{E_f}\right) \quad (6)$$

Since $\sigma_c = \sigma_m = \sigma_f$ then the modulus of elasticity for perpendicular load applied condition is according to relation 7 .

$$\frac{1}{E_{c,\perp}} = \frac{f_m}{E_m} + \frac{f_f}{E_f} \tag{7}$$

Tensile strength of FRP composites (TS_c) depends on the bonding between the fibers and the matrix. However, the rule of mixtures is sometimes used to approximate the tensile strength of a composite containing continuous parallel fibers:

$$TS_c = f_m \sigma_m + f_f TS_f \tag{8}$$

Where TS_f is the tensile strength of the fiber and σ_m is the stress acting on the matrix when the composite is strained to the point where the fiber fractures. Thus, σ_m is not the actual tensile strength of the matrix. Other properties, such as ductility, impact properties, fatigue properties, and creep prosperities, are difficult to predict even for unidirectional aligned fibers.

Many factors must be considered when designing a FRP composite, including the length, diameter, orientation, amount and properties of fibers, matrix and bonding between fibers and matrix.

In the case of fibers length and diameter it should be considered that fibers can be short, long or even continuous. Their dimensions are often characterized by the aspect ratio l/d, where l is the fiber length and d is the diameter. Typical fibers have diameter varying from 10 μm to 150 μm.

The strength of composite improves when the aspect ratio is large. Fibers often fracture because of surface imperfections. Making the diameter as small as possible gives the fiber less surface area and, consequently, fewer flaws that might propagate during processing or under a load. The long fiber is more preferable. The ends of a fiber carry less of the load than the remainder of the fiber, consequently, the fewer the ends, the higher the load ability of the fibers (figure 4).

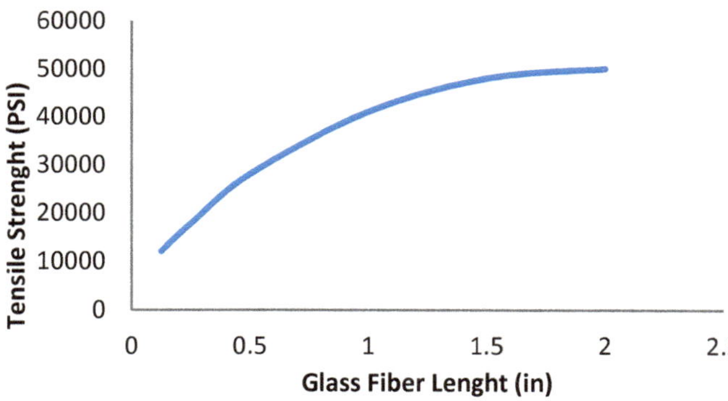

Figure 4. Increasing the length of chopped E-glass fibers in a polyester matrix increases the strength of the composite.

In many FRP systems, discontinuous fibers with an aspect ratio greater than some critical value are used to provide an acceptable compromise between processing ease and properties. A critical fiber length l_c for any given fiber diameter d can be determined by relation 9.

$$l_c = \frac{TS_f d}{2\,\tau_i} \tag{9}$$

Where TS_f is the strength of the fiber and τ_i is related to the strength of the bond between the fiber and the matrix, or the stress at which the matrix begins to deform. If the fiber length l is greater than about $15l_c$, the fiber behaves almost as if it was continuous. The strength of the composite can be estimated from relation 10.

$$\sigma_c = f_f\, TS_f \left(1 - \frac{l_c}{2l}\right) + \mathrm{f_m}\sigma_\mathrm{m} \tag{10}$$

Where σ_m is the stress on the matrix when the fibers break.

Amount of fibers is another factor which controls the strength and stiffness of composite. As we expect from rule of mixture the greater volume fraction of fibers means the more strength and stiffness of the composite. However, the maximum volume fraction is about 80%, beyond which fibers can no longer be completely surrounded by the matrix [1].

Orientation of fibers also could change the amount of strength and stiffness of the composite. The reinforcing fibers may be introduced into the matrix in a number of orientations. Short, randomly oriented fibers having a small aspect ratio are easily introduced into the matrix and give relatively isotropic behavior in the composite.

Long or even continuous unidirectional arrangements of fibers produce anisotropic properties, with particularly good strength and stiffness parallel to the fibers. These fibers are often designated as 0° piles, indicating that all of fibers are aligned with the direction of the applied stress. However, unidirectional orientations provide poor properties if the load is perpendicular to the fibers (figure5).

In most FRP composites the fibers are strong, stiff and lightweight. If the composite is to be used at elevated temperatures, the fiber should also have a high melting temperature. Thus the specific strength and modulus of elasticity of the fiber are important characteristic. In specific characteristic, density has an important role. The lighter the material, the more strength and stiffness. Relation 11 and 12 show the specific strength and modulus. Where ρ is density of the composite.

$$Specific\ Sternght = \frac{TS}{\rho} \tag{11}$$

$$Specific\ Modulus = \frac{E}{\rho} \tag{12}$$

3.2. Flexural properties

Three point flexural tests were conducted to assess the effects of outdoor environments on the flexural strength of FRP composite. The specimens were tested with the exposed surface in compression; in general, however, failure occurred in tension at mid span of specimens. Figure 6 shows flexural strength of ASTM standard, original and acid immersed FRP specimens. The matrix of the specimens is polyester and the reinforcement is E-glass. The specimens were kept for 15 years in storage. Hence, the environmental factors such as humidity, weathering and UV light could lead to aging of them and decrease flexibility of specimens. Furthermore, for increasing the effect of aging the specimens were immersed in acid solution (HCl 30%) at 50°C for interval of 7, 14 and 21 days. The results show that the flexural strength is decreased by aging of composite as well as tensile strength.

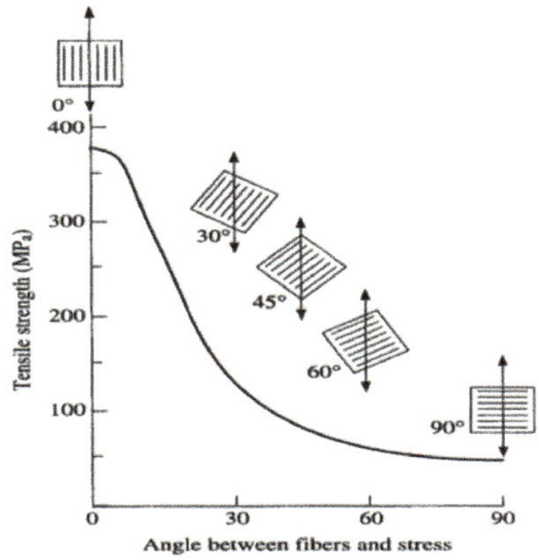

Figure 5. Effect of fiber orientation on the tensile strength of FRP composite. The most strength obtains where the force is applied along the fibers and the least is where the force is applied perpendicular to the fibers.

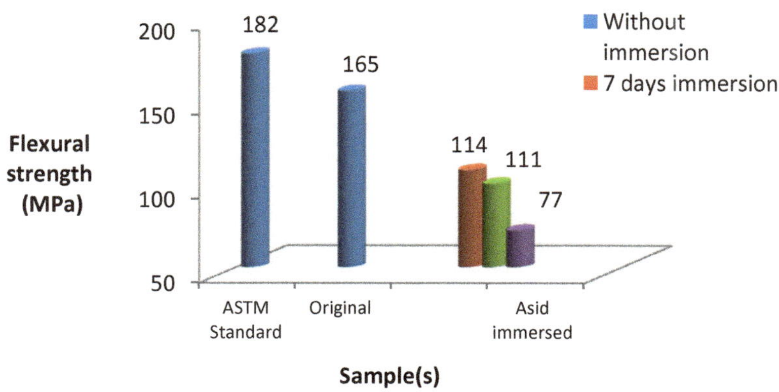

Figure 6. Figure6. Comparison of Flexural strength of FRP specimens (ASTM standard, Original or non immersed and immersed samples). The original samples were stored aged for 15 years. And the immersed samples are the original ones which were immersed at 50°c for different intervals.

By comparison between standard and tested specimens it is revealed that the flexural strength shows 10% decrease after 15 years storage aging. By immersion at 50°C in acid the samples showed gradual deterioration in flexural strength. The results show that storage aging and acid immersion effect on flexural strength is much less than the effect on tensile strength.

The bending displacements corresponding to the flexural strength of the samples (i.e. original, 7, 14, and 21 days immersed samples) at 50°C are also shown in figure 7. Acid immersion caused the gradual decrease of bending displacement similar to the tensile and flexural properties. It seems that penetration of acid made the samples less flexible and more rigid.

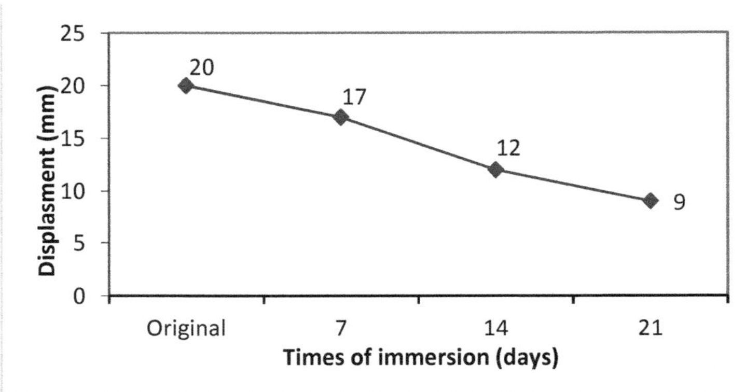

Figure 7. Displacement of original and acid immersed samples due to flexural test. The more immersed samples, the less flexibility and displacement.

When the FRP samples are in contact with the mineral acid solution environment, acid diffuses into the macromolecule of the polymer, degrades the matrix, reinforcements and interfaces. The absorption of acid at different times and temperatures accelerated the rate of aging. Thus reduction of mechanical properties such as tensile strength, flexibility, modulus of elasticity and elongation at break occurred. The effect of this highly destructive process is evidenced by swelling, discoloration and decrease in the mechanical properties of the samples. Chemical attack of polyester matrix by the acid solution environment led to hydrolysis of esters groups of the matrix. Since these groups are located in the chain backbone of the polymer, chain scission occurs. The decrease in molecular weight due to this scission can lead to the reduction in the mechanical properties.

Also the reduction in the flexural strength of the original samples is explained by weathering and thermal degradation. Although the specimens were not suffered by direct exposure of sun light, UV light exists in the environment and could harm the matrix. Weathering is initiated by the absorption of UV radiation by chromophores and in the activation of excited states in the macromolecules. When a polyester composite is exposed to a source of energy (solar or thermal) the absorbed energy by the polymer matrix results in the formation of free radicals. Once free radicals have been produced, reaction with oxygen generates hydroperoxides. These hydroperoxides can dissociate further to produce a series of decomposition products including aldehydes and ketones. Decrease of the flexural properties is due to the formation of these products and scissoring of polymer macromolecules. Regarding to the storage period of specimens, heat, moisture and air-born pollution all influence the mechanism of degradation [6].

3.3. Toughness of FRP

The toughness Gc of a composite (like that of any other material) is a measure of the energy absorbed per unit crack area. If the crack simply propagated straight through the matrix (toughness G_c^m) and fibres (toughness G_c^f), we might expect a simple rule-of-mixtures.

$$G_c = f_f\, G_c^f + f_m\, G_c^m \tag{13}$$

But it does not usually do this. We have already seen that if the length of the fibers is less than l_c , they will not fracture. And if they do not fracture they must instead pull out as the crack opens (Figure 8). This gives a major new contribution to the toughness. Where the matrix shear strength is τ_i, then the work done in pulling a fiber out of the fracture surface is given approximately by relation 14.

$$\int_0^{l/2} F dx = \int_0^{l/2} \pi d\tau_i\, x dx = \pi d\tau_i\, \frac{l^2}{8} \tag{14}$$

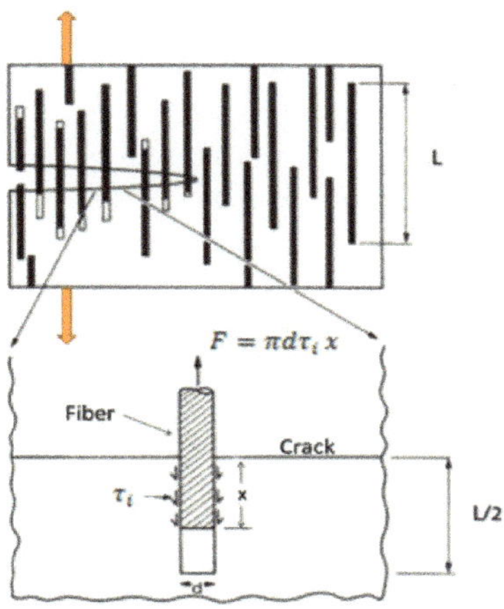

Figure 8. Propagation of crack in a FRP composite.

The number of fibers per unit crack area is $4\, V_f/\pi\, d^2$ (because the volume fraction is the same as the area fraction on a plane perpendicular to the fibers). So the total work done per unit crack area is:

$$G_c = \pi d \tau_i \frac{l^2}{8} \times \frac{4 f_f}{\pi d^2} = \frac{f_f}{2d} \tau_i l^2 \tag{15}$$

This assumes that l is less than the critical length l_c . If l is greater than l_c the fibers will not pull out, but will break instead. Thus optimum toughness is given by setting $l = l_c$ in equation 15 to give :

$$G_c = \frac{f_f}{8} \frac{(TS_f)^2 d}{\tau_i} \tag{16}$$

The equation says that, to get a high toughness, you should use strong fibers in a weak matrix (though of course a weak matrix gives a low strength). This mechanism gives CFRP and GFRP a toughness (50 kJ m^{-2}) far higher than that of either the matrix (5 kJ m^{-2}) or the fibers (0.1 kJ m^{-2}); without it neither would be useful as an engineering material [7].

4. Degradation and aging of FRP composites

The extensive application of composites has seen the emergence of durability problems specific to these materials where durability relates to the long-term performance under adverse conditions, often 20 or even 50 years exposure. These problems are associated with in-service environmental conditions and handling procedures (including maintenance, repair and modifications). Durability is a serious issue from both a health and safety aspect and in terms of economic costs. For large structural applications, such as aircraft, bridge and offshore construction, composite parts are very expensive and due to "parts integration" are often very large. The lack of resistance of composite structures to degradation agents often becomes apparent within a short period. In some circumstances, only a few hours of exposure may lead to catastrophic failure or seriously compromise structural integrity. Irreversible property changes in polyester matrix composites can be induced by any number of degradation agents such as (1) Thermal - static heat ageing, sub-zero exposure or thermal cycling; (2) Humidity (including hot/wet) exposures; (3) Complete immersion in water at ambient and elevated temperatures; (4) Freeze/thaw and dry/wet cyclic conditions; (5) Continuous or intermittent saltwater immersion or spray; (6) Weathering (including rain and sand erosion); (7) Combined load (i.e. stress) and environmental exposures; (8) Chemical (including water, fuel, acids, alkalis, solvents and oxygen); (9) Ultraviolet and high-energy radiation; (10) Electrical stress (e.g. lightning stress and galvanic reactions) and Micro-organisms (e.g. fungi).

In many applications, composite structures will be exposed to a combination of two or more factors, often resulting in complex synergistic degradation of the material. Accelerated degradation may be caused by the combined action of two or more vectors (e.g. temperature and humidity). The relative importance of each agent will depend on the agents present and their levels. Degradation from one agent can also reduce resistance to other agents, similar to biological systems. The two predominant factors in climatic exposure are humidity and temperature. The severity of these two factors will depend on geographical location, and need to be taken into account when designing with these materials.

Failure of polyester composites may occur because of cumulative damage to the matrix, interfacial separation with the fibers, chemical attack of the fibers or a combination of two or more of these processes. The net effect is loss of stiffness and mechanical integrity. This section examines the degradation of composite materials and constituent components (i.e. fiber, matrix, fiber-matrix interface and interphone) as a result of exposure to aggressive environments.

4.1. Thermal ageing

Thermal degradation refers to the chemical and physical processes in polyesters that occur at elevated temperatures. Increased temperature accelerates most of the degradation processes that occur in polyesters such as oxidation, chemical attack and mechanical creep. Oxidation is generally considered to be the most serious problem when using polyesters at elevated temperatures. The influence of temperature on the oxidation processes will depend on the chemical structure of the polyesters. Thermo oxidation is initiated by the reaction of free radicals P° with oxygen to form peroxide radicals:

$$P^{\circ} + O2 \rightarrow PO^{\circ}$$

All polyesters contain these free radicals due to their polymerization and processing history. However, the concentration of free radicals can be significantly increased by interaction with light, ionizing radiation or the presence of transition metals. Once formed the peroxide radicals undergo slower propagation reactions that breakdown the polyester chains. The overall degradation process will normally involve a relatively long induction period during which little degradation is observed. At the end of this period there is a rapid increase in degradation leading to a significant reduction in the mechanical properties of the polyester. This induction period is temperature sensitive and is reduced significantly at elevated temperatures. The induction period of the degradation process can normally be regarded as the serviceable lifetime of the polyester.

Other physical changes can occur in the polyesters at elevated temperatures, one of the most common being thermal expansion. Thermal expansion is reversible and in general does not significant affect the life expectancy of a polymer. However, in polyester composites the mismatch between the thermal expansion of the polyester matrix and the fibers may cause thermo-mechanical degradation during thermal cycling.

4.2. Weathering

Weathering or more specifically photo-oxidation of polyesters refers to the chemical and physical changes that occur when radiation is absorbed by a polymer. Photo-degradation is initiated by solar radiation, which results in the absorption of UV radiation by chromophores and in the activation of excited states in macromolecules. However, other climatic quantities such as heat, moisture and air-born pollution all influence the mechanisms of degradation and the subsequent results of ageing.

When a polymer is exposed to solar radiation the energy absorbed by the polymer results in the formation of free radicals within the polymer by the dissociation of the C-H bonds in the polymer chains.

$$PH \rightarrow P^{\circ} + H^{\circ}$$

The extent of this chemical reaction depends on the radiation exposure that is the quantity of ultraviolet light (<350nm) to which it is exposed. Once free radicals have been produced, reaction with oxygen generates hydro peroxides (POOH).

$$P^{\circ} + O2 \rightarrow POO^{\circ}$$

$$POO^{\circ} + PH \rightarrow POOH + P^{\circ}$$

These hydro peroxides can dissociate further to produce a series of decomposition products including aldehydes and ketones. The presence of these carbonyl groups in a degraded polyesters can be used as a chemical index for degradation. When these free radicals formed, it can continue to react via propagation reactions long after the initial UV exposure has ended. Termination of these free radical reactions is normally achieved through the reaction of pairs of free radicals.

$$P^{\circ} + P^{\circ} \rightarrow P\text{-}P$$

The formation and propagation of free radicals in itself does not serious affect the mechanical properties of the polymer, as they do not significantly alter the long-chain nature of the polymer molecules. Degradation of the mechanical properties occurs because the free radicals produced are highly unstable and readily undergo chain scission reactions. This results in the formation of two smaller polymer chains which each of them is a free radical and they are capable of further reactions. These reactions continuo and go on and on.

Fortunately the intensity of the UV radiation decreases with increasing depth in the material, so that the reaction tends to be a near surface process. Since oxygen is involved in the reaction process, there is an important balance between UV radiation and oxygen diffusion, and of course temperature since that will also determine the kinetics of reaction and the transport of reactive species.

4.3. Chemical degradation

Chemical attack of polyesters involves specific chemical reaction of the polymer with the fluid with the most common mode of failure being hydrolysis by water, acids and alkalis. Esters, and carbonate groups are particularly susceptible. Where these groups are located in the backbone chain rather than the side chain, chain scission ensues. A general hydrolysis scheme can be summarized as follows (figure9):

The reduction in molecular weight consequent upon chain scission can lead to a reduction of toughness and fracture strain. Stress is known to accelerate the chain scission process and also to enhance the rate of fluid uptake.

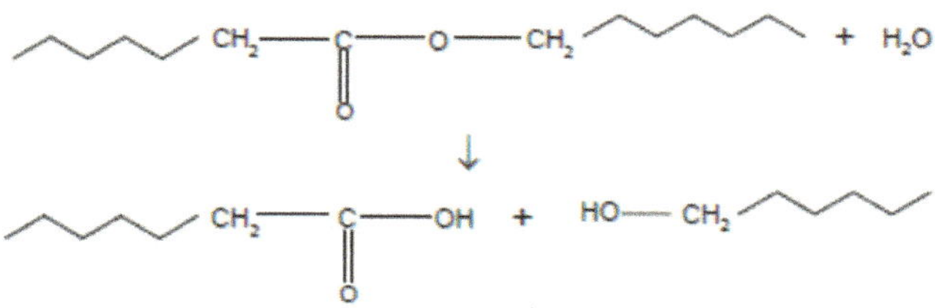

Figure 9. Hydrolysis scheme of polyester due to chemical attack.

4.4. Environmental stress cracking

Environmental stress cracking (ESC) remains one of the most common causes of failure in polyester matrix composites. The main reason for this is the complexity of the phenomenon, the agents such as chemical compatibility, liquid diffusion, craze formation and crack development all involve in this phenomenon. While crystalline and amorphous polymers are both susceptible to ESC, amorphous polymers are particularly susceptible due to their relatively open structure that leads to easy fluid penetration. Once the fluid has penetrated to the polymer it becomes locally dissolved and promotes cracking and crazing in the polymer. Cracking is normally preceded by the formation of crazes initiated at sites of stress concentration or at regions of local micro structural inhomogeneity. Crazes are voids that are held together by highly drawn fibrils, which bridge the void allowing the craze to transmit stress and prevent the craze from propagating. The mechanism of crazing in chemical environments is generally considered to be identical to that in air. In general terms craze initiation is considered to evolve from micro deformation processes in localized regions about 30 nm in diameter.

As the deformation region develops, further localized deformation is induced. The growth and coalescence of such deformed nuclei create a narrow plastic zone. In the presence of dilatational stress, voids develop. This voided structure is considered the precursor of the fibrillated craze structure that ultimately leads to failure. The environment accelerates the craze formation process by local plasticization, for example enhancement of the local relative movement of molecular chains by reduced intermolecular interaction between chains.

4.5. Ionizing radiation

Ionizing radiation covers a wide range of different forms of radiation including x-rays, gamma rays, neutrons, alpha particles and beta particles. When a polyester composite is irradiated by the ionizing radiation it will be degraded by the formation of free radicals or ions in the polyester. These reactive intermediates are capable of initiating chemical reactions which occur by free radical or ionic mechanisms and which result in scission as well as in cross-linking reactions. Free radicals with a long lifetime, which are present in the bulk of the material after irradiation, are responsible for changes in properties even a long time after exposure.

The intensity of ionizing radiation on the earth's surface is not normally high enough to significantly affect most plastics, hence radiation degradation is only occurred in connection with nuclear plants and where radiation is used for applications such as medical x-rays, sterilization or cross-linking.

4.6. Effect of moisture and water on FRP composite performance

Most FRPs will absorb small, but potentially Damaging amounts of moisture from the surrounding environments with the degree of degradation linked directly to the amount of moisture absorbed. The absorbed water may adversely affect the material in a number of ways as like as (1) dimensional changes (swelling); (2) reduction in the glass transition temperature T_g of the resin and (3) reduction in mechanical and physical properties (i.e. stiffness, strength and hardness). In many instances, water reacts with the matrix and causes irreversible chemical

changes and diminished performance. Capillary action along the fibers can account for a significant proportion of initial moisture uptake, although a chemically resistant matrix may protect the fibers. Shrinkage of the resin away from the fibers during curing is a contributing factor to the capillary effect. The effect of moisture is to cause hydrolytic breakdown of the fiber matrix interface resulting in a loss in the efficiency of load transfer between the matrix and the fiber reinforcement.

The moisture absorption kinetics of polymer systems differs widely between resin systems and also changes with chemical ageing. The glass transition temperature for a typical polyester resin decreases by approximately 15-20ºC for a 2% moisture weight gain. This reduction in T_g is induced by softening or plastization of the polymer matrix and in some cases by loss of organic additives through leaching to the surrounding media. It is advised that when using FRP products be sure that the maximum operating temperature is at least 30-40ºC below the T_g of the polyester (taking into account moisture effects). Table 7 shows the variation of T_g with moisture content for a composite E glass / polyester[8].

Material (Exposure Months)	Moisture Content (wt%)	T_g (ºC)
As-received	0.23±0.01	94.7
Dry	0	97.0
1	0.27±0.02	92.2
2	0.34±0.04	95.5
3	0.56±0.04	92.3
4	0.39±0.10	95.2
6	0.54±0.02	96.3

Table 7. Variation of T_g with moisture content in an accelerated weathering condition for E glass/polyester specimens. The increase in moisture content is also generally commensurate with a reduction in T_g. The results shown in this table may be indicative of moisture being mainly limited to the outer layers of the composite.

Although the process of moisture absorption within the surface layers occurs almost immediately on contact with the environment, moisture diffusion into the bulk material is usually a slow process. It may take weeks to months before a substantial amount of moisture has been absorbed by the composite, and considerably longer periods (i.e. 1-2 years) before the material is saturated. The rate of moisture uptake by a FRP composite is dependent on the temperature, relative humidity, exposure time and mechanical load. At elevated temperatures, the rate of moisture uptake and material property degradation is accelerated. Also the presence of tensile loads accelerates moisture uptake by opening existing internal cavities or voids, and by contributing to micro-crack formation. A FRP composite containing micro-cracks will absorb considerably more moisture than an undamaged laminate. Exposing the wet composite to sub-zero temperatures can further exacerbate this process.

4.7. Glass fiber degradation

Fibers degradation is primarily a concerning issue for different fibers which are used in FRP composite in various environments. The tensile strength of freshly drawn E-glass fibers is typically 3.5 GPa. This strength can be fully accomplished, provided the fibers are carefully handled during fabrication to avoid surface damage and are stored in a dry environment. Exposure to humid air will decrease the load bearing capacity of the fibers, resulting in a loss of strength. Carbon fibers are relatively insensitive to moisture, and hence the variability in the tensile breaking stress and strain for carbon fiber tows is noticeably less than for E-glass fiber tows. The loss of tensile strength of E-glass fibers is dependent on exposure time, temperature and degree of humidity. On initial exposure to a humid/water environment, the rate of fiber degradation is relatively rapid, even in safe environments, such as air-conditioned laboratories. The tensile strength is reduced to 3.0 GPa after 3 weeks exposure to standard laboratory conditions (23ºC and 50% relative humidity (RH)). Immersion in water at the same temperature for the same period results in a 20% reduction (~2.5 GPa). Further strength reduction occurs with increasing exposure time. After 100 days, in air and water the strength is 2.6 GPa and 2.1 GPa, respectively. Exposure to boiling water for 24 hours results in a 75% loss of strength. As a consequence of handling and moisture, an intrinsic tensile strength of 2.0 GPa is often assumed for design purposes.

Degradation of E-glass fibers in water can be mainly attributed to leaching of alkali oxides (sodium and potassium oxide) from the fiber surface resulting in the formation of surface micro-cracks, which act as stress concentrators. The loss of strength can be expected to be permanent at all conditioning temperatures and exposure times. The presence of Na^+ ions in solution slows down the exchange of alkali ions, such as $NaOH^-$, and restricts entry of Cl^- ions into the silic acid network. Chloride ions slow the degradation process, although only slightly. Under the influence of humidity or water the fiber forms a water skin in which the alkali ions (e.g. $NaOH^-$) are leached from the fiber surface and replaced by protons (H^+). The thickness of the silic acid structure or skin increases with exposure time and is dependent on the temperature and humidity of the surrounding environment. The water surrounding the glass fibers evolves into an aggressive alkali solution as the alkali ions dissolve out of the glass, slowly decomposing the glass fibers. Increasing the alkali content of the glass tends to reduce environmental attack from water and alkali solutions. Drying of the composite will remove most of the skin of water adjacent to the fiber, but a small permanent layer with retained water will still remain, and the mechanical properties of the fiber will be permanently degraded. Below reaction shows the chemical reaction of water with a fiber glass. Where R is Na, K, Ca, Mg, Al. As with moisture effects, acid and alkali degradation processes are accelerated at elevated temperatures.

$$-Si-O-R+H_2O \rightarrow -Si-O-R^+ + OH^-$$

In the case of acid environment, the acid come into contact with glass fibers, ionic exchange occurs between the metallic cations (e.g. Na+ ions) at the glass surface and the hydrogen ions in the acid solution, resulting in leaching of sodium, potassium, calcium, magnesium, boron and aluminum from the outer layer or sheath of the fiber. The dissolution of the supporting network results in a slight enlargement of the fiber diameter and shortening of the fiber length as longitudinal stresses relax, which is resisted by the unaffected core. As the outer layer becomes depleted, tensile stresses imposed by the core of the fiber build up, which significantly decreases the load capability of the fiber and eventually leads to cracks. Below reaction shows the chemical reaction of acid with a fiber glass [9].

$$-Si-O-Na+H^+ \rightarrow -Si-O-H+Na^+$$

Figure 10(a) shows micrograph of fracture section of the FRP sample before acid immersion at ambient temperature. From this figure one may notice the closely adherence of fibers. A few fibers are pulled out from matrix when tensile stress was applied. Figure 10(b) shows micrograph of fracture section of the FRP sample after 21 days acid immersion at 35°C temperature. It is clear that when tension was applied all the fibers were pulled out of matrix because of the loss bond between matrix and fibers. This figure confirms destructive effect of the acid on the FRP composite structure and shows how acid penetration deteriorates fibers/matrix interfacial bond. During the acid absorption of the interphase, acid solution penetrated into the free space of polymer which induced more new cavities and cracks, and washed out the fibers surface. Thus the interphase gradually is damaged and the cohesive bonds between fiber glass surfaces and the matrix has been retarded. In addition to destructive effect of acid, aging of the specimens intensified the amount of damage imposed to the samples [6].

In the case of alkali attack, the chemical reaction involves a breakdown of the silica network by hydroxide (OH^-) ions and eventually dissolution of all the species in the fiber glass. The glass fibers gradually lose weight and strength when they are in contact with strong alkalis. Immersion in weak caustic solutions at room temperature can result in strength reductions of 30% within 2 weeks. The rate of degradation of glass fibers to alkalis is not determined by the rate of diffusion, but by the active dissolution of the SiO2 network. The loss of mass is proportional to time. Acid corrosion resistant glass is only slight more resistant to strong alkalis. Below reaction shows the chemical reaction of alkali with a fiber glass. Where R is Na, K, Ca, Mg and Al.

Figure 10. Fracture section of the aged FRP sample before acid immersion (a), after 21 days acid immersion at 35°C temperature (b).

$$-\mathrm{Si}-\mathrm{O}-\mathrm{R}+\mathrm{H_2O} \rightarrow -\mathrm{Si}-\mathrm{O}-\mathrm{R}^{+}+\mathrm{OH}^{-}$$

$$-\mathrm{Si}-\mathrm{O}-\mathrm{Si}+\mathrm{OH}^{-} \rightarrow -\mathrm{Si}-\mathrm{OH}^{+}-\mathrm{Si}+\mathrm{O}^{-}$$

It must be mentioned that the resistance of glass fibers can be improved by modifying the chemical composition. Table 8 shows corrosion resistance comparison between different steels and FRP in hostile environments. It is obvious that FRP materials are resistible in all corrosive environments .However, the amount of resistibility depends on the kind of polyester and fiber, method of production and the grade of hostility.

	√= unaffected			•= corroded			
Material	**Dilute H_2SO_4**	**Conc. H_2SO_4**	**Dilute HCl**	**Conc. HCl**	**Dilute HNO_3**	**Chloride Salts**	**Dilute NaOH**
FRP (laminate)	√	√	√	√	√	√	√
Carbon steel (1020)	√	√	•	•	√	•	√
Stain less (316)	√	√	•	•	√	•	√
Hastelloy (C)	√	√	√	√	√	√	√

Table 8. Comparison of corrosion resistance of different materials in corroded environments.

5. Application of FRP composites

FRP composites have properties and capabilities that metals lack and they usually cost less than their metals counterparts: austenitic stainless steels, high nickel content alloys, or titanium. FRP is one-fourth the density of steel, which means that in many instances, equipment can be handled manually instead of renting a crane. FRP is easy to repair and does not necessitate arc welding in hazardous areas.

The dielectric properties of FRP means that it can be used safely where electrical conductivity cannot be tolerated. The anisotropic nature of FRP (different physical properties in different directions) enables the engineer to align the fiber reinforcement with the principal strain field, thus making the equipment stronger and lighter than a corresponding steel fabrication.

The applications of composites are not limited to fiber reinforced plastics, either. In another important application, and by no means the only other example, polymer concrete bridge deck overlays provide chloride protection for bridge deck reinforcement steel while restoring roadway profile, durability, and ride quality. The overlays can be placed under a variety of extreme environmental conditions and, as a result, two to four-hour cure times are achievable allowing a rapid return to public traffic.

Whether it is the material itself or its corrosion resistance, all of these advantages translate into better engineered systems that perform better, last longer, and cost less.

Proper applications of FRP often require an understanding or appreciation of the physical and corrosion resistant properties of metals and other common materials of construction. In the case of metals, the table 9 lists some of typical properties in comparison to those of FRP reinforced composites. The properties of FRP can vary with the type of resin and the construction employed for the reinforcement.

Likewise, metal properties may vary considerably depending on the particular alloy or the manner in which the metals have been annealed or pre-processed.

Reinforced composites are not as stiff as most metals, and obviously can be limited in applications requiring high modulus of elasticity. FRP also does not display other favorable properties of metals such as ductility or malleability. On the other hand, FRP features a low density, which can often give a good strength to weight ratio, which is important in transportation and many structural applications. FRP is also a relatively good thermal as well as an electrical insulator. Fabrication or repairs can be made easily, and without the need for arc welding in hazardous areas.

Table 9 compares the principal material properties of FRP to those of various metals. The light weight (density) and the strength to weight ratio of FRP are clear advantages in transportation and in handling. The thermal conductivity is a clear advantage when storing, using, or transporting fluids at elevated temperature. With a thermal conductivity only 1/187 that of carbon steel and 1/900 that of aluminum, heat loss is much less and the hazard that hot equipment poses for workers is reduced.

	FRP		Carbon Steel	Stainless Steel	Hastelloy	Aluminum	Titanium
Property	With Glass Mat Roving	All Glass Mat	AISI 1020	316L	C	1050-O	Grade 12
Density, lb/in^3	0.065	0.050	0.284	0.286	0.324	0.098	0.163
Tensile Strength, psi x 10^3	12-20	10-20	55	80	80	11	89
Yield Strength, psi x 10^3	10-20	9-15	33	34	51	4	69
Modulus of Elasticity psi x 10^6	0.8-1.5	0.7-1.0	30	30	26	10	14
Coefficient Thermal Expansion In/in°F x 10^{-6}	13	17	7	9	6	13	6
Thermal conductivity, Btu/hr/ft^2/ft/°F	0.15	0.15	3	9	7	135	11

Table 9. Comparison between mechanical and physical properties of different materials with FRPs.

The reduced tensile strength, coefficient of thermal expansion (at twice that of carbon steel), and modulus of elasticity introduce design considerations that may, at first, appear to be disadvantages to design engineers more familiar with steel, but are, in fact, advantages. The coefficient of thermal expansion (at twice that of carbon steel) may be seen as a disadvantage in the case of applying an FRP liner to a steel substrate, e.g. as in relining a steel tank, but with a modulus 1/30th that of steel, resultant thermal forces in a piping system are only 1/15th that of carbon steel (because of the lower modulus of elasticity). An actual calculation is necessary to weigh this advantage for thrust blocks in restrained piping systems, since the wall thickness and pipe size also enter into the calculation of the thrust force. Conversely, the lower modulus of elasticity means that pipe guides will be closer together due to the earlier onset of elastic instability [10].

5.1. Application of FRP composites in aerospace industry

The composite materials used in aircraft industry are generally reinforced fibers or filaments embedded in a resin matrix. The most common fibers are carbon, aramid, glass and their hybrid. The resin matrix is generally polyester or an epoxy. These systems are requiring curing temperatures between 120° and 180°C.

The first structural composite aircraft components, which were introduced during 1950-60, were made from glass fiber reinforced plastics. These components included the fin and the rudder of Grumman E-2A, helicopter canopies, frames, radomes, fairings, rotor blades, etc. Due to high strength and stiffness combined with low density, composites like boron fiber reinforced plastics (BFRP) and carbon fiber reinforced plastics (CFRP) were preferred instead of aluminum for high performance aircraft structures. For lightly loaded structures, aramid fiber reinforced plastics (AFRP) which possess low density, have been used. The use of AFRP continues to be restricted to the lightly loaded structures due to the fact that although these fibers possess high tensile strength, they have very low compressive strength. For light aircraft and lightly loaded structural components, glass fiber reinforced plastics (GFRP) has become one of the standard materials. Over the years, use of composite materials has also increased from few small access panels and canopy frames to almost complete airframe surfaces thereby providing weight savings leading to improved performance, reduced drag and also improved durability and corrosion resistance. Consequently, now-a-days, composite materials like GFRP, CFRP and AFRP have become standard materials for flight control surfaces, engine cowlings, fairings, radomes, landing gear doors, floor panels, fan ducts, etc. in aircraft application.

FRP composite materials are being used for different helicopter components as well. Use of advanced FRP composites in helicopter application started way back in 1959 with the development of Optimum Pitch Blade for the XCH-47 twin rotor helicopter of Vertol Aircraft Corporation. There-after, use of composites in helicopter application has been progressively extended to various parts, which include main & tail rotor blades, stabilizers and fuselage portions. Experience has shown that GFRP main rotor blades have a service life of around 10,000 hours as compared to blades with steel/titanium spars, which have a life of around 1000- 2000 hours.

Flex-beam CFRP rotors are also in use on McDonnell Douglas MD 520N/MD 900 helicopters with NOTAH (No Tail Rotor) system. Composite blades can be designed to be fail-safe and unlike metal blades they do not require frequent inspections for defects. In addition, blade and rotor system efficiencies have been improved due to tailorability of composites. Their longer life and reduced in-service inspection requirements make them very attractive and cost effective.

5.2. GFRP composites developed for encasing electrical equipment

Modular enclosures which aid the assembly of electrical switch gear equipment have been introduced into the UK by the Anglo-Douch Holec organization.

By replacing conventional cast iron or sheet steel box with Halyester (a glass fiber-reinforced polyester) containers, the danger of personal contact with electrical voltage has been considerably reduced. As well as its dielectric properties, Halyester is said to be used both indoors and outdoors.

5.3. Application of FRP composites in off-shore systems

Today's world FRP is widely used as material of construction for chemical process equipments and storage in chemical off-shore industries. It is considered one of the best materials having high resistance against corrosion and has very high strength and light weight and high dielectric properties. It doesn't require any painting and very easy to handle and easily repairable if there is any problem.

In FRP storage tanks generally, the tank shell is filament wounded and top and bottom is hand press molded. Usually tank design is based on international standards like BS-4994, ASTM-D 3299 and ASME section ten.

Although, in FRP storage tanks and vessels selection of materials is depended on parameters like the fluid media and work temperature, generally the reinforcements is E-Glass fibers and the matrix material is isopthalic polyester, bisphenole and epoxy based vinylesters. Some off-shore applications are as follows:

- Composite Grids/ Gratings
- Hand rails & Ladder Components
- Aqueous Piping System
- Water & fuel storage tanks, Vessels
- Low pressure composite valves
- Spoolable type thermosetting tubes
- Sump Caissons and pull tubes
- Modular paneling for partition walls
- High pressure accumulator bottles
- Flexible & Floating Risers, Drill pipe
- Sub sea structural components
- Boxes, housings and shelters
- Fire water pump casing & sea water lift pump casing
- Tendons
- Offshore bride connecting between platforms
- Blast & Fire protection

5.4. Application of FRP composites in constructions

FRP composites have long been used in the construction industry. Applications range from non-structural gratings and claddings to full structural systems for industrial supports, buildings, long span roof structures, tanks, bridge components and complete bridge systems. Their benefits of corrosion resistance and low weight have proven attractive in many low stress applications. An extension to the use of high performance FRP in primary structural applications, however, has been slower to gain acceptance although there is much development activity. Composites present immense opportunities to play increasing role as an alternate material to replace timber, steel, aluminum and concrete in buildings.

Construction holds priority for the adaptation of composites in place of conventional materials being used like doors and windows, paneling, furniture, non-structural gratings, long span roof structures, tanks, bridge components and complete bridge systems and other interiors. Components made of composite materials find extensive applications in shuttering supports, special architectural structures imparting aesthetic appearance, large signages etc.

with the advantages like corrosion resistance, longer life, low maintenance, ease in workability, fire retardancy etc.

Natural fibers, as a substitute for glass fibers in FRP composite components, have gained interest in the last composite ceiling panel decade, especially in the housing sector. Fibers like flax, hemp or jute are cheap, have better stiffness per unit weight and have a lower impact on the environment. Structural applications are rare since existing production techniques are not applicable and availability of semi-finished materials with constant quality is still a problem.

5.5. Application of FRP composite in fan impeller

The axial flow fans are widely used for providing required airflow for heat and mass transfer operations in various industrial equipment and processes. Energy efficient axial flow FRP fans were developed under the advanced FRP composites. The project focused on improving aerodynamic profile of the impellers by selecting appropriate aerofoil and providing composite structural design to suite specific airflow and pressure requirements towards replacing aluminum and mild steel impellers. Hollow FRP blades reduce material and installation costs and possibility of damage to the fan and drive during sudden stops.

Some example of FRP fans are in cooling towers for power plants, mine ventilation for air heat exchanger, diesel locomotive radiator cooling fan and textile mill humidifiers. The performance of all the above fans was tested in actual field conditions with an efficiency differential as high as 25% over conventional fans with aluminum impellers.

Also application of these impellers in corroded environment such as ventilating the acidic gas and fumes is so desirable. Because if the rate of corrosion and cost of FRP impellers are being considered in comparison with stainless steels and titanium impellers, the FRP impellers will be more beneficial than the latter.

5.6. Application of FRP composites in automobile/transportation systems

Despite the potential benefits of lighter weight and durability resulting from corrosion resistance, advanced composites are not recognized as a material of choice in the near term for automotive applications. Significant changes on a broad spectrum would be required to make advanced composites attractive for widespread commercial use in cars and trucks. The principal barrier is the high cost of the raw and fabricated materials when compared to existing options.

However there are opportunities for advanced composites in specific components in the commercial automotive sector. In specialty vehicles of several types, produced in small numbers advanced composite materials have an opportunity to demonstrate their performance benefits, apart from the requirements of the competitive marketplace.

The composite industry worldwide is investing in process improvements for the molding of polymer composites using forms of conventional E-glass in mid-level performance resins, both thermoplastic and thermoset. Automobiles segment of composites accounts for about 50% of the thermoplastic and 24% of the thermoset composite market in the world. Glass-reinforced thermoplastic polymer is a promising material for weight reduction because of the relatively low cost of the fiber, its fast cycle time and its ability to facilitate parts integration. Carbon fiber reinforced polymer is another candidate but will require breakthroughs in cost and manufacturing techniques to be cost effective for high volume production.

5.7. Application of FRP composites in valves and pumps

Filament-wound FRP valves can be employed in corrosive environments. They have advantages over both PVC and stainless steel. FRP valves are competitively priced with stainless steel and are roughly one-third the weight of steel valves. In addition, they are four times stronger than comparable moulded PVC valves. One manufacturer makes filament-wound, fibreglass, wall-ball valve in sizes ranging from 2-10. It can be employed in both high- and low-pressure applications and has found wide acceptance in the CPI, petrochemical, marine and pulp and paper industries. The valve has a temperature service range from -40°C. It is constructed from continuous glass windings impregnated with a thermoset vinyl ester resin. The exterior has a thick resin coating that provides good resistance to chemicals and attack from ultraviolet light. The valve is also available with fire-resistant resin coatings. FRP pumps are available that are used in general industrial, chemical and petroleum products applications. Such a pump is fabricated from 30% fibreglass filled polyester via injection moulding. Pump gaskets are constructed from Viton, capable of withstanding attack from a wide variety of chemicals. Fasteners in the design are stainless steel. Pedestalmounted units and close coupled, electric motor-driven and hydraulicdriven pumps are equipped with- stainless steel shaft sleeves for added shaft protection. These pumps can be used in a wide range of applications including handling water, brine solutions, acids and many organic solvents having a pH less than 10. Advantages are good rugged construction, light weight and resistance to a large number of chemicals to an upper operating temperature of 54.4 °C.

5.8. Application of FRP composites in grating

FRP structural products are used in a wide range of industrial construction applications. They have the advantages of high strength, light weight, ease of installation, corrosion resistance, are nonconductive, non-sparking, fire retardant, have a long service life, and, in the case of fibreglass-reinforced grating, are slip resistant. Fibreglass-reinforced grating has been widely accepted for use in corrosive environments for such applications as flooring, bridges, trench covers, stairs, tower packing supports, walkways, ramps and grilles. It is employed as a fire-retardant grate on tank car and truck loading catwalks in the oil refinery industry and as antiskid trench cover installations in paper mills. In the metal refining industry, it is used for aisles between cells. In food processing operations where acid conditions exist, FRP grating is employed as floor mats. It has also found use in fencing off high-voltage transformers. FRP grating has approximately one-fourth the specific gravity of steel and roughly two-thirds that of aluminium. In general, fibreglass grating has many of the qualities of metal plus several superior advantages. Gratings can be fabricated from a variety of resins on the market. The particular resin used will depend on the specific corrosion environment for which it is intended. Manufacturers will supply samples of grating for immersion tests or test sections can be installed on the site. Fibreglass grating is available in a plain top surface or a surface that has deeply imbedded grit particles. The latter design is employed in applications where maximum non skid features are required. Particles are fabricated into the structure and, as such, are not worn away after constant exposure to corrosive environment and continuous wear. The material is imbedded with angular silica particles during the laying up process. The product is essentially a matrix of polyester resin. Resin wears faster than the silica grit so the grit profile always protrudes above the resin base. This results in very high friction coefficients, even after many years of service.

5.9. Application of FRP composites in tank constructions

Filament-Wound laminate construction of standard vertical FRP tanks is done with automated equipment on a rotating mould called a mandrel. A resin-rich layer of chemically resistant polyester or vinylester is applied uniformly to a specified thickness on an inner surface mat. The mat is usually of a chemically resistant fibreglass reinforcement composition or an organic veil. After the initial layer is applied (approximately 20 millimetre thickness), a second layer of resin-rich, chemically resistant polyester or vinylester is applied uniformly, to roughly a four times greater thickness. The reinforcement usually consists of randomly oriented, chopped strand fibreglass. The next step involves the construction of a filament-wound layer impregnated with isophthalic polyester resin or other suitable resin. The layer thickness depends on the size requirements and the specific gravity design criteria. Generally, the layer composition is comprised of filament winding interspersed with chopped strand fibreglass. The purpose of this third layer is to provide the primary strength characteristics of the vessel. This layer serves to minimize the strain placed on the corrosion barrier to avoid cracking of this layer. The corrosion barrier can undergo cracking to loads considerably below the ultimate strength of the corrosive layer material, resulting in tank failure. As a fourth layer, a resin-rich outer surface mat is applied that is usually of the same thickness as the inner layer. This generally consists of a chemically resistant fibreglass reinforcement or an organic veil that is saturated by spraying with isophthalic polyester resin. The responsibility of the first and second layers is to provide the tank with non-corroding and non-contaminating features. A cross-sectional view of the standard vertical tank wall laminate is illustrated in Figure 11[11].

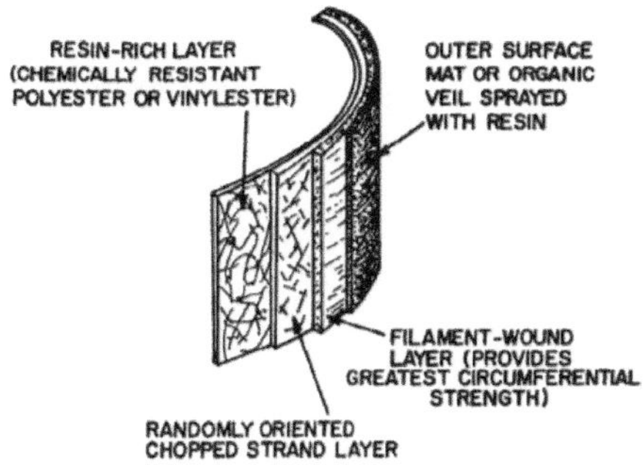

Figure 11. Cross-sectional view of standard vertical tank wall laminate.

6. Conclusion

- FRP composites are composed of two materials or phases joined or connected to each other in such a way to give a combination of properties that cannot be attained otherwise. The properties of the individual components, the relative amount of phases, the orientation of various components, the degree of bonding between the phases, the size, shape and distribution of the discontinuous phase are very important in determining the properties of composite.
- Fiber-reinforced polyester composites provide improvements in strength, stiffness and toughness. They also could have corrosion resistance in different hostile environments. By using different glass fibres electrical resistivity is controllable also. Fibres typically have low densities, giving high specific strength and specific modulus, but they often are brittle. Fibres can be continuous or discontinuous. Discontinuous fibres with high aspect ratio l/d produce better reinforcement.
- Fibres are introduced to the matrix in a variety of orientations. Random orientations and isotropic behaviour are obtained using discontinuous fibres; unidirectional fibres produce composites with anisotropic behaviour, with large improvements in strength and stiffness parallel to the fibre direction. Properties can be tailored to meet the imposed loads by orienting the fibres in multiple directions.
- Although the FRP composites are resistant in hostile environments, such environment as well as acidic and alkali ones could deteriorate the interface of fibres and matrix if they contact with the interface directly. So the mechanical properties could decrease if the specimens are in contact with hostile environments intensively.
- FRP composites have different application in construction, aerospace, automotive, sports and other industries.

Author details

Salar Bagherpour
Islamic Azad University, Department of Materials Science and Engineering, Najafabad-Branch, Iran

7. References

[1] Askland, Donald R. & P.Phule , Pradeep. (fourth edition). (2003). Composites, In: *The Science and Engineering of Materials*, (721-765), Thomson Learning, ISBN 0-534-95373-5, USA

[2] Black, T. & Kosher, R. (tenth edition). (2008). Non Metallic Materials: Plastic, Elastomers, Ceramics and Composites, In : *Materials and Processing in Manufacturing*, (162-194), John Wiley & Sons, Inc, ISBN 978-0470-05512-0, USA

[3] Mathews, F. & Rawlings, R. (1994). Polymer Matrix Composite, In: *Composite Materials: Engineering and sciences*, (168-200), The Alden Press, Oxford, ISBN 0-412-55960-9, UK

[4] Tuttle, M. (2004), Introduction. In: *Structural analysis of Polymeric Composite Materials*, (1-40), University of Washington, ISBN 0-8247-4717-8 USA

[5] Chung, D. (1995). Processing of Carbon Fibres, In: *Carbon fibre Composites*, (13-49), Butter Worth-Heinemann, ISBN 0-7506-9169-7, USA

[6] Bagherpour, S.; Bagheri, R. & Saatchi, A. (2009). *Effect of Concentrated HCl on The Mechanical Properties of Storage Aged Fibre Glass Polyester Composite*, Vol.30, (271-274), ISSN 0261-3069.

[7] Ashby, M.& Jones, D. (second edition). (1999). Polymer and Composite, In : *Engineering Materials 2*, (219-277), Buttere Worth- Heinemann, ISBN 0-7506-4019-7, UK

[8] Broughton, W. (2007). *Towards Accelerated Aging Protocols for Service in Hostile Condition.* 16th International Conference on Composite Materials. Japan, 16 May 2007

[9] Maxwell, A. & et al. (2005). *Review of Accelerated aging Methods and Life Time Techniques for Polymeric Materials*, National Physic Laboratory, ISSN 1744-0270, Scotland

[10] Reich Hold, Inc. (2009). *FRP Material Selection Guide*, Arailble, From: www.Reichhold.Com/Corrosion

[11] Cheremisionoff, N. & Cheremisionoff, P.(1995). *Fibre Glass Reinforced Plastics*, Noyes Publication. ISBN 0-8155-1389-5, USA

Chapter 7

Tribological Properties of Polyester Composites: Effect of Vegetable Oils and Polymer Fibers

Ibrahim Refaay Ahmed and Ali Waheed Yousry

Additional information is available at the end of the chapter

1. Introduction

Short fiber reinforced polymer composites are nowadays used in numerous tribological applications. In spite of this fact, new developments are still under way to explore other fields of application for these materials and to tailor their properties for more extreme loading conditions. [1]. Polyester composites are commonly used nowadays in industrial applications such as bearing materials, brake pads materials, flooring materials and so on. The references given at the end of this chapter describe some of these developments. [2]. Further approaches in designing polymeric composites in order to operate under low friction and low wear against steel counterparts are described. In this work polyester composites was reinforced with graphite or polymer fibers filled with some vegetable oils are proposed to be used as self-lubricated materials, so friction coefficient and wear rates were measured. Due to the great changes in technology that occurred in the last century, a large number of components fabricated in engineering polymers and composites have been used; substituting the most traditional metals in diverse applications, attaining in many cases better advantages as reduction of maintenance costs, save in weight and higher freedom for designing. Some examples of applications can be cited as: self-lubricant bearings, linear guides, mechanical seals, bushings, bearings cages, transporting belts, gears, and pulleys. These components are in turn more required in the aspects of mechanical resistance, fatigue strength and resistance to wear. [3]. Analysis of the wear resistance of polymeric fibres requires a better understanding of both their abrasive scratch behaviour and their frictional response. [4]. Fiber Reinforced Plastics (FRP) are widely used as structural materials in industrial applications, for example, marine boats, automobiles and bathtubs due to their light weight, high degree of rigidity and superior moldability.[5]. Polymers have been favorably introduced as sliding materials in offshore structures for over ten years because of good wear resistance. Mainly under high loads, surface plasticity contributes to low friction, which is favorable for a reduction in dissipated sliding energy. [6]. in industrial applications, the increase in using composite materials means that it is necessary to know their behaviour under working conditions. Wear is an important parameter and its experimental behaviour must be known. [7]. Polymers are frequently used in tribological applications because of their self-lubricating ability and loadability. However, most research on their friction and wear mechanisms is performed on small-scale test samples under relatively low normal loads. [8]. As polymers generally posses good self-lubricating abilities through the formation of a polymer transfer film or 'third body', they are also used in industry as sliding materials in gears, slides and bearings. While polyamides or polyethylene are most commonly used, they should be replaced by polyesters for obtaining higher temperature and fatigue resistance because of the stiffening action of the aromatic phenylene group. [9]. Fibre-Reinforced-Polymer composites are used particularly in the automotive and aircraft industries and the manufacture of spaceships and sea vehicles. [10]. There are the two main characteristics which make these materials attractive compared to conventional metallic designs. They are of relatively low density and they can be tailored to have stacking sequences to provide high strength and stiffness in directions of high loading, [11].Composite materials consist of a resin and reinforcement chosen according to desired mechanical properties and the application, [12].Polyesters are also commonly used as matrix materials, particularly with glass-fibre-reinforcement. Polyester is an economic material that has high chemical resistance and is resistive to environmental effects. It has high dimensional stability and low moisture absorption. Low-volume-fraction glass-fibre/ polyester composites with a wide range of colors have been in use for a long time. The production technologies for thermoset glass/polyester composites are easier and cheaper than those for other glass/resin materials. Glass-fibre-reinforced polymer with thermoset polyester resin is an attractive material that is economically desirable. Its application at low temperatures and under service terms is easy when this material is compared to advanced polymer composites with a complex molecule structure, high strength and working under terms of difficult service, [13].

1.1. Adhesion bonds

When two surfaces are brought into contact, the surface forces of attraction and repulsion act between the atoms and molecules of two approaching surfaces. Due to these forces the bonds formed between the contacting surfaces are followed by junctions developed on the real contact spots. Formation and rupture of the junctions control the adhesion component of friction. For the majority of polymers, the hydrogen bond develops at very short distance in polymers containing the groups OH, COOH, NHCO and others, in which the hydrogen atom is linked with an electronegative atom. Under favorable conditions two approaching atoms are linked together by a common proton providing a strong and stable compound. The junctions sheared under the applied tangential force result in the frictional force. That is, work done by the frictional force results from breakdown of the interfacial bonds. In general case, the interfacial junctions (their formation, growth and fracture) are influenced by nature of the surfaces, surface chemistry and stresses in the surface layers (loading conditions). The interfacial junctions together with products of their fracture and the highly deformed layers where shear deformation is localized, were named by Kragelskii as a 'third body', [14].

1.2. Real contact area

When two surfaces approach each other, their opposing asperities with maximum height come into contact. As the load increases, the new pairs of asperities with lesser height make contact forming individual spots. The overall area of these spots is known as the real contact area (RCA). When simulating the real contact concept with plastics, the temperature and sliding velocity should be taken into account. [15].

1.3. Wear of polymers

1.3.1. Wear modes

The changes in surface layer arise from mechanical stresses, temperature and chemical reactions. Polymers due to their specific structure and mechanical behavior are more sensitive to these factors. The local temperature at the interface may be substantially higher than that of the environment, and may also be enhanced at the asperity contacts by transient 'flashes' or 'hot-spots'. The temperature exerts an influence on wear of polymers. Thus, it was shown that a number of polymers sliding against steel pass a minimum at characteristic temperature. The great diversity of the mechanisms and their interrelation make impossible the rigorous classification of wear processes. It is generally recognized that the most common types of wear of polymers are abrasion, adhesion, and fatigue. [16].

a. Abrasive wear

The key aspect of abrasive wear relates to cutting or plowing of the surface by harder particles or asperities. These cutting points may either be embedded in the counter-face, or loose within the contact zone. The former case is commonly called two-body abrasion, and the latter, three-body abrasion. Abrasion displays scratches, gouges, and scoring marks on the worn surface, and the debris produced by abrasion frequently take on the appearance of fine cutting chips similar to those produced during machining, although at a much finer scale. Most of the models associated with abrasive wear incorporate geometric asperity descriptions, so that wear rates turn out to be quite dependent on the shape and apex angles of the abrasive points moving along the surface. The sources of the abrasive solid are numerous, and the nature of the abrasive wear in a given tribosystem will depend to some extent on the manner in which the abrasives enter the tribosystem: whether they are present in the original microstructure as hard phases, enter the system as contaminants from outside, or generated as debris from the contact surfaces as they wear. There are two distinct modes of deformation when an abrasive particle acts on the plastic material. The first mode is plastic grooving, often referring to as ploughing, in which a prow is pushed ahead of the particle, and material is continually displaced sideways to form ridges adjacent to the developing groove. No material is removed from the surface. The second mode is named cutting, because it is similar to micromachining and all the material displaced by the particle is removed as a chip. In the case of three-body abrasion the free abrasive readily penetrates the polymeric surface which begins to operate as an emery cloth resulting in increasing the wear of counter surface, [17].

b. Adhesion wear and friction transfer

Adhesion wear results from the shear of the friction junctions. The fundamental mechanism of this wear is adhesion, important component of friction outlined above. This wear process evolves in formation of adhesion junction, its growth and fracture. A distinguishing feature of this wear is that transfer of material from one surface to another occurs due to localized bonding between the contacting solid surfaces. It was noted that the transfer of polymer is the most important characteristic of adhesive wear in polymers. It is reasonable that the processes associated with other wear types (fatigue, abrasion and so on) accompany the adhesive wear. The phenomenon of friction transfer is observed for nearly all materials (metals, ceramics, and polymers) and their combinations. The point is that whether the transfer produces an influence on tribological behavior of the friction pair. In this case, the consequences of material transfer may be significantly distinct. If small particles of micrometer size are transferred from one surface to the other, then wear rate varies to only a small extent. Under certain conditions, the situations take place when thin film of soft material is transferred onto the hard mating surface, for example, polymer on metal. [18].

c. Fatigue wear

Fatigue is known to be a change in the material state due to repeated (cyclic) stressing which results in progressive fracture. Its characteristic feature is accumulation of irreversible changes, which give rise to generation, and development of cracks. The similar process takes place at friction accompanying nearly all the wear modes. A friction contact undergoes the cyclic stressing at rolling and reciprocal sliding. In addition, each asperity of friction surface experiences sequential loading from the asperities of counter surface. As a consequence, two varying stress fields are brought about in surface and sub-surface regions with different scales from the diameter of apparent contact area in the first case to that of local contact spot in the second. These fields are responsible for material fatigue in these regions that leads to the generation and propagation of cracks and the formation of wear particles. This process is named friction fatigue. Unlike the bulk fatigue, it spans only surface and sub-surface regions. The loss of material from solid surfaces owing to friction fatigue is referred to as fatigue wear. It has been known that the fatigue cracks are initiated at the points where the maximum tangential stress or the tensile strain takes place. [19].

2. Experimental work

The investigation of wear and friction was carried out to examine the effect of adding vegetable oils and fibers on the tribological behavior of polyester.

2.1. Description of test rig

Experiments were carried out using "pin-on-disc" test rig, figure 1. It consists of a rotary horizontal steel disc of 170 mm diameter and 5 mm thickness driven by variable speed motor, Specimen holder, this holder fastened to the load cell, loading plat at which the normal loads were applied, and digital screen attached to the load cell.

Figure 1. Pin-on-Disk Tribometer

2.2. Test specimens

Test specimens were formed in shape of cylindrical pins of 5 mm diameter and 30mm height, Fig. 2.

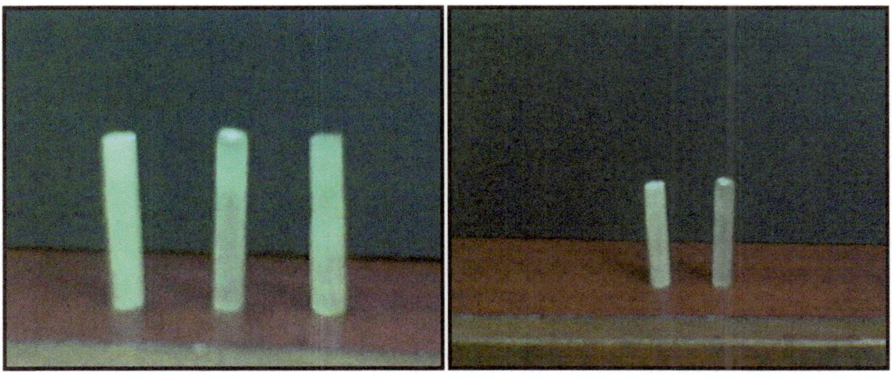

Figure 2. Test Specimens

2.3. Preparation of test specimen

Polyester was mixed with vegetable oils in volumetric ratio from 1 to 10% for unfilled specimens, or filled by some fibers with volumetric ratio from 1 to 5% for filled specimens after well mixing the mixture was molded into a paper mold then left it for two days for solidification before tests.

2.4. Experiments

Tests were carried out at room temperature and normal level of humidity by means of "pin-on-disc" tribometer. Polyester composites were held in specimen holder and loaded against the rotating steel counterface, due to the contact between composites and steel disc a tangential force try to resist the rotating disc, this force depending on the friction coefficient of composites and it was monitored by the digital screen. These forces measured at 190 rpm (0.696 m/s) under variable normal loads 4, 6, 8, and 10 N. friction coefficient calculated by dividing the friction force by the applied load. Wear was determined by measuring the change in composite volume before and after adhesion between the composites and the steel counterface for 5 minutes at 20 N normal load and counterface rotational speed of 100 rpm (0.366m/s). The following equations explain the calculations of wear and friction coefficient.

$$\text{Wear} = \text{Volume loss} = \Delta V = (L_1 - L_2)^* \ A \quad (1)$$

$$\text{Friction coefficient} = \mu = F_f \ / \ F_n \quad (2)$$

Where,

L_1= Length of specimen before test, mm.
L_2= Length of specimen after test, mm.
A= Friction surface area = $(\pi/4)\ d^2$, mm^2.
d= Specimen diameter, mm.
F_f= Friction force N, and
F_n= Normal force N.

3. Results and discussion

Test specimens were divided into two groups, unfilled and filled polyester, the first one divided into three groups according to the type of oil as follows:

1. Polyester impregnated by olives oil.
2. Polyester impregnated by corn oil.
3. Polyester impregnated by sunflower oil.

Each composite divided into two subgroups according to the filler type as follows:

1. Composite filled by polyamide fiber.
2. Composites filled by polytetrafluoroethylene PTFE fiber.

3.1. Unfilled polyester

3.1.1. Frictional behavior of polyester impregnated by olives oil

Friction coefficient decreases to 0.35 with increase of oil contents to 10% by volume of olives oil under low applied loads, it seems that there is an oil film generates at the friction surface and increases with oil content increase, which may be responsible for friction reduction, figure 3.

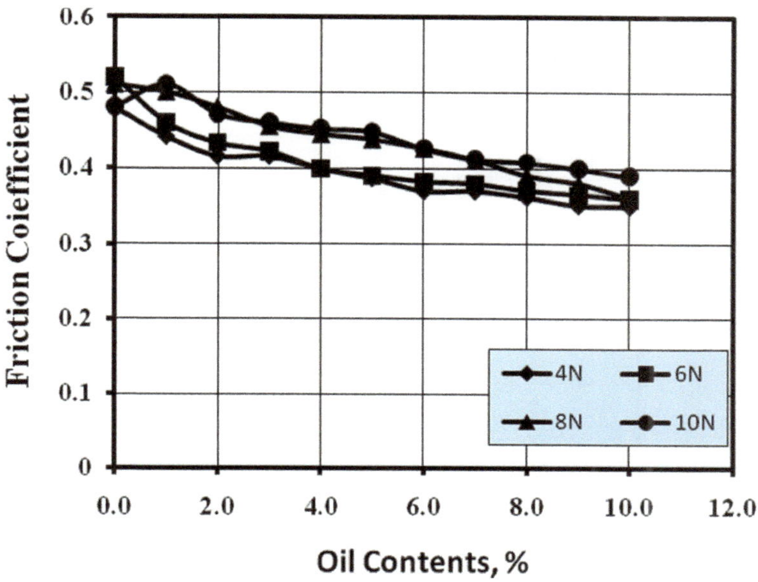

Figure 3. Frictional behavior of polyester impregnated by olives oil

3.1.2. Frictional behavior of polyester impregnated by corn oil

Increases of corn oil in composites decrease friction coefficient to minimum at maximum oil percent and low applied loads, figure 4. It seems that the increase in oil content increases oil film at friction surface which decreased the friction coefficient.

3.1.3. Frictional behavior of polyester impregnated by sunflower oil

Figure 5. shows that the increase of oil contents in composites decreases friction coefficient to 0.33 for composites of 10% by volume of sunflower oil, beside it shows that there is low effect of applied load in friction coefficient; it seems that the increase in sunflower oil contents increases surface hardness of composite which may be responsible for friction reduction.

3.1.4. Wear results of polyester impregnated by vegetable oils

Figure 6. shows that increase of olives, corn or sunflower oils in polyester composites decreased wear drastically to very little values in compare with polyester specimens free of oils, the amount of wear decreased from 3.37 mm^3 for free polyester to 0.3 mm^3 for composite impregnated by 10 vol. % sunflower oils. This can be attributed to the formation of oil film in the contact area which responsible for wear reduction as well as increase of surface hardness for specimens impregnated by sunflower oils.

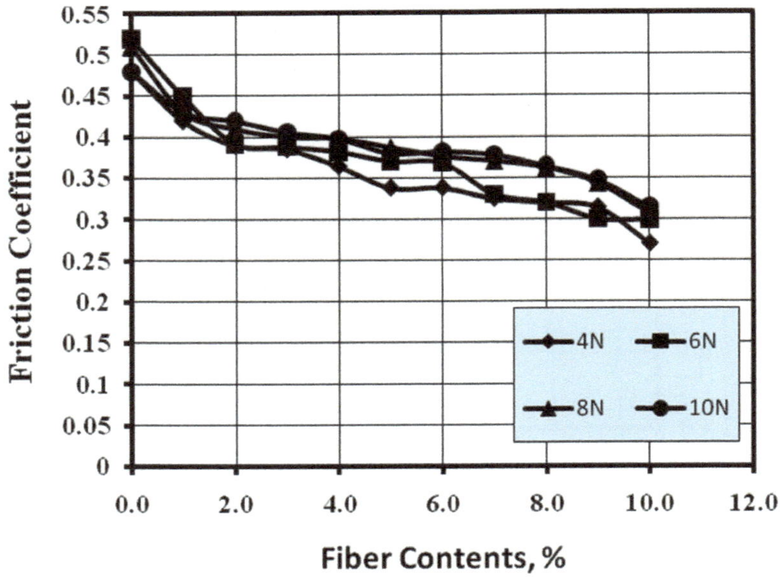

Figure 4. Frictional behavior of polyester impregnated by corn oil

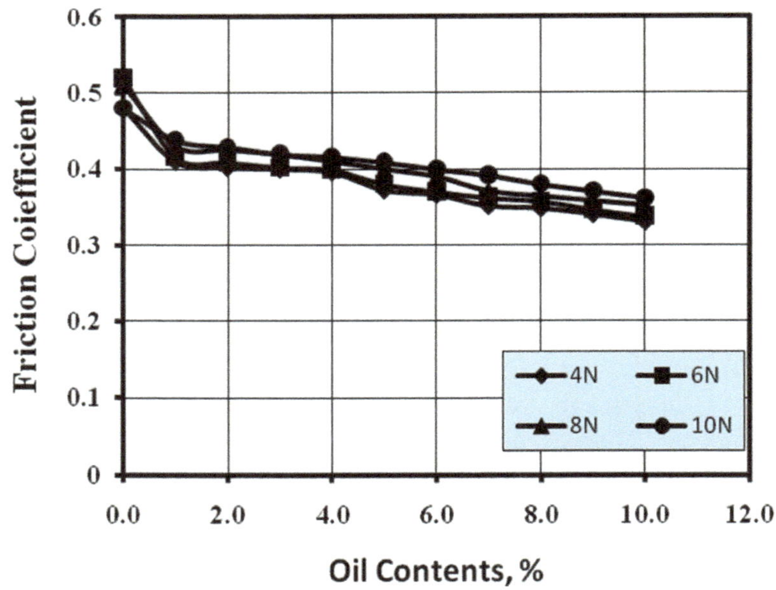

Figure 5. Frictional behavior of polyester impregnated by sunflower oil

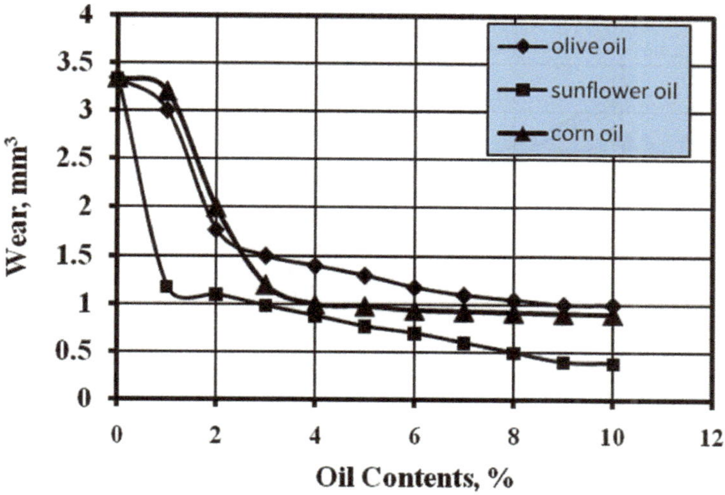

Figure 6. Wear results of polyester impregnated by vegetable oils

3.2. Polyester composites reinforced with polyamide fiber

3.2.1. Frictional behavior of polyester composite free of oil

Figure 7. shows that the increase in fiber content for composite free of oil decreases friction coefficient to 0.22 for composite of 4 % polyamide contents, under 4N. Beside it shows there are great effect for applied loads on friction results. It seems that the increase in fiber content increases the polyamide layer on the friction surface which may be responsible for friction reduction.

3.2.2. Frictional behavior of polyester composites impregnated by olives oil

Friction results of composite impregnated by olive oil shown in figure 8. Increase of fiber contents decreases friction coefficient to 0.183 for composite of 5% fiber content under 6N. This figure shows that there were low effects of applied loads in friction results. The effects of oil increasing in composite shows outstanding friction results, friction coefficient decreases to 0.153 for composite of 1 % fiber content and 10% by volume of olives oil. This recommended these composite as good bearing material under high loads, fig. 9.

3.2.3. Wear results of polyester composites impregnated by olives oil

Wear results of composite impregnated by 5 and 10 % oil content shows remarkably decreases in wear of composite contain 10 % oil content , it reached to 0.196 mm^3 for composite of 4 % fiber content, fig. 10., it may be attributed to the ability of oil film at frictional surfaces to reduce polyester transfer to counterface. That reduce composites wear.

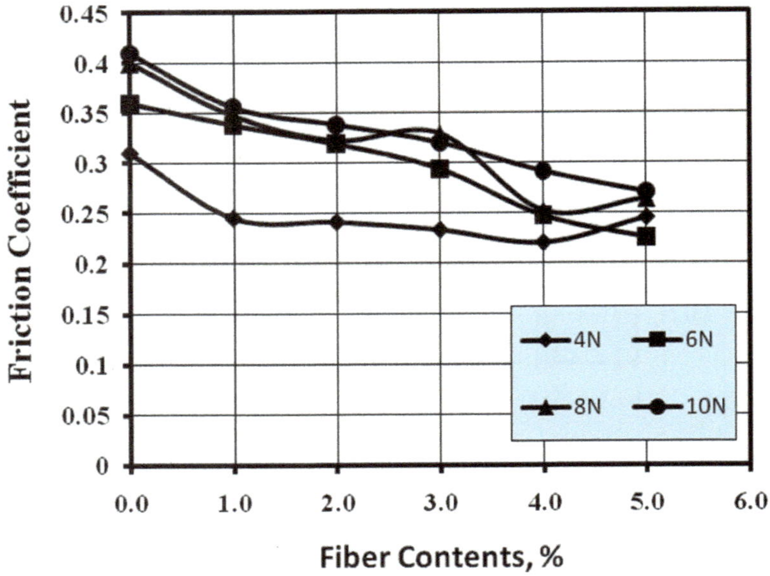

Figure 7. Frictional behavior of polyester composite free of oil reinforced by polyamide fiber

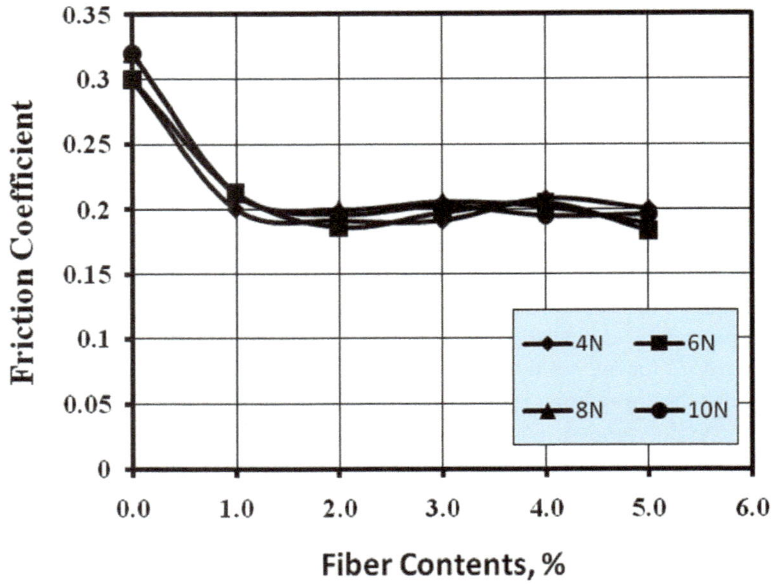

Figure 8. Frictional behavior of polyester composite impregnated with 5.0 vol. % olives oil

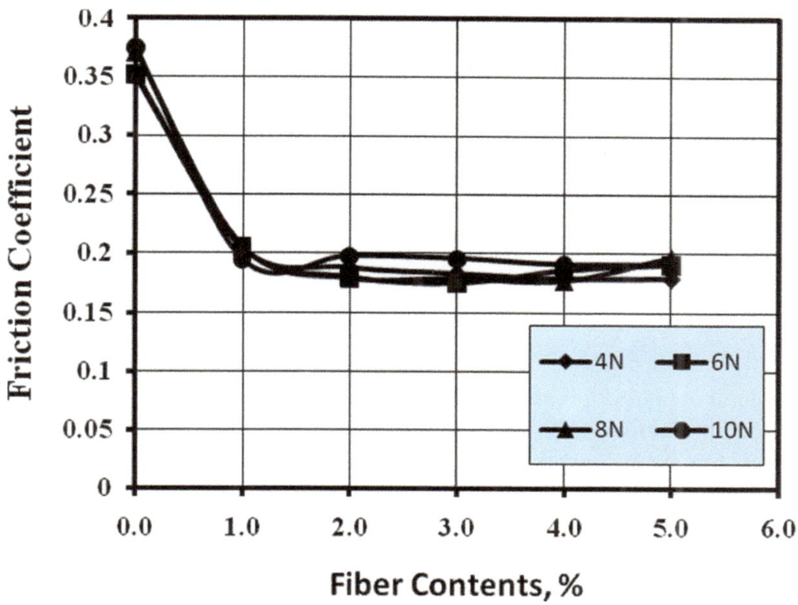

Figure 9. Frictional behavior of polyester composite impregnated with 10 vol. % olives oil

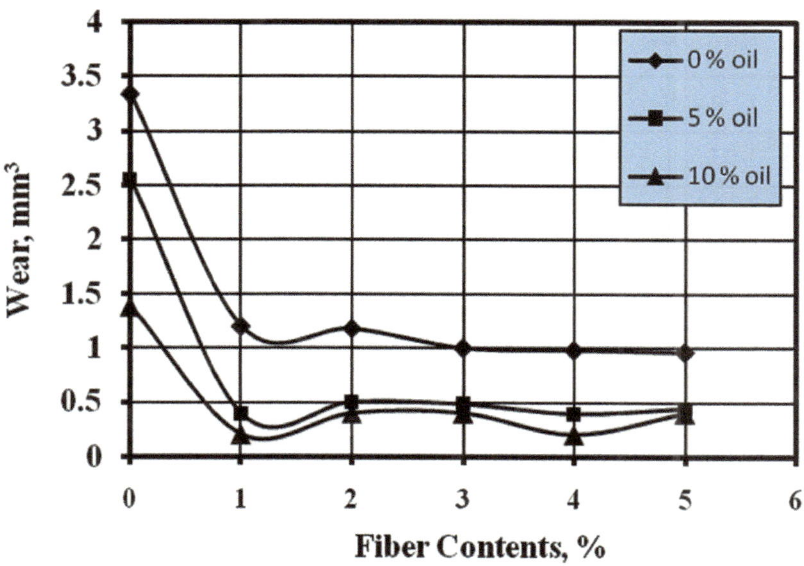

Figure 10. Wear of polyester composite impregnated by olives oil

3.2.4. *Frictional behavior of polyester composites impregnated by corn oil*

Increase of corn oil in composite shows great effect especially with fiber content increasing this was represented in figure 11. It shows that the friction coefficient was decreased to 0.187 for composite of 4 % fiber content. It may be attributed to the increase of oil layer at friction surface. Furthermore, it could be stated that the effect of applied load was decreased in high fiber composite contents. Composite impregnated by 10% by volume of corn oil shows good friction results as fiber content increase, friction coefficient decreases to 0.17 for composite of 5 % fiber content under 10N, fig. 12., beside it show that there is low effect of applied load on friction results. These results recommended those composite as good bearing material.

3.2.5. *Wear results of polyester composites impregnated by corn oil*

Wear results of composite free of oil shows decrease trend with increasing of fiber content, it may be attributed to the ability of polyamide layer for decreases polyester transfer to counterface. Composite impregnated by 5 and 10% by volume of oil content shows significantly decreases in wear to 0.215 mm^3 for composite of 5 % fiber content and 10% oil content, it seems that the increase in oil content increases the presence oil layer which may be responsible for wear reduction, fig. 13.

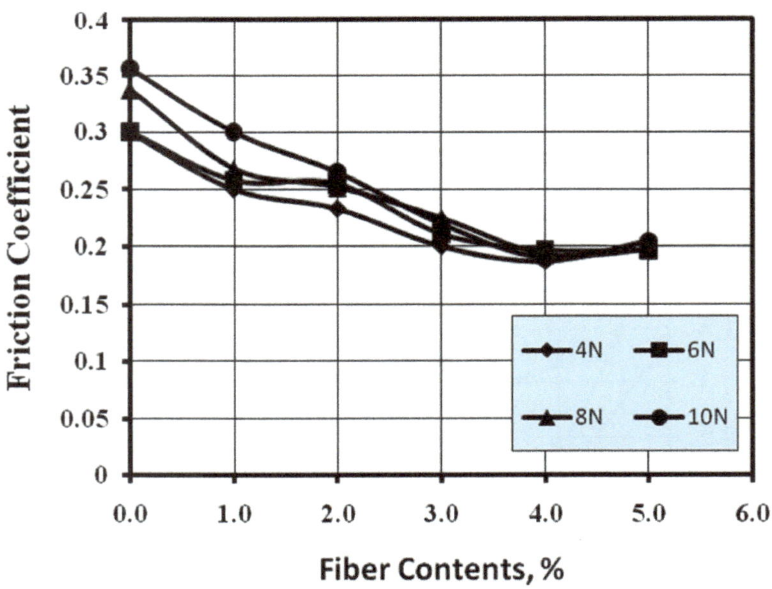

Figure 11. Frictional behavior of polyester composite impregnated with 5.0 vol. % corn oil

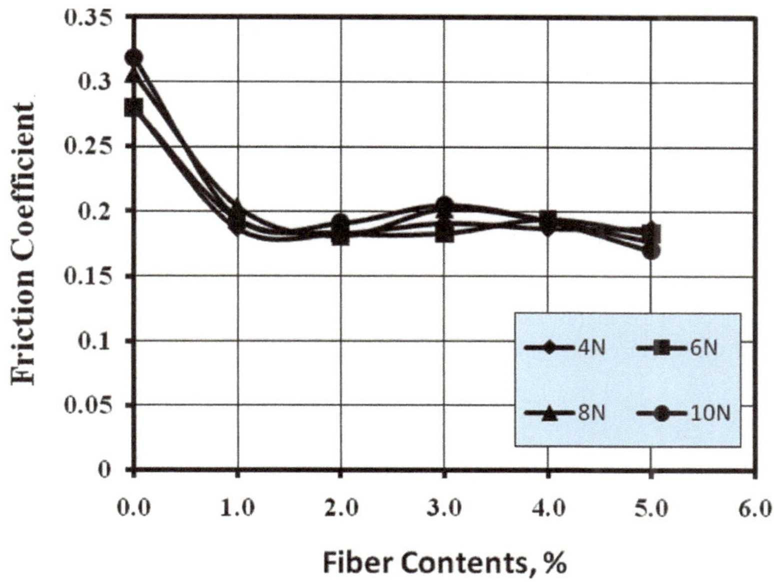

Figure 12. Frictional behavior of polyester composite impregnated with 10 vol. % corn oil

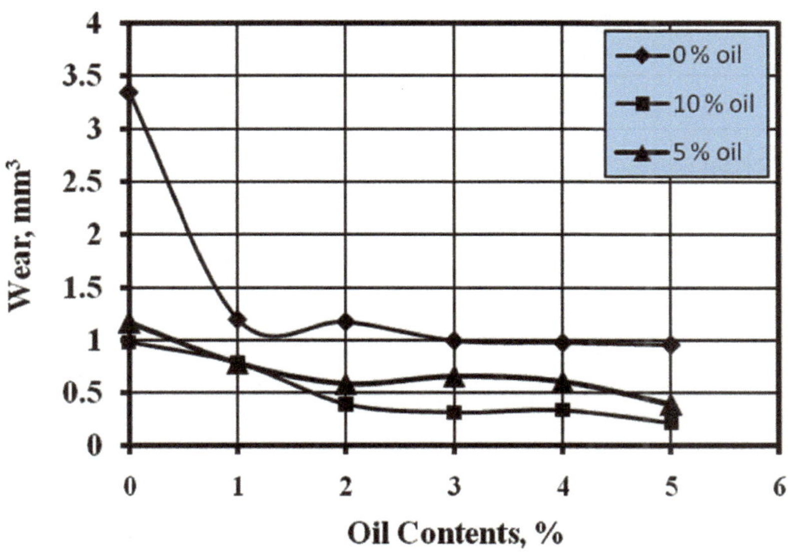

Figure 13. Wear of polyester composite impregnated by corn oil

3.2.6. Frictional behavior of polyester composites impregnated by sunflower oil

Friction coefficient of 4 % fiber composite content impregnated by 5 % sunflower oil under load of 8N was decreased to 0.177, figure 14. Composite impregnated by 10 % sunflower oil content shows good friction results, friction coefficient decreases to 0.166 for composite of 4 % fiber content under 4N, fig 15. beside low effects of applied load in friction coefficient. It seems that increase of sunflower oil in polyester composite increased its hardness and decreased friction coefficient.

3.2.7. Wear results of polyester composites impregnated by sunflower oil

Wear results for composite impregnated by 5 and 10 % sunflower oil were shown in figure 16. it could be stated that increasing the oil content decreased wear value to 0.196 mm^3 for composite of 10 % oil content and 5 % fiber content.

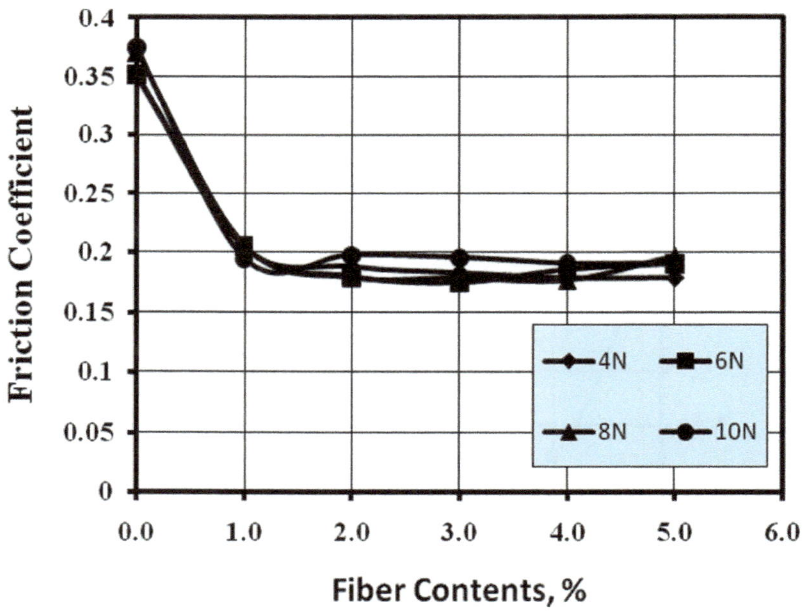

Figure 14. Frictional behavior of polyester composite impregnated with 5.0 vol. % sunflower oil

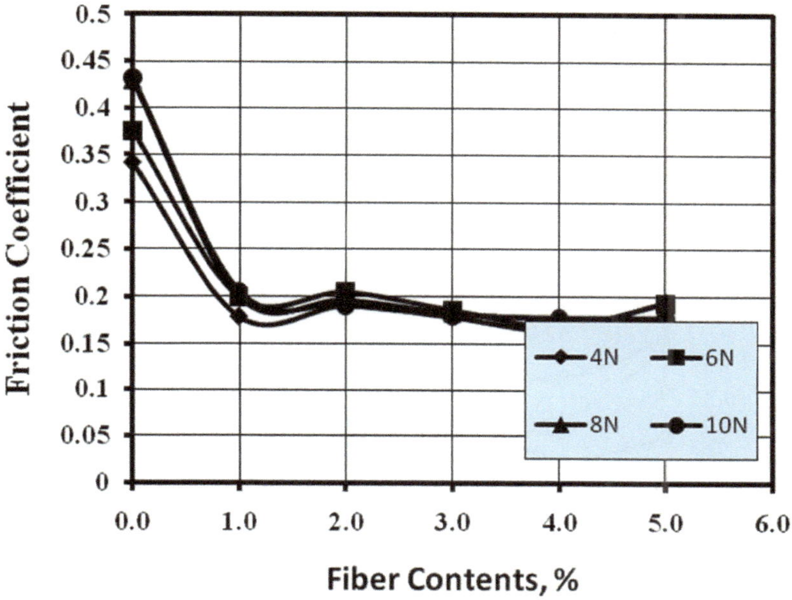

Figure 15. Frictional behavior of polyester composite impregnated with 10 vol. % sunflower oil

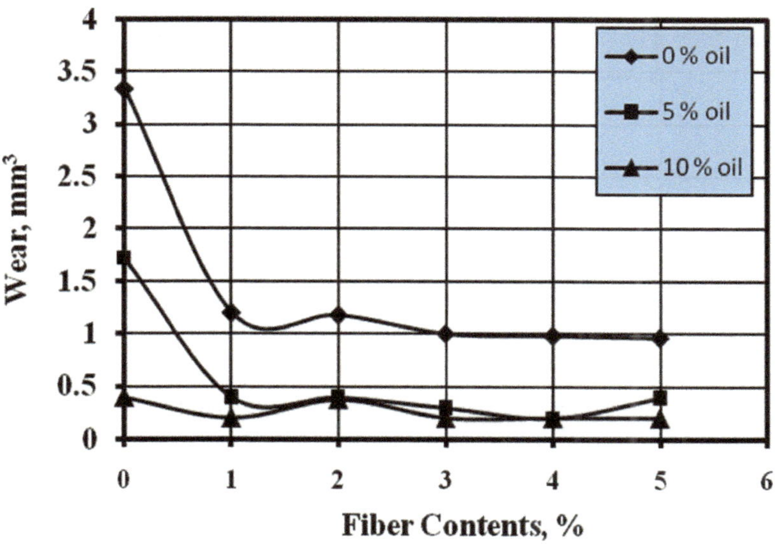

Figure 16. Wear of polyester composite impregnated by sunflower oil

3.3. Polyester composites reinforced by PTFE fiber

3.3.1. Frictional behavior of polyester composite free of oil

Frictional behavior of composite free from oil was shown in figure 17. It could be noted that the increasing of fiber content decreases friction coefficient to 0.257 for composite of 5 % PTFE fiber content under 4N. It may be interpreted by the increase of PTFE layer on the contact area which behaved as solid lubricant layer and reduced friction coefficient.

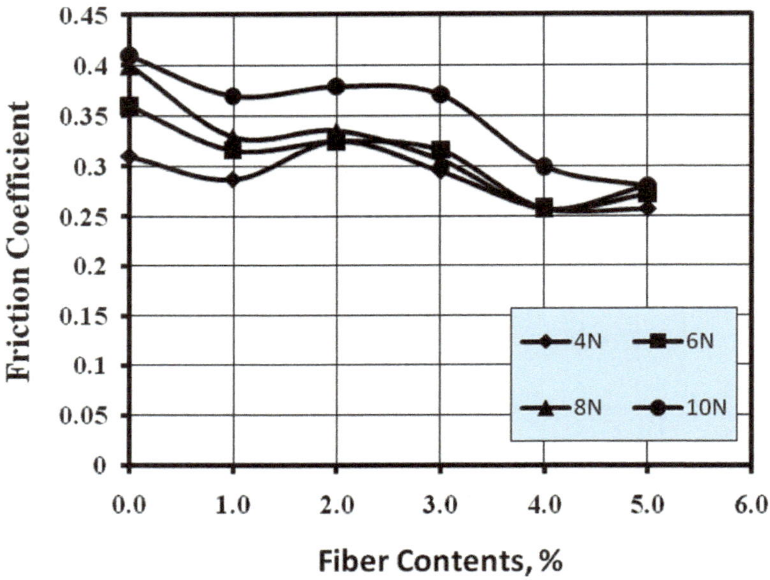

Figure 17. Frictional behavior of polyester composite free of oil reinforced by PTFE fiber

3.3.2. Frictional behavior of polyester composites impregnated by olives oil

Friction results for composite impregnated by 5 % olive oil was shown in figure 18., it shows that the increase in fiber content decreases friction coefficient to 0.165 for composite of 5 % fiber content under 4N. Figure 19. shows that increase of olive oil contents to 10 vol. % decreases friction coefficient remarkably to 0.152 for composite of 5 % fiber content under 4N. It seems that the increase in oil content increased the presence of oil layer at friction surface which may be responsible for friction reduction. These results recommended those composite as good bearing material.

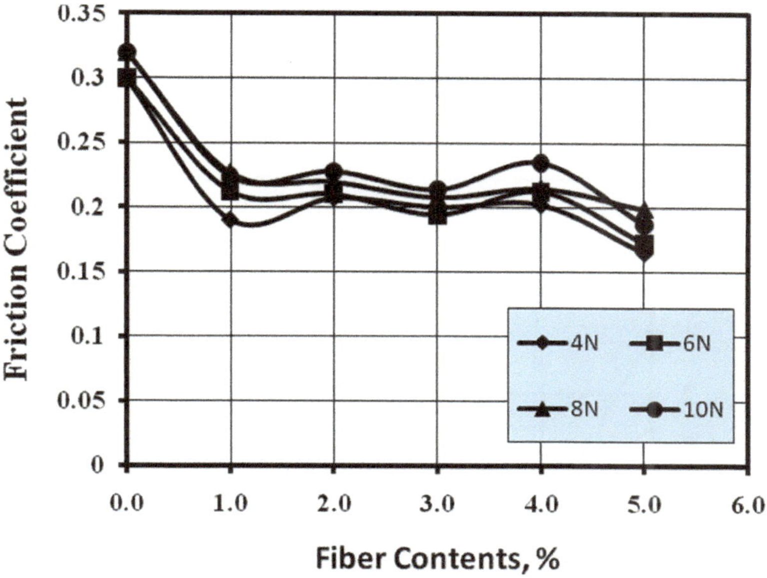

Figure 18. Frictional behavior of polyester composite impregnated by 5.0 vol. % olives oil

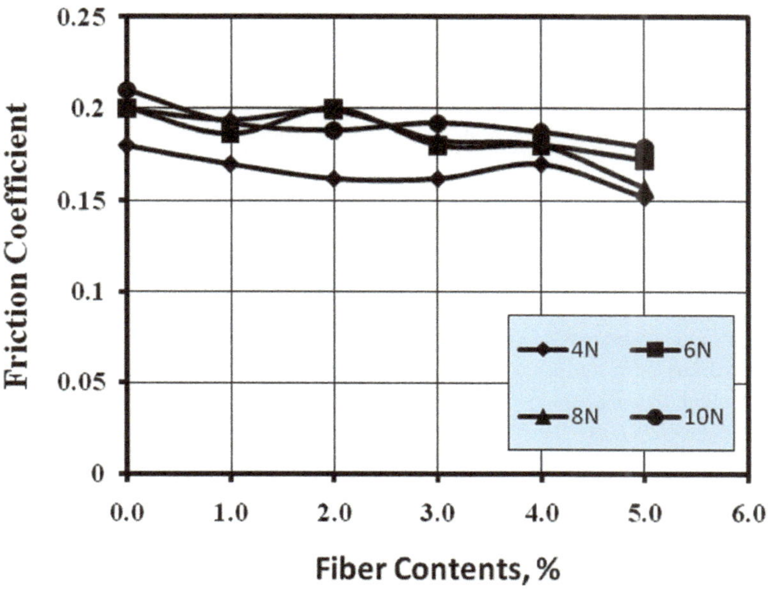

Figure 19. Frictional behavior of polyester composite impregnated by 10.0 vol. % olives oil

3.3.3. Wear results of polyester composites impregnated by olives oil

Figure 20. shows that the increase in oil content decreased the wear value to 0.315 mm^3 for composite of 3 % PTFE fiber content and 10 % oil content. It may be interpreted by increase of oil contents decreased polyester transfer and reduce wear.

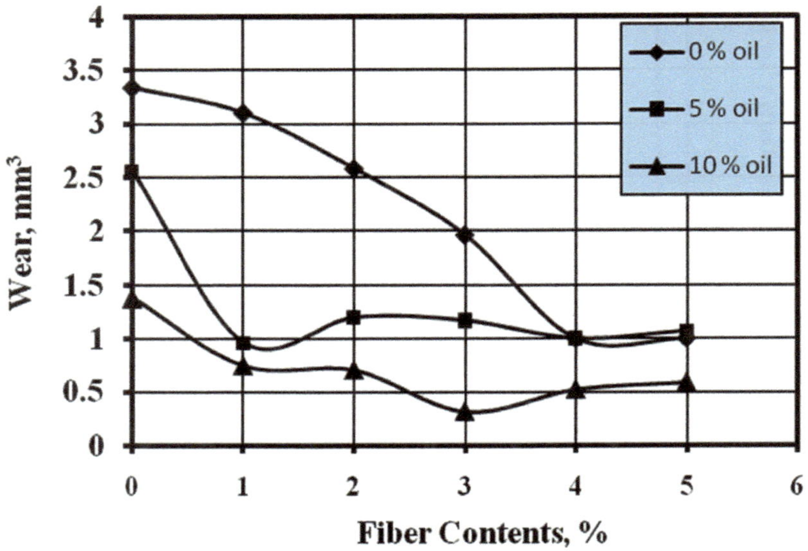

Figure 20. Wear of polyester composite impregnated by olives oil

3.3.4. Frictional behavior of polyester composites impregnated by corn oil

Composite impregnated by 5 % corn oil shows remarkably decreases in friction coefficient with increases of PTFE content; it reached to 0.172 for composite of 5 % fiber content under 6N figure 21. It may be attributed to the ability of corn oil film to reduce adhesion between composite and counterface. This may be responsible for friction reduction. Friction coefficient deceases in composite impregnated by 10 % oil content to 0.182 for composite of 5 % fiber content, figure 22.

3.3.5. Wear results of polyester composites impregnated by sunflower oil

Wear results decreases with increases of oil content it reaches to 0.283 mm^3 for composite of 2 % fiber content and 10 % oil content. It may be attributed to the ability of PTFE layer and the presence oil film to reduce friction as well as polyester transfer to counterface. Figure 23.

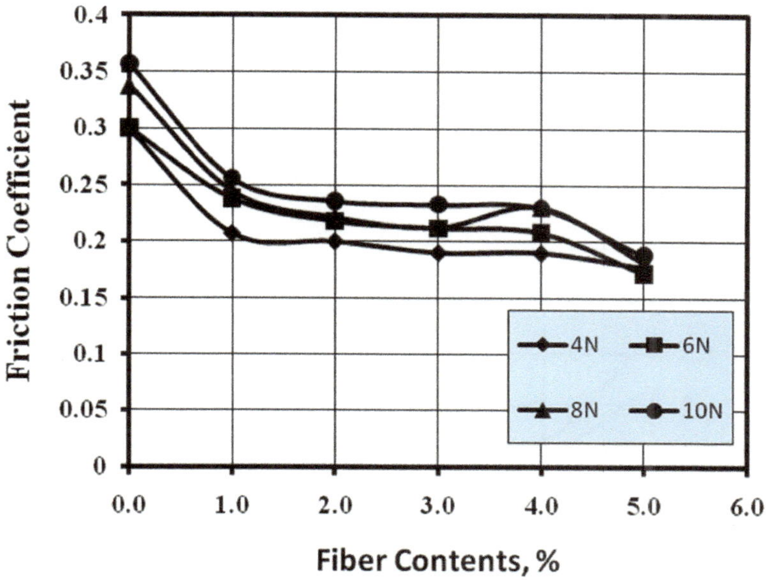

Figure 21. Frictional behavior of polyester composite impregnated by 5.0 vol. % corn oil

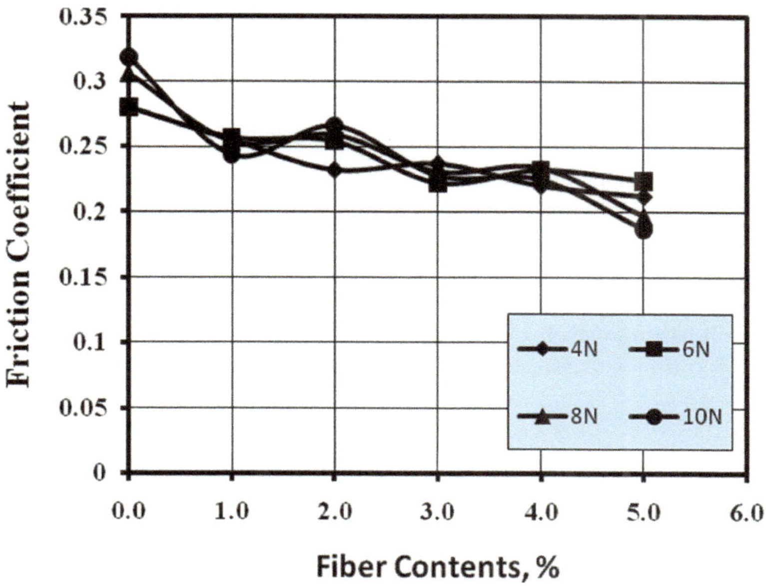

Figure 22. Frictional behavior of polyester composite impregnated by 10.0 vol. % corn oil

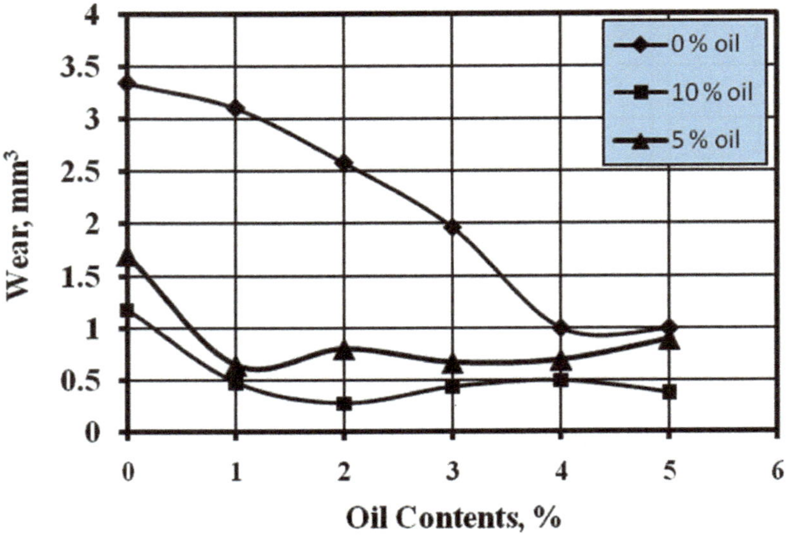

Figure 23. Wear of polyester composite impregnated by corn oil

3.3.6. Frictional behavior of polyester composites impregnated by sunflower oil

Friction results of composites impregnated by 5 % sunflower shows decreases in friction coefficient to 0.185 with increase of PTFE content to 5 % under 10N, figure 24. The reached results showed that there were low effect of applied loads on friction coefficient. It seems that the increase of sunflower oil increases composite hardness. This may be responsible for friction reduction. Composite impregnated by 10 % sunflower oil shows slightly increases in friction coefficient in comparing with composite impregnated by 5 % sunflower. But friction coefficient decreases with increasing of fiber content to 0.2 for composite of 5 % fiber content under 10N figure 25.

3.3.7. Wear results of polyester composites impregnated by sunflower oil

Composite of fiber content impregnated by 5 and 10% of sunflower oil decreased the value of wear to 0.347 mm^3. Especially for composite of 10 % oil content and 5 % fiber content. It may be attributed to the ability of sunflower oil to reduce polyester transfer to counterface and reduce wear. Figure 26.

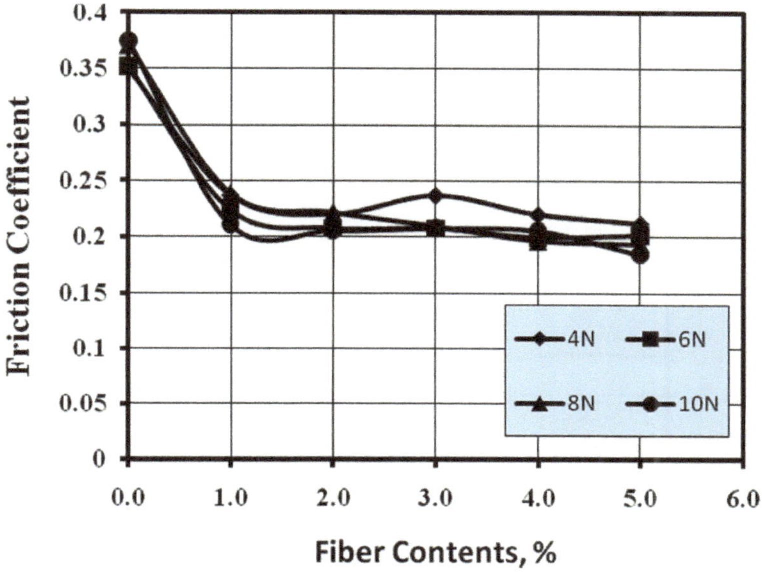

Figure 24. Frictional behavior of polyester composite impregnated by 5.0 vol. % sunflower oil

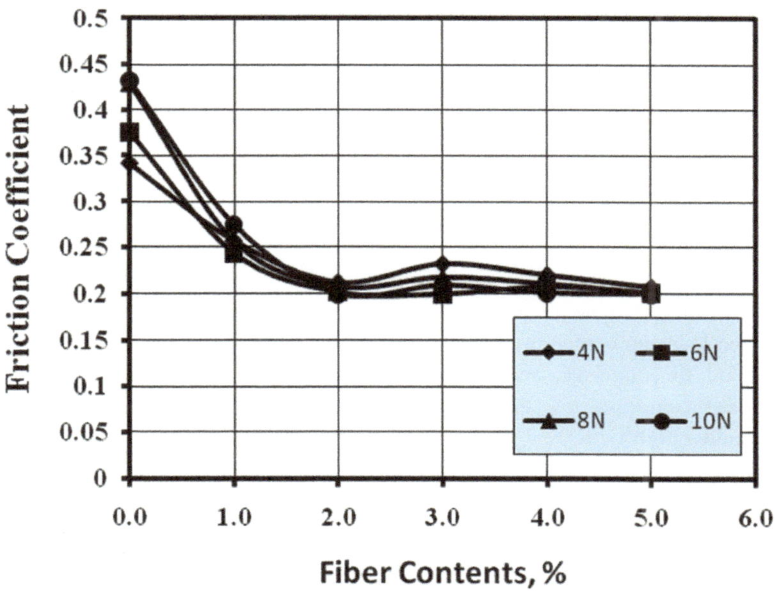

Figure 25. Frictional behavior of polyester composite impregnated by 10.0 vol. % sunflower oil

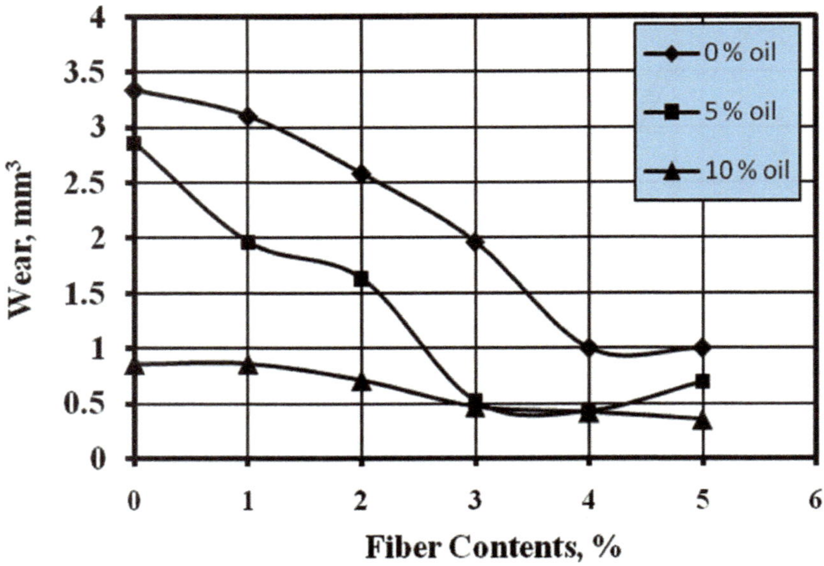

Figure 26. Wear of polyester composite impregnated by sunflower oil

Composite	VHR
Free polyester without oils	11.28×10^6 N/mm^2
Composite impregnated by corn oil	11.43×10^6 N/mm^2
Composite impregnated by glycerin oil	13.77×10^6 N/mm^2
Composite impregnated by olive oil	13.92×10^6 N/mm^2
Composite impregnated by paraffin oil	09.68×10^6 N/mm^2
Composite impregnated by sunflower oil	24.22×10^6 N/mm^2

Table 1. Micro hardness for test specimens

These results confirmed the previous results of friction and wear. The consistency of friction with applied load can be interpreted on the basis that hardness increases with addition of sunflower oil to composite, and consequently the plastic deformation decreases, this recommended composite impregnated by sunflower oil as very good bearing material especially under high loads.

4. Conclusions

Polyester composites filled with vegetable oils and reinforced by polymer fiber shows low friction coefficients and low wear rates, which recommended these composites for industrial applications such as self-lubricated materials.

5. Future work

Polyester composites reinforced with some agricultural wastes are proposed to be used in industrial applications.

Author details

Ibrahim Refaay Ahmed
Faculty of engineering, Beni-Suif University, Egypt

Ali Waheed Yousry
Faculty of engineering, Minia University, Egypt

6. References

[1] Klaus F., Zhong Z. , Alois K., "Effects of various fillers on the sliding wear of polymer composites", Composites Science and Technology 65 (2005) pp 2329–2343.

[2] Karger-Kocsis J., Zhang Z., "Structure–property relationships in nanoparticle/semi-crystalline thermoplastic composites". Mechanical properties of polymers based on nanostructure and morphology, New York: CRC Press; (2005) pp. 547–596.

[3] Hutchings I. M., "Wear-resistant materials: into the next century", J. Mat. Sci. Eng., (1994) pp.185–195.

[4] H. Czichos, D. Klaffke, E. Santner, M. Woydt, Advances in tribology: the materials point of view, Wear 190 (1995) 155–161.

[5] Gauthier C, R. Schirrer. Time and temperature dependence of the scratch properties of poly(methylmethacrylate) surfaces. Journal of Materials Science 2000;35(9):2121–30.

[6] Inoue, Kaoru Miyake, Yuochi, Development of the chemical recycling technology of glass fiber reinforced PA6 Parts, SAE Paper, No 01-0694 2001 (in English).

[7] J.R.T. Branco, S.V. Campos, Wear behaviour of thermally sprayed PET, Surf. Coat. Technol. 120–121 (1999) 476.

[8] Samyn P., Baets P., "Friction of polyoxymethylene homopolymer in highly loaded applications extrapolated from small-scale testing", Tribol. Lett. 19 (2005) pp. 177–189.

[9] Zheng S. F., Weng G.J., "A new constitutive eq. for the long-term creep of polymers based on physical aging", Eur. J. Mech. A21 (2002) pp. 411–421.

[10] Beake B. D., Leggett G. J., Shipway P. H., "Nanotribology of biaxially oriented poly(ethylene terephthalate) film", Polymer 42 (2001) p. 7025.

[11] Neogi S., Hashmi S. A. R., Chand N., "Role of PET in improving wear properties of PP in dry sliding condition", Bull. Mater. Sci. 26 (2003) p.579.

[12] Franklin S. E., "Wear experiments with selected engineering polymers and polymer composites under dry reciprocating sliding conditions", Wear 251 (2001) 1591.

[13] Hashim P., Nihat T., "Investigation of the wear behaviour of a glass-fibre-reinforced composite and plain polyester resin", Composites Science and Technology 62 (2002) pp. 367–370

[14] Tsukruk V., Nguen T., Lemieux M., Hazel J., Weber W., Shevchenko V., et al., "Tribological properties of modified MEMS surfaces. In: Bhushan B, editor. Tribology issues and opportunities in MEMS. Dodrecht: Kluwer Academic Publishers; (1998) pp. 607–614.
[15] Dutta N., Karak N., Dolui S. K., "Synthesis and characterization of polyester resins based on Nahar seed oil", Progress in Organic Coatings 49 (2004) pp. 146–152
[16] Liu Y., Schaefer J. A., "The sliding friction of thermoplastic polymer composites tested at low speeds", Wear 261 (2006) pp. 568–577.
[17] Liu Y., Hild W., Kitsche M., Doering S., Lasse S., Hungenbach G., Scherge M., Schaefer J. A., "Tribological performance of selected bearings and bearing materials used for nanopositioning, in: D., Dowson, et al., Life Cycle Tribology: 31st Lyon-Leeds Symposium on Tribology, Tribology and Interface Engineering Series No. 48 (2005) Elsevier, pp. 739–750.
[18] Sawyer W. G., Freudenberg K. D., Bhimaraj P., Schadler L. S., "A study on the friction and wear behavior of PTFE filled with alumina nanoparticles", Wear 254 (2003) p. 573.
[19] McCook N. L., Burris D. L., Bourne G. R., Steffens J., Hanrahan J. R., Sawyer W. G., "Wear resistant solid lubricant coating made from PTFE and epoxy", Trib. Let. 18 (2005) p. 119.

Chapter 8

Unsaturated Polyester Resin for Specialty Applications

Bharat Dholakiya

Additional information is available at the end of the chapter

1. Introduction

Polymers are substances whose molecules consist of a large number of units of a few types: the units themselves, consisting of a number of atoms, are usually referred to as the segments of the polymer. In the polymerization of a mixture of two monomers, the structure of each macromolecule contains units of both monomers. Such a polymer is called copolymer and the process of its synthesis is called copolymerization.

Polyesters [1-3] are one of the most versatile synthetic copolymers. Polyesters are produced in high volume that exceeds 30 billion pounds a year worldwide [3]. They are widely used commercially as fibres, plastics, composites and for coatings applications too [4-6]. They are heterochain macromolecules that possess carboxylate ester groups as an integral component of their polymer backbones. Polyesters have received a great deal of attention since the early work of Carothers, who initiated study on many step-growth polymerizations [7]. His work involved A-B ω-hydroxy acids, the polymerization of certain lactones and the esterification of A-A linear diols with B-B terminal aliphatic dicarboxylic acids. The resulting copolymers were of low molecular weight (8,000-10,000 g/mol), hard, crystalline solids and susceptible to conversion from the molten state to filaments which could be stretched below their melting point with an ultimate increase in strength.

Carothers worked with aliphatic straight-chain polyesters, which were soluble in organic liquids, low melting and had poor resistance to hydrolysis. These polyesters were not used as textile fibres [5]. The extension of these concepts later led to the discovery of nylon-6,6 in 1935 and Whinfield and Dickson developed poly(ethylene terephthalate) (PET) in 1941[8]. A partially aromatic organic structure was necessary to increase melting temperature (T_m) above 250°C. A large number of polyester structures have found use in industry today which displays a wide variety of properties and applications. More detailed discussion may be found in a number of excellent books and reviews [3-4&6].

1.1. Types of polyester resin

1.1.1. Saturated polyester resin [9]

These are the reaction products of dibasic acids or dibasic acid chlorides with diols and largely used in textile industries e.g. Polyethylene terephthalate.

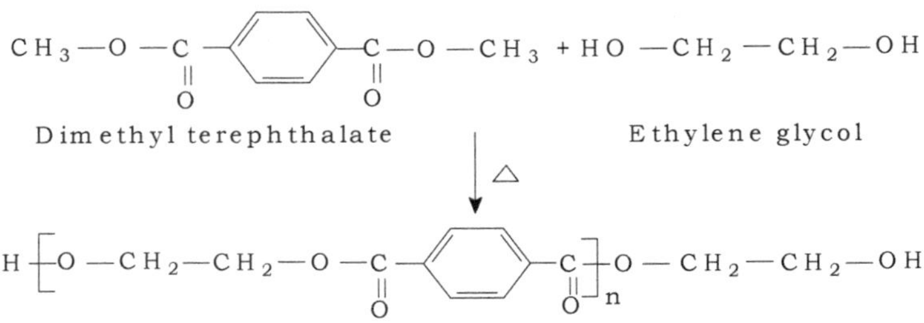

1.1.2. Alkyd resin [9]

Alkyds are the reaction products of polyhydric alcohols with fatty acids followed by reaction with dibasic acids. They are largely used in paint and printing ink industries [10-14].

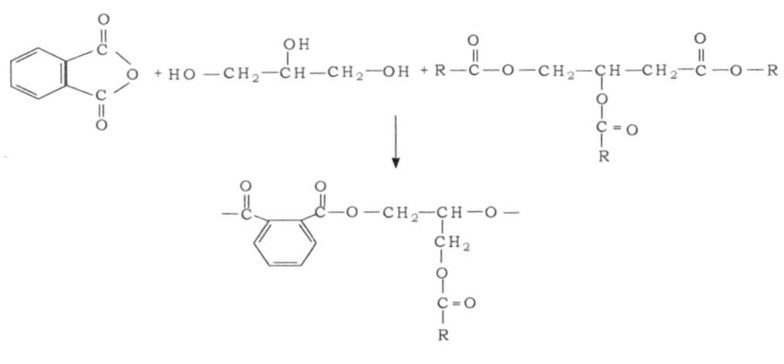

1.1.3. Vinyl ester resin [4]

Vinyl ester resins are becoming increasingly important in new industrial applications such as coating, printed circuit boards, metal foil laminates, building materials, automotive parts, rigid foams and fibre reinforced composites [15-19]. A conventional vinyl ester resin can be prepared by end capping various epoxy resins with unsaturated mono-carboxylic acid [20-22]. They combine the excellent mechanical, chemical and solvent resistance properties of epoxy resins with the properties found in the unsaturated polyester resins. The cured vinyl ester resin has physical properties superior to cured conventional polyester resin, particularly corrosion resistance. Vinyl ester resins were developed in the late 1950's and early 1960's i.e. just after IInd world war. The resins were prepared by reacting glycidyl acrylate and glycidyl methacrylate with bisphenol-A. These resins are very reactive and have very short life. A general structure of vinyl ester obtained by reacting epoxy resin with acrylic acid is as shown below.

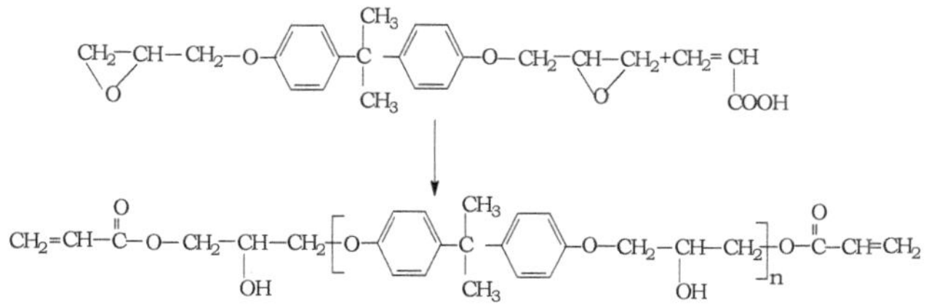

1.1.4. *Unsaturated polyester resin*

Unsaturated polyester resins are the condensation products of unsaturated acids or anhydrides and diols with/without diacids. The unsaturation present in this type of polyesters provides a site for subsequent cross-linking [23-24]. Since 1930, unsaturated polyester resins have been used remarkably for wide range of applications making them a thermosetting system of major importance [25-26]. These resins are compounded with varied fillers, reinforcements and cured by using free radical initiators to yield thermoset articles having a wide range of chemical and mechanical properties depending upon the choice of diacids, diols, cross- linking agents, initiators and other additives [27]. This versatility in the properties of the final thermoset product associated with comparatively low cost has renewed the interest in these resins as an important matrix material for wide range of applications. In 1929, Arvin and Carothers [28] developed unsaturated polyester resins from maleic acid and ethylene glycol reacted at temperature 175^0-185^0C. Ford Motor Co. Ltd.[29] synthesized unsaturated polyester resin by reacting maleic anhydride and phthalic anhydride with propylene glycol at 100^0C and then at 250^0C till the acid number diminished to the value less than 50 (mg of KOH per gm of sample). Corrado and his assistants [30] synthesized low viscosity unsaturated polyester resins by reacting maleic anhydride, phthalic anhydride and dipropylene glycol at 200^0C. Ochsenbein and Olliver [31] synthesized storage stable unsaturated polyester resin by reacting maleic anhydride, propylene glycol and dipropylene glycol at 185^0C under inert atmosphere. General purpose unsaturated polyester resins were prepared by using maleic anhydride, phthalic anhydride and propylene glycol with the molar ratio of phthalic anhydride: maleic anhydride ranging from 1:2 to 2:1[32-33]. For thermoset products, the resultant resin was blended with styrene for cross-linking and small amount of peroxide as initiator. These types of resins are useful in making trays, shower stalls, boats, swimming pool, water tanks etc.

Acrylic modified unsaturated polyester resin having good cracking resistance, flexibility and thixotropic property were synthesized using 65 parts 0.15:1.0:0.3:0.7 molar ratio of dicyclopentadiene: fumaric acid: ethylene glycol : diethylene glycol copolymer and 35 parts of 2-hydroxy ethylmethacrylate [34]. Chemical resistant thermosetting unsaturated polyester resins for laminates having low content of volatile organic components were prepared from isophthalic acid, maleic anhydride and propylene glycol with 35% styrene as reactive diluents [35]. Unsaturated polyester resin prepared from isophthalic acid, maleic anhydride and neopentyl glycol was mixed with dicyclopentadiene resin and styrene to give dielectric compound for electric machines [36]. The effect of styrene content on non-exponential and non-Arrhenius behaviour of the α-relaxation of cured unsaturated polyester resin was investigated by dynamic mechanical analysis [37].

A review on recent research progress in modification methods of unsaturated polyester resin, with an emphasis on the toughening and reinforcement, flame proofing, improving heat resistance and resistance to environmental medium, reducing contraction percentage of cured unsaturated polyester resin and air drying properties was done by Q. Jun-min et. al. [38]. The research development on unsaturated polyester resins during the year 2000-2001 was reviewed with references [39]. The review on the development trends in the production of unsaturated polyester resins was done by Penezek [40]. Literature survey reveals that there are many reports about the synthesis and modification methods of unsaturated polyester resin [41-49].

1.2. Fire resistant polyester resin [48]

Fire resistant polyester resins can be prepared by using halogenated dibasic acids e.g. tetrachlorophthalic anhydride(I), tetrabromophthalic anhydride(II) [49], dibromoneopentyl glycol (III) [50] or tetrabromo bisphenol-A (IV) [51] in place of phthalic anhydride or propylene glycol. Fire resistance of polyester resins can be further improved by blending with flame retardant additives such as triphenyl phosphate and antimony trioxide [52]. Such types of resins are generally used in fume hoods, electrical equipments, building panels and navy boats.

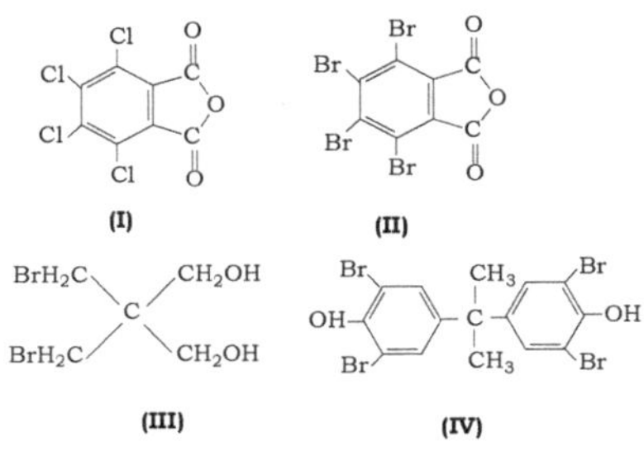

1.3. Electrical resistant polyester resin [48]

Electrical resistant polyester resins can be prepared by using dibasic acids e.g. isophthalic acid, maleic anhydride and neopentyl glycol [53] or tetrabromobisphenol-A [51] in place of phthalic anhydride or propylene glycol. Electrical resistance of polyester resins can be further improved by blending with various additives such as antimony trioxide [54], kaolin [55], mica [56] and calcium carbonate [57]. Such types of resins are generally used in printed circuit board, electrical appliances and electronic equipments.

1.4. Low styrene emission polyester resin

The unsaturated polyester normally blended with styrene to a reactive resin solution. The styrene acts both as a cross-linking agent and as a viscosity reducer so that the resin can be processed. In conventional unsaturated polyesters the styrene content varies between 35-45wt %. The styrene monomer in the unsaturated polyester resin is an environmental and occupational health problem due to evaporation and emissions, which occur during the processing of the resin. During the end of the 1980.s the environmental problems caused by styrene emission from the processing of unsaturated polyesters were debated a lot in the society, especially in the Nordic countries. The producers of unsaturated polyesters started therefore to search actively for alternatives to styrene and started to develop additives inhibiting styrene emission.

2. Unsaturated polyester based fire resistance composites

2.1. Introduction

The fire behavior of materials is an important parameter to be considered for the safety of individuals and it must be taken into account at the beginning of equipment development. The use of composite materials for building equipment is becoming increasingly important and it has been clearly demonstrated by tests on structures that composite materials with an organic matrix behave well than traditional materials (woods in the naval area, aluminium in the aeronautic sector), in the case of a fire. In the naval sector, it is not the case that there are more fires on ships with composite structures than on ships with metallic structures. For several years the use of composites in interiors of cabin and even for structures is common in the aeronautic sector because of weight gains and there have not been more incidents observed on equipment due to the potentially greater fire risk [58-60]. Since many years, unsaturated polyester composites have been used in very varied technology fields as naval constructions, offshore applications, water piping, building construction and automotive applications. In spite of hydrolytic instability problems of these materials, mainly due to the fact of their pre-polymer ester functions [61-62], interest in unsaturated polyester resin comes from their low cost and easy processing. Due to their stable structure, they display good mechanical properties and furthermore, unsaturated polyester resin has good thermal stability. However, the use of these materials is limited due to their poor fire performance. There are some solutions [63]. Survey of literature revels that there are many reports about the fire resistance behavior of the glass reinforced composites [64-88].

The aim of this work was to prepare halogen free fire resistance composites by using non-traditional fillers like fly ash, zinc borate and hydroxyapatite in combination with traditional filler antimony trioxide in different amounts in prepolymer polyester resin and study the thermal and the fire behavior with and without fillers in order to investigate the role played by fillers in the thermal and fire degradation reactions.

2.2. Synthesis of Unsaturated Polyester Resin (UPR-1)

Unsaturated polyester resin (figure-1) was prepared in the present work using technique reported by T Sunemi Hidenari [89]. A mixture of 1.25 mol Propylene glycol (PG), 0.5 mol Phthalic anhydride (PA), 0.2% p- Toluene sulfonic acid (PTSA) and Xylene as solvent was charged in a three-neck reaction kettle equipped with stirrer, thermometer, nitrogen-gas introducing tube, separator and water condenser. The mixture was mechanically stirred and heated at 120⁰C under nitrogen gas stream. When reaction mass becomes clear, it was allowed to cool to 80⁰C and then 0.5 mol Maleic anhydride (MA) was added and continues heating at 150-200⁰C until an acid number of 20 was reached. During esterification reaction, water formed as by product and was continuously removed from the reaction mass as it inhibits the rate of reaction. The Xylene was completely distilled out and reaction product was allowed to cool. When the temperature reached to 160⁰C, 20 mg of hydroquinone was added as inhibitor and when resin temperature dropped below the boiling point of reactive diluent (i.e. Styrene), the polyester resin were mixed with styrene by 35 weight percent of resin [90]. The polymer obtained was characterized Gel Permeation chromatography (GPC) and IR spectroscopy.

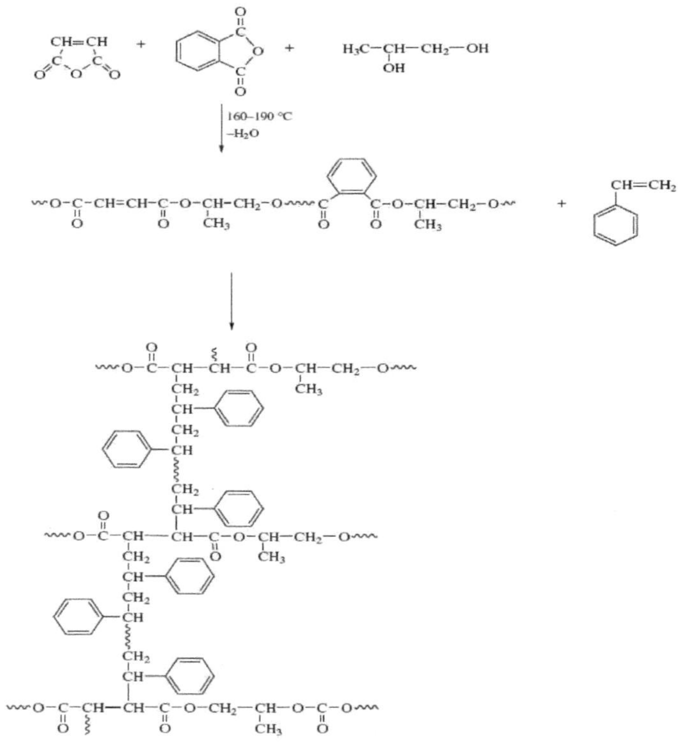

Figure 1. Polyester Resin (UPR-1) Scheme with Cross linking [92]

2.3. Fire resistance formulation and composite preparation

2.3.1. Fillers for fire resistance formulation

Some non-traditional filler like hydroxyapatite, zinc borate and fly ash were used in combination with traditional filler antimony trioxide for fire resistance formulation. The details about fillers are given as below:

2.3.1.1. Hydroxyapatite

The formula of hydroxyapatite is $Ca_{10}(PO_4)_6(OH)_2$. It forms part of the crystallographic family of apatite, isomorphic compounds with the same hexagonal structure. Hydroxyapatite can be found in nature or synthesized by precipitation method using chemical reagents. It is also known as calcium phosphate. This compound is most commonly used for biomaterial and fire resistant applications. This compound can be physically blended in polymers to suppress, reduce, delay or modify the propagation of flame through polymeric materials. The fire retardant property of hydroxyapatite is due to the presence of phosphate. The technical specifications of hydroxyapatite used in this work are given as below:

1. Ca/P molar ratio: 1.61-1.71.
2. Crystalline hydroxyapatite content: 95%.
3. Grain size: 1500-2000 mesh
4. Moisture: 1% max.

2.3.1.2. Zinc borate

Zinc borate, $2ZnO_3B_2O_3$ $3.5H_2O$, is a white crystal powder, which does not dissolve in water and other organic solvent but it does dissolve in hydrochloric and sulphuric acid. It has good thermal stability, lower density and good miscibility with most of the polymers. Zinc borate shows excellent fire resistance performance and can be therefore widely used in polymer, rubber, fibre, paint and ceramic industries. The technical specifications of zinc borate used in this work are given below:

1. B_2O_3: 48%,
2. ZnO: 37%
3. Particle size: 1250 mesh – 1850 mesh
4. Loss on ignition: 13.50 - 15.50%
5. Moisture: 1.0% max.

2.3.1.3. Fly ash

Fly ash is the finely divided mineral residue resulting from the combustion of coal in electric generating plant. Fly ash consists of inorganic matter present in the coal that has been fused during coal combustion. This material is solidified while suspended in the exhaust gases and is collected from the exhaust gases by electrostatic precipitators and is usually of silt sizes (0.074 - 0.005mm). Fly ash is a pozzolanic material and has been classified into two classes F and C based on the chemical composition of fly ash. In the present study class C fly ash was used. Class C fly ash is produced normally from lignite and sub-bituminous coals. The technical specifications of fly ash used in this work are given below:

1. Silicon dioxide (SiO_2) plus Al_2O_3 plus Fe_2O_3: 55.0%
2. Sulfur trioxide (SO_3): 5.0%
3. Moisture content: 3.0% max.
4. Loss on ignition: 6.0%

2.3.1.4. Antimony trioxide

The formula of antimony trioxide is Sb_2O_3. It is a solid white powder and is used mainly as fire retardant additive and as pigment. The substance comes in various grades, depending on particle size. The breakdown of particle size affects both physical properties and pigment property. The finer fraction is used for pigmentation. The flame-retardant property, however, is not affected by particle size. This property is based on antimony trioxide reacting in the event of fire with the surrounding material, with the result that a protective layer of various antimony compounds covers the combustible materials. Since the formation of these compounds consumes heat and because they are generally difficult to ignite, they protect the underlying material from ignition. The technical specifications of antimony trioxide used in this work are given as below:

1. Total antimony oxide content: 99.50%
2. Bulk density: 0.3 - 1.0gms/cc
3. Average particle size: 1.1μm

Fire resistance formulations were prepared by adding non-traditional fillers like fly ash, zinc borate and hydroxyapatite in combination with traditional filler antimony trioxide in different amounts (10 to 30 wt %) in prepolymer polyester resin (table-1). Several combinations were tested to determine effects of each additive in the polymer formulations. Additives in different amount were mixed with prepolymer polyester resin in a round bottom flask equipped with high speed agitator. The resultant formulations were compounded with glass fibre as reinforcing materials and benzoyl peroxide as initiator for composite preparation. Composites were prepared by hand lay-up technique using about 50 weight percent of glass fibre and filler. After application of resin on glass fibre the composites were cured by compression moulding at 7 bar pressure and 120°C temperature for 30 minutes. Two percent benzoyl peroxide (BPO) was used as initiator. The composites obtained from fire resistance formulations were characterized by Limiting oxygen index, (LOI) according to ASTM D-2863 [91], thermo gravimetric analysis (TGA) and IR spectroscopy.

2.4. Characterization

Prepolymer polyester resin synthesized by above procedure was analyzed by GPC, FTIR and TGA.

Composite code	UPR-1 Composition in Wt %	Reinforcement in Wt %	Fillers in Wt %				Limiting Oxygen Index (LOI) in %
			Sb_2O_3	Fly Ash	Hydroxy-apatite	Zinc borate	
FRC-1	100	100	-	-	-	-	21
FRC-2	100	70	30	-	-	-	25
FRC-3	100	70	20	10	-	-	27
FRC-4	100	70	10	20	-	-	29
FRC-5	100	70	-	30	-	-	31
FRC-6	100	70	20	-	10	-	28
FRC-7	100	70	10	-	20	-	30
FRC-8	100	70	-	-	30	-	32
FRC-9	100	70	20	-	-	10	32
FRC-10	100	70	10	-	-	20	34
FRC-11	100	70	-	-	-	30	37

* FRC=Fire resistance composites of UPR-1

Table 1. Fire Resistance Formulation and LOI [92].

2.4.1. Gel Permeation Chromatography (GPC)

GPC is one of the most powerful and versatile analytical technique available for understanding and predicting polymer performance. It is the only proven technique for characterizing a polymer by its complete molecular weight distribution. GPC of polyester resin was done by using Perkin Elmer 200 GPC instrument. The HPLC grade Tetrahydrofuran(THF) was used as a mobile phase at a flow rate of 1.0 ml/min. The GPC system was equipped with two ultrastyragel columns packed with styrene-DVB copolymers of 10^3 and 10^6 A^0 porosity connected in series to cover an exclusion limit of 200 to 10×10^6. Refractive Index detector was used with internal temperature of 35^0C for peak detection. The GPC system was calibrated with eight different polystyrene standards having molecular weight ranging from 2000 to 2.5×10^6. The number average (M_n) of polyester resin sample is 1554, weight average (M_w) is 3576 and polydispersity (M_w/M_n) is 2.301. Gel permeation chromatography of polyester resin is shown in figure -2.

2.4.2. Spectral analysis of polyester resin

An IR spectrum of prepolymer polyester resin is shown in figure-3. A Nicolet Impact 400D FT-IR Spectrophotometer was employed for the measurements. The spectrum was run by applying resin sample on KBr cell covering the range of frequencies from 4000-400 cm^{-1} with scanning period of 20 seconds. The crushed powder sample (2-3mg) in the form of a pellet was prepared by mixing with dry KBr (1g). The spectrophotometer was set at 100% transmittance with pure KBr pellet. A strong absorption band at 755 cm^{-1} and a weak band at 1004 cm^{-1} can be attributed to -C-H bending arising from 1 and 3 position in benzene ring and –C=CH bending arising from isomerisation of maleic anhydride to fumarate during polymerization. A broad-spectrum absorption bend at 1145 cm^{-1} confirms the presence of –C-O-C- of ester linkage. A strong absorption peak appearing at 1306 cm^{-1} was assigned to –C=C– group of polyester. A medium absorption band at 1461 cm^{-1} can be attributed to –C-H bending. The presence of –C=O and symmetric –CH stretching was confirmed by the presence of strong bend at 1736 cm^{-1} and 2985 cm^{-1} respectively. The band at 1736 cm^{-1} confirms the presence of –C=O ester group and also confirms the formation of polyester resin. In addition to this, disappearance of anhydride peak at 1755 cm^{-1} in maleic anhydride & phthalic anhydride and hydroxyl peak at 1375 cm^{-1} proplyene glycol confirms the formation of polyester.

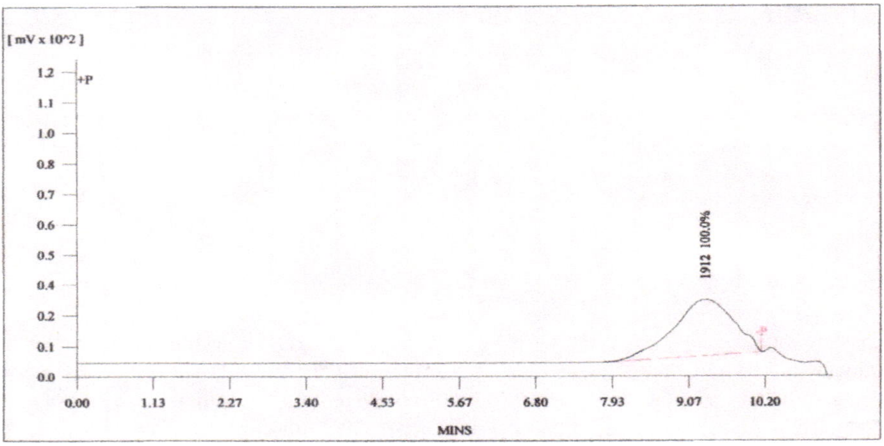

Figure 2. GPC of Polyester Resin

An IR spectrum of cured polyester resin is shown in figure-4. The band at around 2985 cm^{-1} became sharper and bands due to –CH=CH- group almost disappears in composite and a new sharp band at 1447 cm^{-1} becomes visible in the IR spectrum of composite. This indicates the presence of alkane group, which clearly can be attribute to participation of –CH=CH- group in curing process and conversion of this group to alkane during cross linking process.

2.4.3. *Thermogravimetric analysis (TGA)*

TGA of the cured composites was carried out on TA Instruments, USA model 5000/2960. This instrument measures the weight loss as well as the rate of weight loss of a material continuously as a function of increase in temperature. The present study concerns the percentage weight loss of the material at different temperature at a heating rate of 10°C/min. The sample for analysis was kept in an oven at 35°C under vacuum overnight. The cured sample (about 8 to 12 mg) was placed in the sample container and suspended on the quartz rod in an atmosphere of nitrogen gas. The weight of the sample was noted on the TGA balance. The whole assembly was introduced into the furnace and the experiment started by heating the system at constant and definite rate throughout the experiment. Simultaneous change in weight was recorded automatically with the time while temperature was increased at a known uniform rate (10°C/min). The experiment was stopped when no further change in weight could be observed on further heating. Proper knowledge of the thermal stability of polymers is essential for their appropriate applications [93-95]. The thermal behaviour of polymers with reference to their thermal stability is of paramount importance.

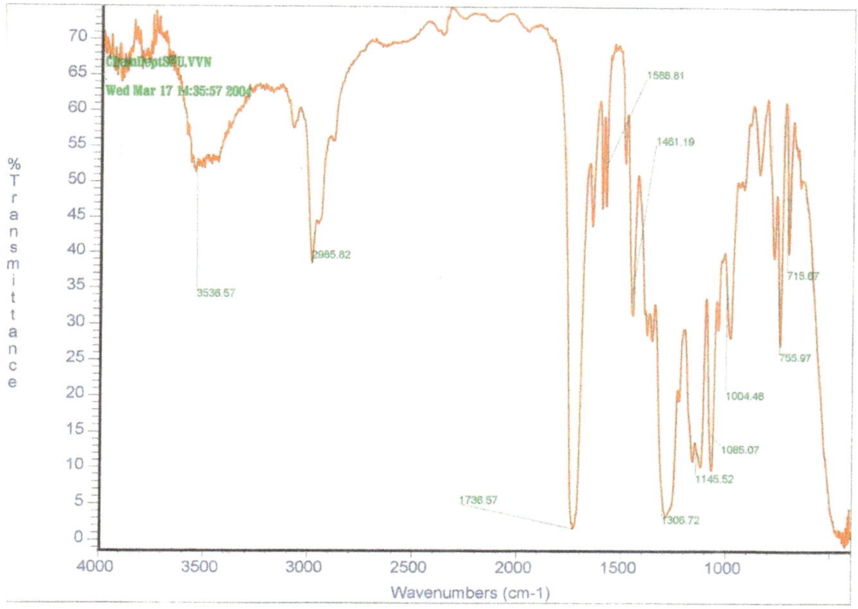

Figure 3. IR Spectra of UPR-1

TGA thermograms of all the composites indicate a similar decomposition pattern of two steps degradation. The first step can be assigned to degradation of resin part and the second one to that of filler part. The initial decomposition in almost all composites start at around 200°C with only 3-7 % loss which clearly shows very good thermal stability of these composites up to 200°C. In unfilled composite (figure-5) loss of weight at 100, 200, 300, 400 and 450 °C is -3.09, -11.37, -34.71, -82.60 and -83.60% respectively. In filled composite (figure-6) loss of weight at 100, 200, 300, 400 and 450 °C is -0.96, -10.30, -28.58, -65.00 and -66.02% respectively. Filled composites show very good thermal stability than unfilled composites.

2.5. Fire resistance properties of composites

In this work, fire behaviour of composites was evaluated. Many procedures exist to evaluate fire behaviour of the composite materials [96-97].The key fire resistance properties of interest for polymers to be used in fire resistance applications are ease of ignition, flame spread, ease of extinction, smoke obscuration, smoke toxicity, heat release rate and limiting oxygen index. In the present study limiting oxygen index (ASTM D-2863) was determined to characterize the composites.

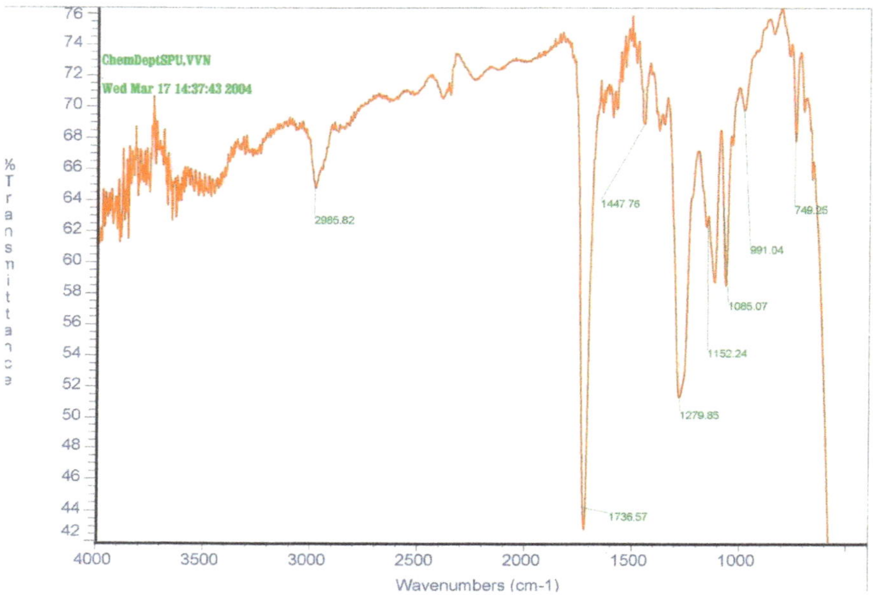

Figure 4. IR Spectra of FRC-1

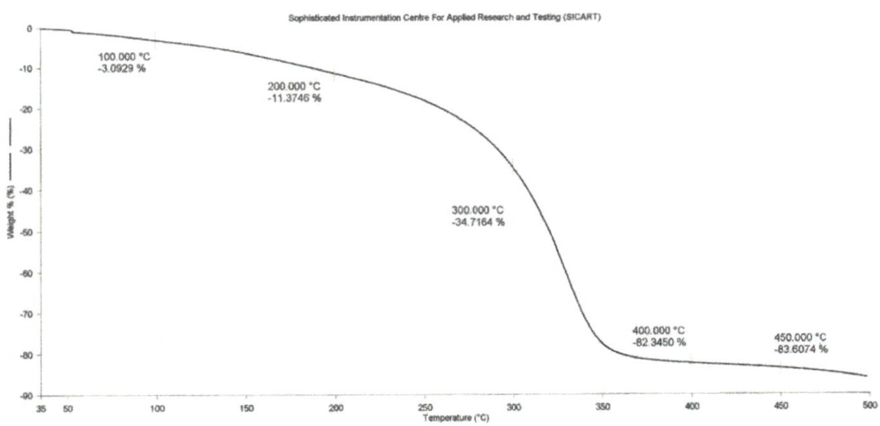

Figure 5. Thermogram of FRC-1

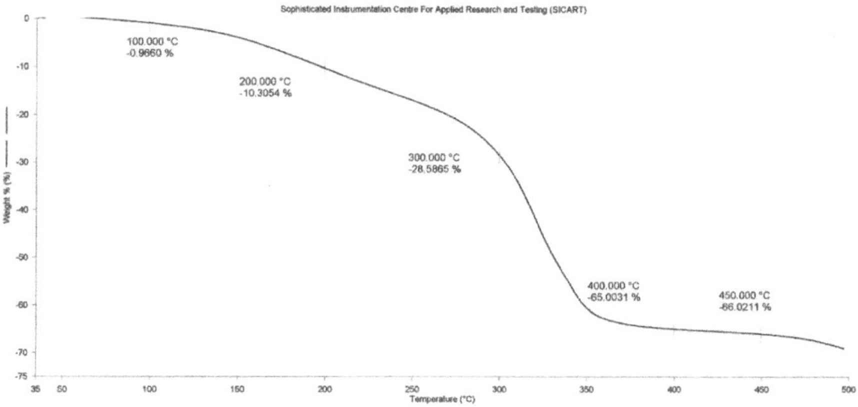

Figure 6. Thermogram of FRC-5

2.5.1. Limiting Oxygen Index (LOI) ASTM D-2863 [91]

Fennimore and Martin developed this method [98]. LOI is defined as measure of minimum amount of oxygen in an environment ($O_2 + N_2$) necessary to initiate and support the burning (flame) under specified conditions. ASTM D2863 is a method to determine the minimum concentration of oxygen in an oxygen/nitrogen mixture that will support a burning of test specimen. The specimen used for LOI test having size150 mm length, 10 mm wide and 4 mm thick. The apparatus used for LOI was Stanton Redcroft FTA flammability unit. The test sample is positioned vertically in a glass chimney and an oxygen/nitrogen environment is established with a flow from the bottom of the chimney. The top edge of the test sample is ignited and the oxygen concentration in the flow is decreased until the flame is no longer supported. Oxygen Index is calculated from the final oxygen concentrations tested. This is very simple and advantageous method for assessing fire resistance of polymers. The critical amount of oxygen was measured. LOI is expressed in percentage as:

$$\text{LOI}=\frac{[O_2]}{[O_2]+[N_2]}\times 100$$

Table-1 represents limiting oxygen index of filled and unfilled composites. In the present study values of limiting oxygen index of filled composites are in the range of 25-37. In practice material is often called the self extinguish materials if LOI is greater than 26 [99]. Thus the results of present study are in good agreement with the reported values. The results clearly indicate that the composites of present study can be used in all fire resistance applications. Composites prepared using fillers give better fire resistance than unfilled composites and also there is increase in fire resistance with increase in amount of filler.

2.6. Mechanical properties of fire resistance composites

Mechanical properties measured in this study include Izod impact strength, Rockwell Hardness and Flexural strength.

2.6.1. Izod impact strength

The impact properties of the polymeric materials are directly related to the overall toughness of the material. The objective of the Izod impact test is to measure the relative susceptibility of a standard test specimen to the pendulum type impact load. The results are expressed in terms of kinetic energy consumed by the pendulum in order to break the specimen. A study of Izod impact strength was carried out in terms of resistance to breakage under high velocity impact conditions, according to ASTM D-256 [100].

Zwick model No. 8900 Impact Machine was used for the present study. For the measurement, a specimen was cut from the fabricated composite (6.4 x 1.27 cm^2) and V notched (2.5 mm) at the middle. The test specimen was clamped into position so that notched end of the specimen remained facing the striking edge of pendulum. The impact test indicates the energy to break standard test specimen of specified size under the stipulated conditions of specimen mounting and pendulum velocity at impact. All the measurements were carried out at room temperature.

2.6.2. Rockwell hardness

Hardness of a material is defined as the resistance to deformation, particularly permanent deformation, indentation or scratching. The Rockwell hardness test measures the net increase in depth impression as the load on an indentor is increased from a fixed minor to a major load and then returned to minor load. Rockwell Hardness study was carried out at room temperature according to standard method of testing ASTM D-785 [101].

In the present study, Rockwell hardness tester model RAS/Saroj Engg. Udyog Pvt. Ltd., Jaysingpur was used. Load of 100kgf was applied for each measurement. The specimen with parallel flat surfaces was placed on the avail of the apparatus and minor load (10kgf) was applied by lowering the steel ball onto the surface of the specimen. The dial was adjusted to zero on the scale under minor load and the major load (100kgf) was immediately applied by releasing the trip lever. After 15 second the major load was removed and the specimen was allowed to recover for 15 second. Rockwell hardness was read directly on the dial.

2.6.3. Flexural strength

It is one of the most important mechanical properties of interest for any comparison of rigid materials or modulus of rupture. Flexural strength is the ability of the material to withstand bending forces applied perpendicular to its longitudinal axis. The test specimens were conditioned in accordance with the ASTM D 790 [102].

A Dutron's Tensile Tester Model No. 130 was used in the present study. For the measurement, a strip of the dimensions 8.0 x 1.25 cm^2 was cut from the composites. It suited the gauge length of 7.5 cm i.e. the experiment was carried out a distance between two jaws to be 7.5 cm apart. The results were recorded with the chart speed of 2 mm/min at room temperature. The test was initiated by applying the load to the specimen at the specified crosshead rate. The deflection was measured by gauge under the specimen in contact with it in the centre of the support span. Table–2 represents Izod impact strength, Rockwell hardness and Flexural strength.

Composite code	Izod Impact in Jcm^{-1}	Flexural Strength in MPa	Rockwell Hardness(M)
FRC-1	5.09	238.85	59
FRC-2	4.15	230.85	57
FRC-3	5.00	238.75	59
FRC-4	4.00	230.15	57
FRC-5	3.50	225.00	50
FRC-6	7.00	240.00	60
FRC-7	10.05	245.85	62
FRC-8	11.50	247.90	64
FRC-9	7.25	242.00	61
FRC-10	10.10	246.15	63
FRC-11	11.80	247.90	66

Table 2. Mechanical Properties of Fire resistance Composites [92]

The mechanical properties of composites improve remarkably with increase in the filler content except in antimony trioxide and fly ash. Increase in mechanical properties with increase in filler content is due to the reinforcement action provided by the filler particles to the polymer matrix [103-104]. Composites prepared using non-traditional fillers in combination with traditional filler gives better performance than unfilled composites and also there is increase in mechanical properties with increase in amount of fillers. Antimony trioxide gives poor results compared to other fillers and also mechanical performance decreases with increase in amount. For antimony trioxide and fly ash best results obtained for 20:10 % weight of antimony trioxide to fly ash in composites.

The results also revealed that the fire retardancy of the polyester based composites can be improved by using non traditional fillers like zinc borate, hydroxyapatite and fly ash. They have an advantage over a traditional filler antimony oxide to increase the fire retardancy without decreasing mechanical and thermal properties of the composites. There is however, considerable variation in the efficiency of these fillers. The adequacy of the fire resistance performance of these filled composites is dependent on both types of filler and incorporation level of the fillers. The use of antimony oxide and fly ash increases the fire resistance behaviour but there is decrease in mechanical properties and thermal stability. The optimum result obtained was with 30% zinc borate as filler with good improvement in fire resistance with considerable increases in mechanical properties and thermal stability. The use of such fillers can also solve the problems of toxic emissions of halogenated fire retardants and also it lowers the cost of polyester resin. The composites can be used for facade elements, dome light crowns, in the transportation sector, in the electrical industry, e.g. for cable distribution cupboards, for boats and shipbuilding, tanks, tubes, vessels and others electrical, electronic and electro technical applications like circuit breakers, switch board cabinets, automotive distributor caps, printed circuit board etc.

3. Unsaturated polyester resin based electrical laminates

3.1. Introduction

The unbeatable combination of characteristics such as ease of fabrication, low cost, light weight and excellent insulation properties have made composite materials to be one of the most desirable materials for electrical applications. The electrical properties of polymers are related to the behaviour of polymers as an electrical insulating material for the general interest. The purpose of electrical insulation may generally be described as the prevention of unwanted contacts between the electrical conductors at different potentials. A simple example such as two mounted terminals can usefully serve as an illustration. The insulation must maintain the separation under a wide variety of environmental hazards which include humidity, temperature, vibration, radiation, presence of gases, moisture and other contamination. It must be sufficiently strong physically to withstand the mechanical forces which may be exerted by the conductors and at the same time must prevent any significant flow of current between the conductors. No insulator is perfect, but what constituents a significant flow of current depends upon the application. A few micro amps in an electrical supply circuit would be negligible. In every case the current must be sufficiently small to eliminate secondary effects such as temperature rise, mechanical stress and electrochemical action.

The primary function of thermosetting plastics in electrical applications has been that of an insulator. Thermosetting plastics not act as effective insulators but also provide mechanical support for field carrying conductors. For this reason the mechanical properties of thermosetting plastics materials used as insulators become very important.

The use of glass-reinforced laminates in the field of dielectric materials has been of prime importance in the engineering materials. The choice of dielectric material depends upon its dielectric and other properties over a wide range of temperature and dielectric field frequencies. Investigation of dielectric properties is one of the most convenient and sensitive methods in studying structural property relationship of the glass reinforced for engineering purpose. Typical electrical application of fibre reinforced composite include terminals, connectors, industrial and house hold plugs, switches and printed circuit boards. Now a day unidirectional glass reinforced polymer composites have become a popular alternative to porcelain in the manufacture of high voltage insulators [105-127]. Several investigations on the effects of electrical properties and mineral fillers have been reported [128-129]. Survey of literature revels that there are some reports about the dielectric behaviour of the glass reinforced laminates [130-149]. The glass fibre reinforced laminates were studied for the electrical properties. There are several experimental and numerical procedures which can be used to evaluate the laminates in terms of electrical properties [150-165].

The aim of this work was to prepare electrical resistance composites and study the thermal and electrical behavior with and without electrical resistance additives in order to investigate the role played by additives in the electrical resistance. This chapter discusses the electrical resistance composites which were prepared by using unsaturated polyester resin and fillers (up to 40% by wt.) like mica, raw kaolin and surface treated kaolin with glass fiber and were evaluated for arc resistance, surface resistivity and volume resistivity according to ASTM standards.

3.2. Synthesis of Unsaturated Polyester Resin (UPR-2)

UPR was prepared in the present work using technique reported by T Sunemi Hidenari [89]. A mixture of 42.8 gm (1.3 mol) Propylene glycol (PG), 35.96 gm (0.5 mol) Isophthalic acid, 0.2 gm (0.2%) p- Toluene sulfonic acid (PTSA) and Xylene as distillating solvent was charged in a three - neck reaction kettle equipped with stirrer, thermometer, nitrogen-gas introducing tube, Dean & Stark apparatus and water condenser. The mixture was mechanically stirred and heated at 120 ^{0}C under nitrogen gas stream and esterification was carried out while removing water formed by the reaction from the reaction system. When reaction mass becomes clear, it was allowed to cool. When the temperature was dropped to 80^{0}C, 21.24 gm (0.5 mol) Maleic anhydride (MA) was added and continues heating at 200-225^{0}C until an acid number of 25 were reached. The Xylene was completely distilled out and reaction product was allowed to cool. When the temperature reached to 160^{0}C, 20 mg of hydroquinone was added as inhibitor and when resin temperature dropped below the boiling point of reactive diluent (i.e. Styrene), the polyester resin was mixed with styrene by 35 weight percent by resin [90].

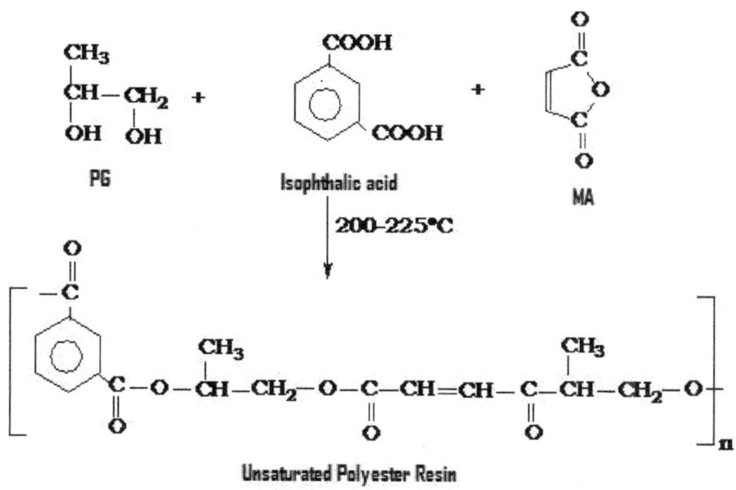

Figure 7. Reaction mechanism for UPR-2

3.3. Electrical resistance formulation and composite preparation

3.3.1. *Fillers for electrical resistance formulation*

3.3.1.1. *Raw kaolin and surface treated kaolin*

Raw Kaolin is the common name for the mineral products comprised totally or substantially of the aluminium-silicate clay. Kaolin is an inherently fine particle-size, platy, chemically inert mineral. The work reported here seeks to improve the mechanical performance of composites by using surface treated kaolin. Surface treated kaolin was prepared by treating raw kaolin with surface active coupling agent. Commercially available silane was chosen as a surface active coupling agent because these materials have a history of applications in other polymer composites [166].

3.3.1.2. *Mica*

Mica is phyllosilicate mineral that have a layered or platy texture. The value of mica is in its unique physical properties. The crystalline structure of mica forms layers that can be split or delaminated into thin sheets. Mica is also chemically inert and is stable when exposed to electricity, light, moisture and extreme temperatures.

Electrical resistance formulations were prepared by adding various additives viz: raw kaolin, surface treated kaolin and mica in different amounts (10 to 40 wt %) in unsaturated polyester resin (table-3). Several combinations were tested to determine effects of each additive in the polymer formulations. Additives in different amount were mixed with unsaturated polyester resin in a round bottom flask equipped with high speed agitator. The resultant formulations were compounded with glass fibre as reinforcing materials and benzoyl peroxide as curing catalyst for composite preparation. Composites were prepared by hand lay-up technique using about 50 weight percent of glass fibre and filler. After application of resin on glass fibre the composites were cured by compression moulding at 7 bar pressure and 120°C temperature for 30 minutes. Two percent benzoyl peroxide (BPO) was used as initiator. The composites obtained from electrical resistance formulations were characterized by surface resistivity, volume resistivity, arc resistance and IR.

3.4. Characterization

3.4.1. *Gel Permeation Chromatography (GPC)*

Gel permeation chromatography of unsaturated polyester resin is shown in figure-2. Gel permeation chromatography of polyester resin was done by using Perkin Elmer 200 GPC instrument. PL GEL mixed B type of column, Tetrahydrofuran as solvent and refractive index detector were used in this analysis. Volume of sample injected was 20μl/min. with retention time 9.309 minutes. The number average (M_n) of polyester resin sample is 1554, weight average (M_w) is 3576 and polydispersity (M_w/M_n) is 2.301.

Composite Code	UPR-2 Composition in Wt %	Reinforcement in Wt %	Fillers in Wt %		
			Mica	Raw Kaolin	Surface Treated Kaolin
ERC-1	100	100	-	-	-
ERC-2	100	70	10%	-	-
ERC-3	100	70	20%	-	-
ERC-4	100	70	30%	-	-
ERC-5	100	70	-	10%	-
ERC-6	100	70	-	20%	-
ERC-7	100	70	-	30%	-
ERC-8	100	70	-	40%	-
ERC-9	100	70	-	-	10%
ERC-10	100	70	-	-	20%
ERC-11	100	70	-	-	30%
ERC-12	100	70	-	-	40%

ERC: Electrical Resistance Composites of UPR-2

Table 3. Electrical Resistance Composite Formulation

3.4.2. Spectral analysis of polyester resin

An IR spectrum of unsaturated polyester resin (UPR-2) is shown in figure-8. A Nicolet Impact 400D FT-IR Spectrophotometer was employed for the measurements. The spectrum was run by applying resin sample on KBr cell covering the range of frequencies from 4000-400 cm^{-1}. A strong absorption band at 742.54 cm^{-1} and a weak band at 1031.34 cm^{-1} can be attributed to –C-H bending arising from 1 and 3 position in benzene ring (isophthalic acid moieties) and –C=CH bending arising from isomerization of maleic anhydride to fumarate during polymerization. A broad-spectrum absorption bend at 1105.22 cm^{-1} confirms the presence of –C-O-C- of ester linkage. A strong absorption peak appearing at 1246.2 cm^{-1} was assigned to –C=C– group of polyester. A medium absorption band at 1447.76 cm^{-1} can be attributed to –C-H bending. The presence of –C=O and symmetric –CH stretching were confirmed by the presence of strong bend at 1736.57 cm^{-1} and 2979.1 cm^{-1} respectively. An IR spectrum of cured polyester resin is shown in figure-9.

3.5. Electrical resistance properties of composites

The key electrical properties of interest for plastics to be used in electrical applications are dielectric strength, dielectric constant, dissipation factor, volume resistivity, surface resistivity and arc resistance. In the present study, volume resistivity, surface resistivity and arc resistance were determined to characterize the composites.

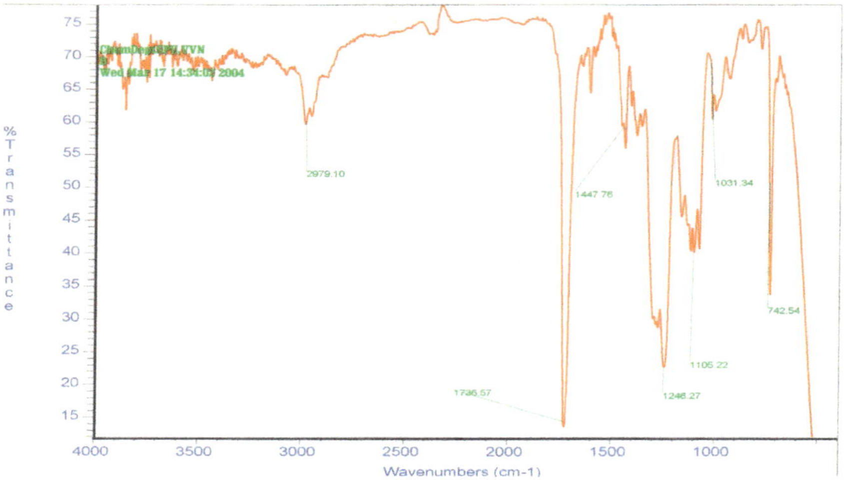

Figure 8. IR Spectra of UPR-2

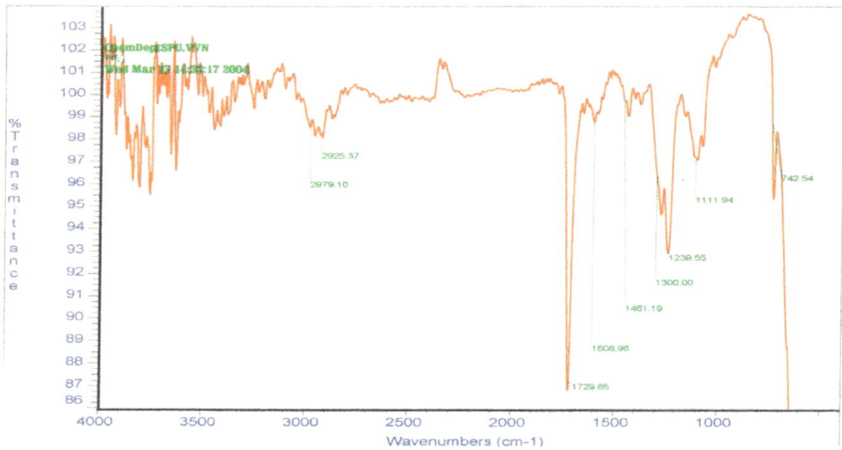

Figure 9. IR Spectra of ERCM-1

3.5.1. Volume resistivity and surface resistivity

The volume resistivity and surface resistivity were measured in terms of resistance of composites by a Hewlett Packard Model No. 4329 High resistance meter according to the standard method of testing ASTM D-257 (1978) [167]. A sample sheet was cut from the composite sheet and all dimensions measured as the average of several measurements of length, breadth and thickness.

For volume resistivity, the specimen was placed between two cylindrical copper electrodes. The electrodes were cleaned before each experiment. The entire rod of the electrode was covered with Teflon and one end of each rod was used for the connection with resistance meter. Volume resistivity was calculated using the following formula.

$$Volume\ \mathrm{Re}sistivity\ \rho\,(\Omega \cdot cm) = \frac{A}{t}(Rv)$$

Where, A = Area of the specimen, cm^2, t = Thickness of specimen, cm, Rv = Volume resistance (Ω)

For surface resistivity, the electrodes were placed on the same surface of the specimen at a fixed distance and direct voltage was applied between the two electrodes. The resulting current between the electrodes after a given fixed time of application of the voltage was measured ignoring the possible polarization effects. Surface resistivity was calculated using the following formula

$$\text{Surface resistivity} = \frac{\text{normal voltage gradient}}{\text{current per unit width of current path}}$$

The tests were carried out with all the prepared composite sheets and the data are reported in the Table-4.

3.5.2. Arc resistance

The arc resistance of composites was measured according to standard method of testing ASTM D- 495 (1989) [168], which is high voltage, low current dry arc resistance of solid electrical insulating Material. Figure 10 Illustrates a typical setup for an arc resistance test.

Voltage was applied to test sample and its severity increased in steps until the failure occurred. Arc resistance was measured as the time (sec) at which the surface between two electrodes becomes conductive. The results of arc resistance of all composites under study has been obtained and reported in Table-4.

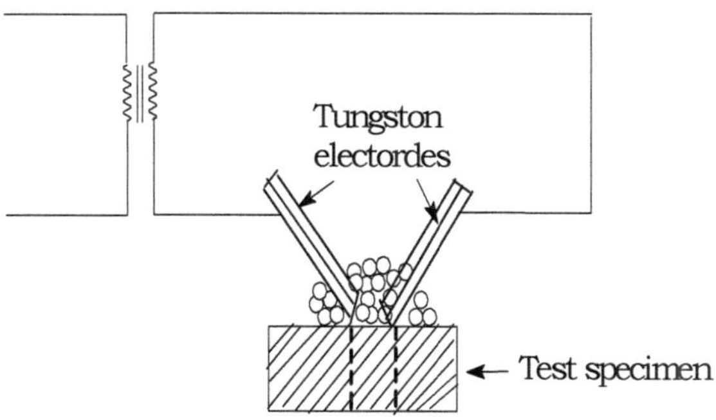

Figure 10. Arc Resistance Test

Table-4 represents arc resistance, surface resistivity and volume resistivity of filled and unfilled composites. In the present study, the values of surface resistivity and volume resistivity are in the range 1.8×10^{18}-10.9×10^{17}. The reported values for asbestos filled melamine are 1.2 $\times10^{12}$. Urea-formaldehyde resins show the values in the range of 10^{12}-10^{13}. Thus the result of present study are in good agreement with the reported value of PF, MF and UF resins. The results clearly indicate that the composites of present study can be used in all electrical applications where PF, MF and UF resins are used. Plastics such as epoxies, polyethylenes and fluorocarbons are excellent arc-resistant materials with arc resistance of 120-200 sec and above. It has been reported that arc resistance of plastics can be improved substantially by the addition of fillers such as glass, mineral, wood flour, asbestos and other inorganic fillers. In the present study all the composites show arc resistance in the range of 158-169 sec indicating very good arc resistance.

Sample	Arc Resistance in Sec.	Surface Resistivity in $\Omega \cdot 10^{17}$	Volume Resistivity in Ω Cm $\cdot 10^{18}$	Izod Impact in J/cm	Flexural Strength in MPa	Rockwell Hardness (M)
ERC-1	150	2.1	1.3	5.44	239.89	60
ERC-2	162	7.9	2.0	8.35	238.90	65
ERC-3	164	8.7	3.3	11.00	240.00	69
ERC-4	167	9.9	3.5	12.25	241.15	72
ERC-5	158	7.8	1.8	5.67	241.86	61
ERC-6	161	8.5	2.6	6.95	244.98	63
ERC-7	163	8.9	3.0	7.38	246.68	64
ERC-8	Leaching of filler	Leaching of filler	Leaching of filler	Leaching of filler	Leaching of filler	Leaching of filler
ERC-9	161	8.9	2.1	7.15	242.91	63
ERC-10	165	9.9	3.6	10.10	246.84	66
ERC-11	168	10.8	4.5	11.73	248.88	68
ERC-12	169	11.2	4.9	12.20	249.92	69

Table 4. Electrical and Mechanical Properties of Composites [169-170]

3.6. Mechanical properties of electrical resistance composites

Mechanical properties measured in this study include Izod impact strength, Rockwell Hardness and Flexural strength (table-4). The unbeatable combination of characteristic such as ease of fabrication, low cost, light weight and excellent insulation properties have made composite materials to be one of the most desirable materials for electrical application such as insulators, printed circuit boards, circuit breakers, cabinets etc. Arc resistance, surface resistivity and volume resistivity were evaluated for all the composites filled with different weight percentages of fillers. The arc resistance, surface resistivity and volume resistivity increases with increase in amount of filler because kaolin, surface treated kaolin and mica are electrically inert materials whose increased amount in laminates improves electrical behaviour. Surface treatment of fillers improves the mechanical as well as electrical performances because surface treatment of kaolin with the coupling agents will bridged polymer molecules and mineral particles and have the unique ability to substantially improve the bond strength between organic polymer and kaolin particles. Due to these fillers can be easily and uniformly dispersed throughout the polymer matrix and leaching of filler from the polymer matrix is prevented. This will ultimately improve the performance of composites. Composites prepared using fillers give better electrical performance than unfilled composites and also there is increase in electrical resistance with increase in amount of filler, but when amount of filler exceeds 30 weight percent, untreated fillers i. e. raw kaolin and mica leaches out from the polymer matrix. While surface treated kaolin doesn't leach out of the polymer matrix. The effects of concentration of filler and surface treatment of filler on mechanical properties have been examined. The results revealed that concentration of filler and surface treatment of filler remarkably influence the mechanical and electrical behaviour of composites [103-104]. Also cost-effective electrical composites can be formulated with competitive performance by using cheaper and widely available fillers. The composites can be used for electrical, electronic and electro technical applications like circuit breakers, switch board cabinets, automotive distributor caps, printed circuit board etc.

4. Acrylic modified low styrene emission unsaturated polyester resin having improved fire resistance and mechanical properties

4.1. Introduction

Unsaturated polyester resins are one of the most important matrix resins for commodity glass fibre reinforced composites [171]. They are obtained in a two step process; first unsaturated and saturated acids or anhydrides are reacted with diols in a polycondensation reaction, secondly the resulting linear polyester prepolymer is dissolved in styrene into syrup-like resin [172-173]. The styrene acts both as a cross-linking agent and as a viscosity reducer so that the resin can be processed. In conventional unsaturated polyesters the styrene content varies between 35- 45wt %. The styrene monomer in the unsaturated polyester resin is an environmental and occupational health problem due to evaporation and emissions, which occur during the processing of the resin. In 1992 it was found that about 75% of the workers in the reinforced plastic industry in Finland are exposed to styrene concentrations exceeding the current occupational exposure limit of 20ppm [174]. Much improvement of the work environment has been made by using proper ventilation systems together with styrene absorbing collection systems, by improvement of the work practice and the housekeeping, by using personal respiratory protection and by using low styrene emission resins in the industry. During the end of the 1980.s the environmental problems caused by styrene emission from the processing of unsaturated polyesters were debated a lot in the society, especially in the Nordic countries. The producers of unsaturated polyesters started therefore to search actively for alternatives to styrene and started to develop additives inhibiting styrene emission.

This work deals with the development of low styrene emission unsaturated polyester resins with improved fire resistance and mechanical properties. These studies showed clearly that the styrene emission cannot be avoided as long as the basic chemical composition is kept unchanged. Structural modifications are therefore needed. The fire behaviour of composites can also be improved by using a halogenated resin [175] in place of a traditional resin. The main disadvantage of this solution is that a fire could create toxic and possibly corrosive smoke [176]. Recently, environmental problems relating to brominated flame retardants (BFRs) have become a matter of greater concern than ever before, because of the recent marked increase in levels of polybrominated diphenyl ethers (PBDEs) found in human milk in Sweden and North America [177]. It is shown in this work that non-halogen flame retardants are very effective and does not create any toxic and corrosive smoke in unsaturated polyester resins. Their advantages lie in their high effectiveness, which enables very low concentrations to be used, while at the same time meeting the most stringent requirements.

The experimental work in this article deals synthetic modifications of unsaturated polyester resin in order to achieve better fire resistance and mechanical properties, lower styrene emission, better styrene solubility and lower processing viscosity. These new ecological unsaturated polyester resins will be used for marine, automotive, electric and electronic, building construction, sport and leisure, domestic and sanitary appliances, furniture as well as military applications. Two concepts were used for development of these new ecological unsaturated polyester resins; Introduction of acrylic monomers viz; Methyl methacrylate (MMA), Butyl methacrylate (BMA) and Acrylonitrile (AN) into the unsaturated polyester resins to reduce styrene emission and use of halogen free fire resistance fillers (Zinc borate and antimony trioxide) in place of halogenated fillers. The main disadvantage of halogenated fillers is that a fire could create toxic and possibly corrosive smoke. Liquid crystalline unsaturated polyester resins were possible to synthesize and it was also possible to prepare solutions of unsaturated polyesters in styrene. By modification with acrylic monomers it is possible to prepare resins with a styrene content as low as 20 wt %. The mechanical and fire resistance properties of the cured samples depending on the type of acrylic monomer and amount and type of filler. Modification of unsaturated polyesters in this way is a possibility to get unsaturated polyester resins with low styrene emission, better fire resistance and mechanical properties, lower shrinkage, reduced brittleness and increased flexibility.

4.2. Synthesis of acrylic modified low styrene emission unsaturated polyester resin

Unsaturated polyester resin was prepared in the present work by the technique reported by B.Parkyn [171]. A mixture of 1.25mol Propylene glycol (PG), 0.5mol Phthalic anhydride, 0.2% p-Toluene sulfonic acid (PTSA) and Xylene as distillating solvent was charged in a three-neck reaction kettle equipped with stirrer, thermometer, nitrogen-gas introducing tube, Dean & Stark apparatus, and water condenser. The mixture was mechanically stirred and heated at 120^{0}C under nitrogen gas stream, and esterification was carried out while removing water formed by the reaction from the reaction system. When reaction mass becomes clear, it was allowed to cool. When the temperature was dropped to 80^{0}C, 0.5mol

Maleic anhydride (MA) was added and continues heating at 150-200^{0}C until an acid number of 20 were reached. The Xylene was completely distilled out and reaction product was allowed to cool. When the temperature was dropped to 160^{0}C, 20mg of hydroquinone was added. When the temperature was dropped to below 100^{0}C, an amount of styrene was added to a styrene content of 20%.

Unsaturated polyester resin was modified by replacing 15% styrene (table-5) by different acrylic monomers viz. Methyl methacrylate (MMA), Butyl methacrylate (BMA) and Acrylonitrile (AN). Modification of unsaturated polyester resin was carried out in three neck flask equipped with stirrer. Firstly required quantity of styrene diluted unsaturated polyester resin (20% styrene) was added and then 15% of acrylic monomers were added with constant stirring at room temperature. The resultant modified unsaturated polyester resin was named as low styrene emission unsaturated polyester resin (LSEUPR).

Low Styrene Emission Unsaturated Polyester Resin	Styrene in Wt %	Acrylic Monomer in Wt %		
		MMA	BMA	AN
LSEUPR-1	20	15	-	-
LSEUPR-2	20	15	-	-
LSEUPR-3	20	15	-	-
LSEUPR-4	20	15	-	-
LSEUPR-5	20	15	-	-
LSEUPR-6	20	-	15	-
LSEUPR-7	20	-	15	-
LSEUPR-8	20	-	15	-
LSEUPR-9	20	-	15	-
LSEUPR-10	20	-	15	-
LSEUPR-11	20	-	-	15
LSEUPR-12	20	-	-	15
LSEUPR-13	20	-	-	15
LSEUPR-14	20	-	-	15
LSEUPR-15	20	-	-	15

Table 5. Low Styrene Emission Unsaturated Polyester Resin [178]

4.3. Fire resistance formulation and composite preparation

In present study low styrene emission unsaturated polyester resins were used with antimony trioxide and zinc borate for composite preparation (table-6). Several combinations were tested to determine effects of additives in the polymer formulations. Fire resistance formulations were prepared by adding these additives in different amounts (10 to 30 wt %) in low styrene emission unsaturated polyester resin. Additives in different amount were mixed with low styrene emission unsaturated polyester resin in a round bottom flask equipped with high speed agitator by means of agitation. The resultant formulations were compounded with glass fibre as reinforcing materials and benzoyl peroxide as curing catalyst for composite preparation.

Composites were prepared by hand lay-up technique using about 50 weight percent of glass fibre and filler. After application of resin on glass fibre, the composites were cured by compression moulding technique at 100 *psi* pressure and 120°C temperature for 30minutes. 1% Benzoyl peroxide (BPO) was used as an initiator. The composites obtained from fire resistance formulations were characterized by Limiting oxygen index according to ASTM D-2863[91].

Composite Code	Low Styrene Emission Unsaturated Polyester Resin Composition 100 Wt %	Reinforcement Wt %	Fillers in Wt %		Catalyst in Wt %	Limiting Oxygen Index (LOI) in %	Izod Impact in J/cm	Flexural Strength in MPa	Rockwell Hardness (M)
			Sb_2O_3	Zinc borate					
LSEUPRC-1	LSEUPR-1 (100)	100	-	-	2	21	5.09	238.85	59
LSEUPRC-2	LSEUPR-2 (100)	70	30	-	2	25	4.15	230.85	57
LSEUPRC-3	LSEUPR-3 (100)	70	20	10	2	28	7.25	242.00	61
LSEUPRC-4	LSEUPR-4 (100)	70	10	20	2	30	10.10	246.15	63
LSEUPRC-5	LSEUPR-5 (100)	70	-	30	2	32	11.80	247.90	66
LSEUPRC-6	LSEUPR-6 (100)	100	-	-	2	22	4.90	240.00	58
LSEUPRC-7	LSEUPR-7 (100)	70	30	-	2	26	4.00	231.85	56
LSEUPRC-8	LSEUPR-8 (100)	70	20	10	2	29	7.15	242.25	60
LSEUPRC-9	LSEUPR-9 (100)	70	10	20	2	31	10.00	246.55	62
LSEUPRC-10	LSEUPR-10 (100)	70	-	30	2	34	11.55	248.10	65
LSEUPRC-11	LSEUPR-11 (100)	100	-	-	2	21	5.65	235.90	60
LSEUPRC-12	LSEUPR-12 (100)	70	30	-	2	25	4.85	228.15	58
LSEUPRC-13	LSEUPR-13 (100)	70	20	10	2	27	7.65	240.00	63
LSEUPRC-14	LSEUPR-14 (100)	70	10	20	2	29	10.20	244.35	66
LSEUPRC-15	LSEUPR-15 (100)	70	-	30	2	31	11.85	246.15	68

LSEUPRC: Low Styrene Emission Unsaturated Polyester Resin Composites

Table 6. Fire Resistance Formulation and Properties with Mechanical Properties [178]

4.4. Fire resistance properties of composites

In this work, fire behaviour of composites was evaluated. Many procedures exist to evaluate fire behaviour of the composite materials [96-97].The key fire resistance properties of interest for polymers to be used in fire resistance applications are ease of ignition, flame spread, ease of extinction, smoke obscuration, smoke toxicity, heat release rate and limiting oxygen index. In the present study limiting oxygen index (ASTM D-2863) was determined to characterize the composites. Table-6 represents limiting oxygen index of filled and unfilled composites. In the present study values of limiting oxygen index of filled composites are in the range of 25-34. Material is normally considered as flame resistance if LOI is greater than 26[99]. Thus the results of present study are in good agreement with the reported values. The results clearly indicate that the composites of present study can be used in all fire resistance applications. Composites prepared using fillers give better fire resistance than unfilled composites and also there is increase in fire resistance with increase in amount of filler.

4.5. Mechanical properties of composites

Mechanical properties measured in this study include Izod impact, Rockwell Hardness and Flexural strength. Table-6 represents Izod impact strength, Rockwell hardness and flexural strength. Mechanical properties of laminates improve remarkably with introduction of acrylic monomers and fillers. Increase in strength with increase in filler content is due to the reinforcement action provided by the filler particles to the polymer matrix. Laminates prepared using LSEUPR and fillers give better mechanical performance than traditional unfilled laminates and also there is increase in mechanical properties with increase in amount of filler. Also acrylic modification by using various acrylic monomers viz; methyl methacrylate, butyl methacrylate and acrylonitrile of unsaturated polyester resins greatly influences mechanical properties and also reduced styrene emission. Butyl methacrylate modified LSEUPRC gives better flexural strength than methyl methacrylate and Acrylonitrile modified LSEUPRC. Methyl methacrylate modified LSEUPRC gives better flexural strength than acrylonitrile modified LSEUPRC but lower than butyl methacrylate modified unsaturated polyester resin. Acrylonitrile modified LSEUPRC shows better Izod impact strength and Rockwell hardness than methyl methacrylate and butyl methacrylate modified LSEUPRC.

Mechanical properties of composites improve remarkably with increase in the filler content except in antimony trioxide. Increase in mechanical properties with increase in filler content is due to the reinforcement action provided by the filler particles to the polymer matrix [103-104]. Composites prepared using fillers antimony trioxide and zinc borate gives better performance than unfilled composites and also there is increase in mechanical properties with increase in amount of fillers. Antimony trioxide gives poor results compared to zinc borate and also mechanical performance decreases with increase in amount as antimony trioxide is poor in mechanical performance. The results revealed that concentration and type of filler remarkably influence the mechanical and fire behaviour of composites. The study reveals that acrylic modification of unsaturated polyester resin by partially replacing styrene monomer drastically reduces styrene emission and greatly influences mechanical properties.

5. Conclusion

Composites prepared using fillers antimony trioxide, fly ash, zinc borate, hydroxyapatite, mica, surface treated kaolin and raw kaolin gives better performance than unfilled composites and also there is increase in mechanical properties with increase in amount of fillers, but leaching of fillers observed in mica and raw kaolin when amount exceeds more than 30 weight %, while the surface treated kaolin filled composites doesn't show such problem. Antimony trioxide and fly ash gives poor results compared to other fillers and also mechanical performance decreases with increase in amount as antimony trioxide has poor binding property. Also composites prepared using surface treated kaolin gives better performance than composites prepared from antimony trioxide, fly ash, zinc borate, hydroxyapatite, mica and raw kaolin due to surface treatment.

The results also revealed that the fire retardancy of the polyester based composites can be improved by using non traditional fillers like zinc borate, hydroxyapatite and fly ash. They have an advantage over a traditional filler antimony oxide to increase the fire retardancy without decreasing mechanical and thermal properties of the composites. There is however, considerable variation in the efficiency of these fillers. The adequacy of the fire resistance performance of these filled composites is dependent on both types of filler and incorporation level of the fillers. Hydroxyapatite and zinc borate gives good results as compared to other fillers. Tests show that the fire results are good with a decrease in heat and smoke production during combustion and also decomposition does not release additional toxic gases.

Composites prepared using fillers give better electrical performance than unfilled composites and also there is increase in electrical resistance with increase in amount of filler. Surface treatment of fillers improves the mechanical as well as electrical performances because surface treatment of kaolin with the coupling agents will bridged polymer molecules and mineral particles and have the unique ability to substantially improve the bond strength between organic polymer and kaolin particles. Due to this filler can be easily and uniformly dispersed throughout the polymer matrix and leaching of filler from the polymer matrix is prevented. This will ultimately improve the performance of composites.

The study reveals that acrylic modification of unsaturated polyester resin by partially replacing styrene monomer drastically reduces styrene emission and greatly influences mechanical properties. From overall study we conclude that composites prepared from unsaturated polyester resin using additives like hydroxyapatite, zinc borate, fly ash, antimony trioxide, mica, surface treated kaolin and raw kaolin are of better/upgraded in fire resistance and electrical resistance with good mechanical properties and acrylic modification by partially replacing styrene monomer reduces styrene emission.

Author details

Bharat Dholakiya
Applied Chemistry Department, Sardar Vallabhbhai National Institute of Technology (SVNIT), Gujarat, India

6. References

[1] Bjorksten Research Laboratories; Polyesters and Their Applications; Reinhold; New York; 1956.

[2] V. V. Karshak and S. V. Vinogradova; Polyesters; Pergamon; New York; 1965.
[3] I. Goodman; Encyclopedia of Polymer Science and Engineering; 2nd ed.; 12; Wiley; New York; 1988.
[4] I. Goodman; J. A. Rhys; Polyesters; Saturated Polymers; Iliffe Books; London; 1; 1965.
[5] I. Goodman; 2nd ed.; 16: 159; Wiley: New York, 1968.
[6] P. W. Morgan; Condensation Polymers; By Interfacial and Solution Methods; Inerscience Publishers; New York; 1965.
[7] W. H. Carothers; J. Am. Chem. Soc.; 51; 2548; 1929.
[8] J. R. Whinfield; Nature; 158; 930; 1946.
[9] M. P. Stevens; Polymer Chemistry; 2nd ed.; Oxford University Press. Inc.; pp 338-358; 1999.
[10] H. Makoto; M. Nobuo and M. Masatoshi; Jpn. Pat. JP 2001 262, 031; 2000; CA. 135 243881y; 2001.
[11] L. Geng; J. A. Baghdachi and J. L. Massingill; Jr. Polym. Material Sci. Eng. 85; 135-136; 2001.
[12] C. Eddy; Verfkroniek; 74(4); 13-15; CA. 135 290199s; 2001.
[13] T. Hiroyuki; O. Sumio and F. Tadashi; Jpn Pat. JP 2001 279, 150; CA. 135 290293t; 2001; 2002.
[14] H. Norimitsu; T. Shinichi; T. Toshiyuki and S. Isao; Shikizai Kyokaishi; 74(2); 57-62; CA. 135 290170a; 2001.
[15] K. M. England; J. W. Gillespie; W. John, Jr. and B. K. Fink; J. Compos. Mater.; 35(15); 1392-1414; 2001.
[16] M. Koichi and I. Hidemi; Jpn. Pat. JP 2001 293, 732; C.A. 135 304974e; 2002.
[17] G. Jianjun and J. Wugang; Gongcheng suliao Ying Young; 28(3); CA. 135 332164j; 2000.
[18] M. Fumio; M. Katsuhisa; H. Yoshitaka; M. Takashi; S. Junnosuke; S. Fumiaki; H. Nobuhito and N. Takeshi; Jpn. Pat. JP 2001 287; 313; CA..135 304831f; 2002.
[19] H. Yasumochi and K. Kazuo; Jpn; Pat. JP. 2001 294; 688; CA.135 319323r; 2000.
[20] H. Takashi; Jpn. Pat JP. 77 00; 509; CA. 87 109445w; 1975.
[21] R. E. Young; "Unsaturated Polyester Technology"; P.E. Bruins; Ed. Garden and Breach; New York; 315; 1976.
[22] S. H. Ridrr and E. E. Hardy; "Polymerization and Polycondensation Processes"; N.A.J.Platzers Ed; Advances in Chemistry Series no. 34; American Chemical Society; Washington Dc.; 173; 1962.
[23] H. V. Boenig; "Unsaturated Polyesters"; Structure and Properties; Elsevier; Amsterdam; 1964.
[24] B. Parkyn; F. Lamb and B. V. Clinton; Polyesters; Unsaturated Polyesters and Polyester Plasticizers; Elsevier; New York; 2; 1967.
[25] A. Frodet; and P. Arland; "Unsaturated Polyesters"; Comprehensive Polymer Science; Pargamon Press, New York, Ch. 19, 331; 1989.
[26] H. F. Mark; N. G. Gayord and N. M. Bikales; Encyclopedia of Polymer Science and Technology; John Wiley and Sons; New York; 1970.
[27] H. V. Boenig; Unsaturated Polyesters"; Elsevier Publishing Co.; Amsterdam; 1964.
[28] J. A. Arvin and W. H. Carothers; J.Am. Chem. Soc.; 51; 2560; C.A. 23, 4438; 1929.

[29] Ford Motor Co. Ltd.; Brit.; 991, 561; C.A. 63, 7208d; 1965.
[30] G. Corrado; E. Bertotti and B. Sopino; Ger. Often.; 2; 952, 679; C.A. 93, 16916p; 1980.
[31] M. Ochsenbein and J. P. Olliver; Ger. Often.; 3; 033; 063; C.A. 94; 193123q; 1981.
[32] K. H. Schmidt; A. Weber; D. Rowe and B. Brinker; US; 3,418,286; C.A. 70, 38474r; 1968.
[33] T. Sarai; H. Miyashita and Y. Sano; Japan, 69 31,832; C.A. 72, 112258; 1970.
[34] M. Hitoshi; S. Hatsuhiko and A. Mamoru.; JP 2003183489; Jpn. Kokai Tokkyo Koho; 2003.
[35] Y. Howard B.; V. James F. and M. Karen E.; EP 1132429; Eur. Pat. Appl.; 2001.
[36] O. Toshishide; JP 2002294057; Jpn. Kokai Tokkyo Koho; 2002.
[37] N. T. Qazvini and N. Mohammadi; Polymer; 46(21); 9088-9096; 2005.
[38] Q. Jun-min and J. Zhi-hao; School of Materials Science and Engineering; Xi'an Jiaotong University; Xi'an, Peop. Rep. China. Reguxing Shuzhi; 17(4); 31-35; 2002.
[39] Li, Tong.; Tianjin Synthetic Material Res. Inst., Tianjin; Peop. Rep. China.; Reguxing Shuzhi; 17(3); 31-35; 2002.
[40] P. Penezek; Polymery; 17(5); 248; C.A. 77; 127278s; 1972.
[41] W. Freitag; W. Sarfert and W. Lohs; Ger (East). DD; 260; 834; C.A. 111, 24521g; 1989.
[42] J. Korbar; J. Golob and A. Sebenik; Polym. Eng. Sci.; 33(18); 1212; C.A. 119, 250590a; 1993.
[43] Owens Corning Fibre Glass Corp.; Belg.; 867; 256; C.A. 90, 104895v; 1979.
[44] S. Miyamato; S. Shimazu; F. Sato; Y. Higuchi; T. Abe and K. Yamato; Ger. Often; 3; 019; 859; C.A. 94, 192925r; 1981.
[45] K. H. Jin; P. G. Guk and S. C. Dong.; KR 2000061809; Repub. Korean Kongkae Taeho-Kongbo; 2000.
[46] B. Michael A.; and C. Richard.; US 6472069, U.S.; 2002.
[47] E. K. Gamstedt; M. Skrifvars; T. K. Jacobsen; Composites; Part A: Applied Science and Manufacturing; 33A (9); 1239-1252; 2002.
[48] M. P. Stevens; Polymer Chemistry; 2nd ed.; Oxford University Press. Inc.; 106-120; 1999.
[49] M R.C. Nametz; J.D. Pietro and I.N. Einhorn; Amer. Chem. Soc.; 28(1); 204-24; 1968.
[50] Tanaka and E. Iwami; Jpn Pat. JP 62 201; 326; CA. 108 22799t; 1986.
[51] N.Yoshihiro and I. Yasuaki; Jpn Pat. JP 11 60, 858; CA. 130 223995g; 1997.
[52] N. Kazunri and K. Haruyuki; Jpn Pat JP 11 116, 779; CA. 130 353416e; 1997.
[53] O. Toshishide; JP 2002294057; 7pp; Jpn. Kokai Tokkyo Koho; 2002.
[54] W. Shoichi; N. Seiichi; K. Koichi and U. Tomoyuki; JP 11140287; 8pp; Jpn. Kokai Tokkyo Koho; 1999
[55] B. Jaljakumari; K. G. K. Warrier; K. G. Satyanarayana and C. Pavithran; Journal of Reinforced Plastics and Composites; 7(5); 402-412; 1998.
[56] D. Baral; P. P. De and G. B. Nando; Polymer Degradation and Stability; 65; 47-51; 1999.
[57] A. L. G. Saad and A. F. Younan; Polymer Degradation and Stability; 50; 133-140; 1995.
[58] D. K. Brown and D. W. Chalmers; The management of safety of warships in the UK; Transaction of the Royal Institution of Naval Architects; 132; 1990.
[59] 59C. M. Sprague; P. E. Robert and M. A. Blanchard; A methodology for evaluation of ship fire safety; Naval Engineers; 104; 1992.

[60] Fire safety on ships; Development into the 21st century; The Institute of Marine Engineers; London; 106; 1994.
[61] K. H. G. Ashbee and R. C. Wyatt; Proc. Royal Soc.; London; A312; 553; 1969.
[62] B. Mortaigne; V. Bellenger and J. Verdu; Polymer Networks and Blends; 2(4); 187; 1992.
[63] R. D. Deanin; Flame Retardants for Polyolefins; Polymeric Materials Science and Engineering; 66; 320–322; 1989.
[64] K. Yashuiro; I. Hiroyuki and T. Mitsuo; Jpn Kokai Tokkyo Koho; JP 2002188015; 11pp; 2002.
[65] L. In-Tae; S. Jae-Kyong and K. Kyo-Myong; Repub. Korea; KR 9509156; 1995.
[66] W. Shoichi; N. Seiichi; K. Koicji and U. Tomoyuki; Jpn Kokai Tokkyo Koho; JP 11140287; 8pp; 1999.
[67] O. Yasuo; Jpn Kokai Tokkyo Koho; JP 10095910; 10pp; 1998.
[68] T. Susumu; T. Yuichi and I. Yuji; Jpn Kokai Tokkyo Koho; JP 10087977; 8pp; 1998.
[69] K. Hideyuki; O. Toshe; M. Masaaki and F. Kenju; Jpn Kokai Tokkyo Koho; JP 07173362; 11pp; 1995.
[70] Y. Yasushi; M. Haruhiko; N. Asaichi and K. Tsukasa; Jpn Kokai Tokkyo Koho; JP 02283442; 4pp; 1990.
[71] V. Eduard; CS 216617; 5pp; 1984.
[72] K. Erich; US 4139577; 7pp; 1979.
[73] O. Takao and M. Masahiko; Jpn Kokai Tokkyo Koho; JP 46006865; 7pp; 1971.
[74] A. R. Jack and J. A. Clive; GB 1222738; 5pp; 1971.
[75] M. Shunli; H. Silie; L. Shuhong; X. Zhijun and X. Bingshi; Faming Zhuanli Shenqing Gongkai Shoumingshu; CN 1379078; 6pp; 2002.
[76] W. Shui-Ping; G. Wen-Hua and Z. Jia; Journal of Wuhan University of Technology; Material Science Edition; 17(4); 86-88; 2002.
[77] A. D. La Rosa; O. Motta and A. Recca; Journal of Polymer Engineering; 22(5); 341-352; 2002.
[78] Q. Jun-Min and J. Zhi-hao; Reguxing Shuzhi; 17(4); 31-35; 2002.
[79] L. Tong; Reguxing Shuzhi; 17(3); 31-35; 2002.
[80] B. K. Kandola; O. R. Horrocks; P. Myler and D. Blair; Composites Part-A; Applied Science And Manufacturing; 33A(6); 805-817; 2002.
[81] S. Yeng-Fong and J. Ru-Jong; Polymer Degradation and Stability; 77(1); 67-76; 2002.
[82] Z. Liming and G. Wenfeng; Wuhan Gongye Daxue Xuebao; 23(2); 15-17; 2001.
[83] B. Gaoliang; Faming Zhuanli Shenqing Gongkai Shuomingshu; CN 1268753; 7pp; 2000.
[84] S. Yasutoshi; Jpn Kokai Tokkyo Koho; JP 46040423; 5pp; 1971.
[85] L. B. Manfredi; E. S. Rodríguez; M. W. Przybylak and A. Vázquez; Polymer Degradation and Stability; 91(2); 255-261; 2006.
[86] B. K. Kandola; M. H. Akonda and A. R. Horrocks; Polymer Degradation and Stability; 88(1); 123-129; 2005.
[87] F. Le Lay and J. Gutierrez; Polymer Degradation and Stability; 44(3); 323-333; 1994.
[88] S. Horold; Polymer Degradation and Stability; 64(3); 419-425; 1999.

[89] Sunemi, H. T., Fushiki, Y., Nishimura, A., Kawai, Y.: Prepolymer polyester resin compositions and electrical laminates made therefrom, Kanegafuchi Chemical Ind., CA 1337915, 1996

[90] Kim, J., Jeong, D., Son, C., Lee, Y., Kim, E., Moom, I.: Synthesis & Application of Unsaturated Polyester Resins based on PET Waste, Korean J. Chem. Eng., 24(6), (2007), 1076-1083.

[91] ASTM D – 2963: MODERN Plastics Encyclopedia, McGraw-Hill, New York, 1967, 238.

[92] Bharatkumar Z. Dholakiya: Use of non-traditional fillers to reduce flammability of polyester resin composites; polimeri; 30 (1), (2009), 10-17.

[93] Edith, A.: Thermal Characterization of Polymeric Materials, Turi Acadamic press, 1981.

[94] Wendlandt, W. W., Collins, L. W.: Thermal Analysis, series two, Benchmark papers in Analytical Chemistry, Dowden, Hutchinson and Ross, Inc.,Pennsylvania, 1976.

[95] Turi, E. A,, khanna, Y. P., Taylor, T. J.: A guide to materials, characterization and chemical analysis, Ed.by J.P.sibilia,VHC publishers, 1988, 205.

[96] Tewarson, A.: Flammability Parameters of Materials: Ignition, Combustion, and Fire Propagation, J Fire Sci., 12(1994)4, 329-356.

[97] Tewarson, A., Pion, R. F.: Flammability of plastics—I. Burning intensity, Combustion and Flame, 26(1976), 85-103.

[98] Fennimore, C. P., Martin, F. J.: Flammability and sensitivity of materials in oxygen-enriched atmospheres, Mod. Plast., 44(1966)3, 141-148.

[99] Van Krevelen, D. W.: Properties of Polymers, Elservier/North-Holland, New York, 1975.

[100] ASTM D -256: Impact Resistance of Plastics and Electrical Insulating Materials, 1984.

[101] ASTM D-785: Rockwell Hardness of Plastics and Electrical Insulating Materials, 1965.

[102] ASTM D-790: Flexural Properties of Plastics and Electrical Insulating Materials, 1984.

[103] T.K. Jayasree et. al., Effect of Fillers on Mechanical Properties of Dynamically Cross linked Styrene Butadiene Rubber/High Density Polyethylene Blends, Journal of Elastomers and Plastics, 40 (2), 127-146, 2008.

[104] Azhar Iqbal et. el., , The effect of filler concentration on the electrical, thermal, and mechanical properties of carbon particle and carbon fiber-reinforced poly(styrene-co-acrylonitrile) composites, Polymer Composites, 28(2), 186–197, 2007.

[105] S. J. Harris; B. Noble and M. J. Owen; Metallographic investigation of the damage caused to GRP by the combined action of electrical, mechanical, and chemical environments; J Mater Sci.; 19; 1596–1604; 1984.

[106] A. Akhtar; J. S. Nadeau; J. Y. Wang; D. P. Romily and C. Taggart; Brittle fracture of non-ceramic insulators; Report for the Canadian Electrical Association (186 T 350); Prepared by the British Columbia Hydro and Power Authority; September 1986.

[107] M. J. Owen; S. J. Harris and B. Noble; Failure of high voltage electrical insulators with pultruded glass fibre-reinforced plastic cores; Composites; 17; 217–226; 1986.

[108] A. Akhtar and J. Y. Wong; Failure analysis of brittle fracture in nonceramic insulators; J Compos Technol Res.; 9; 95–100; 1987.

[109] M. Kumosa; Micro-fracture mechanisms in glass polymer insulator materials under combined effects of electrical, mechanical and environmental stresses; Final Report to the Bonneville Power Administration; Electric Power Research Institute and the

Western Area Power Administration; Oregon Graduate Institute, Portland; Oregon; July 1994.

[110] M. Kumosa; Q. Qiu; E. Bennett; C. Ek; T. S. McQuarrie and J. M. Braun; Brittle fracture of non-ceramic insulators; Proceedings of the Fracture Mechanics for Hydroelectric Power Systems Symposium '94; Canadian Committee for Research on the Strength and Fracture of Materials (CSFM); BC Hydro; 235–54; 1994.

[111] Q. Qiu; Brittle fracture mechanisms of glass fibre reinforced polymer insulators; Ph.D. Thesis; Oregon Graduate Institute of Science and Technology; Portland; Oregon; October 1995.

[112] M. Kumosa and Q. Qiu; Failure analysis of composite insulators (failure investigation of 500 kV non-ceramic insulators for pacific gas and electric company); Final Report to the Pacific Gas and Electric Company; Department of Engineering; University of Denver; May 1996.

[113] M. Kumosa; H. Shankara Narayan; Q. Qiu and A. Bansal; Brittle fracture of non-ceramic suspension insulators with epoxy cone end-fittings; Compos Sci Technol.; 57; 739–751; 1997.

[114] Interview with Maciej Kumosa; Research of brittle fractures in composite insulators; Insulator News and Market Report; 46-51; July/August 1997.

[115] J.T. Burnham and R J.Waidelich; Gunshot damage to ceramic and nonceramic insulators; IEEE Trans Power Deliv.;12(4);1651–1656; 1997.

[116] M. Kumosa et al.; Micro-fracture mechanisms in glass/polymer insulator materials under the combined effect of mechanical, electrical and environmental stresses; Final report to BPA,APA,PG and E,WAPA and NRECA, University of Denver; Denver; Colorado; December 1998.

[117] A. R. Chughtai; D. M. Smith and M. Kumosa; Chemical analysis of a field-failed composite suspension insulator; Compos Sci Technol.; 58; 1641–1647; 1998.

[118] D. Armentrout; T. Ely; S. Carpenter and M. Kumosa; An investigation of brittle fracture in composite materials used for high voltage insulators; J Acous Emission; 16(1–4); S10–S18; 1998.

[119] T. S. McQuarrie; Improved dielectric and brittle fracture resistant core rods for non-ceramic insulators; Insulator World Congress on Insulator Technologies for the Year 2000 and Beyond, Barcelona; Spain; 14–17; 1999.

[120] S. H. Carpenter and M. Kumosa; An investigation of brittle fracture of composite insulator rods in an acidic environment with static or cyclic loading; J Mater Sci.; 35(17); 4465–4476; 2000.

[121] C. de Tourreil; L. Pargamin; G. Thevenet and S. Prat; Brittle fracture of composite insulators: Why and how they occur; Power Engineering Society Summer Meeting 2000; 4; 2569–2574; 2000.

[122] M. Kumosa et al.; Fracture analysis of composite insulators; EPRI; Palo Alto; CA; 1006293; 2001.

[123] J. T. Burnham; T. Baker; A. Bernstorf; C. de Tourreil; J. M. George; R. Gorur; R. Hartings; B. Hill; A. Jagtiani; T. S. McQuarrie; D. Mitchell; D. Ruff; H. Schneider; D.

Shaffner; J. Yu and J. Varner; IEEE Task Force Report: Brittle Fracture in Non-Ceramic Insulators; IEEE Trans Power Deliv.; ;17(3); 2002.
[124] M. Kumosa et al.; Failure analysis of composite high voltage insulators. EPRI; Palo Alto; CA 1007464; 2002.
[125] M. Kumosa; Y. Han and L. Kumosa; Analyses of composite insulators with crimped end-fittings; part I-non-linear finite element computations; Compos Sci Technol.; 62;1191–1207; 2002.
[126] M. Kumosa; D. Armentrout; L. Kumosa; Y. Han and S. H. Carpenter; Analyses of composite insulators with crimped end-fittings; part II-suitable crimping conditions. Compos Sci Technol.; 62; 1209–1221; 2002.
[127] D. Armentrout; M. Kumosa and T. S. McQuarrie; Boron free fibres for prevention of acid induced brittle fracture of composite insulator GRP rods; IEEE Trans Power Deliv.; 18(3); 684–693; 2003.
[128] F. F. Hanna; A. A. Yehia; I. K. Hakim; A. M. Bishai and A. L. G. Saad; Br. Polym. J.; 15; 154; 1983.
[129] F. F. Hanna; K. N. Abdel-Nour and S. L. Abdel Messieh; Polym. Deg. Stab.; 35; 49; 1992.
[130] V. Mentilik; Thermochim. Acta.; 93; 353; 1985; CA 103 216292g; 1985.
[131] P. Penezek and W. Tokarski; Polymer; 16(9-10); 475; 1971; CA 76 113977m; 1971.
[132] Matsushita Electric Work Ltd.; Japan Kokai Tokkyo Koho; JP 60 84 350; 1985; CA 103 88677e; 1986.
[133] Toray Industries Inc.; Japan Kokai Tokkyo Koho; JP 59 119 618; 1984; CA 101 172676v; 1984.
[134] K. Shigeru; Jpn. Kokai Tokkyo Koho; JP 2002245856; 6pp; 2002.
[135] O. Kenji; Jpn. Kokai Tokkyo Koho; JP 2002088127; 4pp; 2002.
[136] M. Testsu; M. Yuichiro; K. Takao and Y. Minoru; Jpn. Kokai Tokkyo Koho; JP 2001131397; 11pp; 2001.
[137] K. Shigeru; Jpn. Kokai Tokkyo Koho; JP 2000154312; 5pp; 2000.
[138] W. Shoichi; N. Seiichi; K. Koichi and U. Tomoyuki; Jpn. Kokai Tokkyo Koho; JP 11140287; 8pp; 1999.
[139] O. Yasuo; Jpn. Kokai Tokkyo Koho; JP 10095910; 10pp; 1998.
[140] O. Toshishide; Jpn. Kokai Tokkyo Koho; JP 2002294057; 7pp; 2002.
[141] O. Kenji; Jpn. Kokai Tokkyo Koho; JP 2002088127; 4pp; 2002.
[142] A. Zheglova; Fachtagung Ueber Verarbeitung Und Anwendung Von Polymeren, 17th Chemnitz; Germany; 701-706; 2001.
[143] B. Jalajakumari; K. G. K. Warrier; K. G. Satyanarayana and C. Pavithran; Journal of Reinforced Plastics and Composites; 7(5); 402-412; 1988.
[144] R. Schmidtschneider; Gummi; Fasern; Knustst; 47; 304; 1994; CA. 123 58005e; 1995.
[145] Toho Belson Co. Ltd.; Jpn. Pat.; JP 82 49,612; 1982; CA. 97 56696b; 1982.
[146] S. Murakami; M. Takezawa; M. Miyazaki; Y. Ishida; and H. Inone; Jpn. Pat. JP 62 292,839; 1987; CA 108 187922r; 1988.
[147] W. Tao; H. Yao; Y. Chu; T.Chao and S. Li; KO Fen Tzu Tung Hsun; 6(5); 337; 1964; CA 64 6825g; 1966.

[148] L. T. Kravchenko; and Y. Y. Zherdev; Plasticheskic Massy; 4; 47; 1964; CA 64 2419d; 1966.
[149] E. Shiratsuchi; I. Ogura; S. Kobayashi and K. Tomioka; Jpn. Pat. JP 72 16,657; 1973; CA 77 165590V; 1973.
[150] G. S. Springer and S. W. Tsai; Thermal conductivities of unidirectional materials. J Compos Mater.; I(2); 166–173; 1967.
[151] J. Halpin and S. W. Tsai; Effects of environmental factors of composite materials; AFML-TR; 67-423; 57–64; 1969.
[152] C. H. Shen and G. S. Springer; Moisture absorption and desorption of composite materials; J Compos Mater.; 10; 2–20; 1976.
[153] G. S. Springer; Moisture content of composites under transient conditions; J Compos Mater.; 10; 107; 1976.
[154] C. D. Shirell and J. Halpin; Moisture absorption and desorption in epoxy composite laminates; Composite Materials: Testing and Design (Fourth Conference); ASTM STP 617; 514–28; 1977.
[155] K. Kondo and T. Taki; Moisture diffusivity of unidirectional composites; Environmental effects on composite materials; Lancaster; PA: Technomic Publishing Company; 1; 288–299; 1981.
[156] G. S. Springer; editor. Environmental effects on composite materials; Lancaster; PA: Technomic Publishing Company; 1; 1981.
[157] G. S. Springer; editor. Environmental effects on composite materials; Lancaster; PA: Technomic Publishing Company; 2; 1984.
[158] M. Woo and M. R. Piggott; J Compos Technol Res.; 10; 20–24; 1988.
[159] ASTM standard D5229/D5229M-92; Standard test methodology for moisture absorption properties and equilibrium conditioning of polymer matrix composite materials. American Society for Testing and Materials; 1992.
[160] L. R. Bao and A. F. Yee; Moisture diffusion and hygrothermal aging in bismaleimide matrix carbon fibre composites: part I-uni-weave composites; Compos Sci Technol.; 62; 2099–2110; 2002.
[161] H. G. Carter and K. G. Kibler; Langmuir-type model for anomalous moisture diffusion in composite resins; J Compos Mater. 12; 118–131; 1978.
[162] M. E. Gurtin and C. Yatomi; On a model for two phase diffusion in composite materials; J Compos Mater.; 13; 126–130; 1979.
[163] J. P. Lucas and J. Zhou; The effects of sorbed moisture on resin–matrix composites; JOM;37–40; 1993.
[164] L. W. Cai and Y. Weitsman; Non-fickian moisture diffusion in polymeric composites; J Compos Mater.; 28; 130–154; 1994.
[165] J. Zhou and J. P. Lucas; The effects of a water environment on anomalous absorption behavior in graphite/epoxy composites; Compos Sci Technol.; 1995.
[166] M. W. Ranney; S. E. Berger and J. G. Marsden; 27th Ann. Technol. Conf. Reinfor. Plastics/Compos.; Inst. Soc. Plas. Ind. 21D; 1-22; 1972.
[167] ASTM D -495, Test Method for High- Voltage, Low-Current, Dry Arc Resistance of Solid Electrical Insulation, 1984.

[168] ASTM D-257, Tests for DC Resistance or Conductance of Insulating Materials, 1978.
[169] B. Z. Dholakiya et. al., Kaolin filled unsaturated polyester resin based electrical laminates in Chemical Engineering World, Vol. 40(7), 80-85, July 2000.
[170] B. Z. Dholakiya et. al., Reinforced polymer composites based on acrylic modified unsaturated polyester resin-mica having improved electrical and mechanical properties in Ultra Science Vol. 19(3), 397-404, 2007.
[171] O.C.Zaske, S.H.Goodman; .Unsaturated Polyester, Vinyl Ester Resins., In S.H.Goodman; (Ed.); .Handbook of Thermoset Plastics., 2nd Ed.; Noyes Publication, Westwood, 97, (1998).
[172] H.V.Boenig; .Unsaturated Polyesters., Structure and Properties, Elsevier, Amsterdam, (1964).
[173] B.Parkyn; F.Lamb, B.V.Clinton; Polyesters, Unsaturated Polyesters and Polyester Plasticizers, Elsevier, New York, 2, (1967).
[174] A.Saamanen; Methods to Control Styrene Exposure in the Reinforced Plastics Industry, Doctoral Thesis, University of Kuopio, 29 August, (1998).
[175] W.Shoichi, N.Seiichi, K.Koicji, U.Tomoyuki; Jpn Kokai Tokkyo Koho, JP 11140287; 8 (1999).
[176] Isao Watanabea, Shin-ichi Sakaib; Environment International, 29, 665-682 (2003).
[177] Cynthia A.de Wit; Chemosphere, 46, 583-624 (2002).
[178] Synthesis, Characterization and Glass reinforced Composites of Low Styrene Emission Unsaturated Polyester Resin having improved fire resistance and mechanical properties in Macromolecule, MMAIJ, 3(4), 169-175, 2007.

Chapter 9

Time Dependent Behavior of Polymer Concrete Using Unsaturated Polyester Resin

Ghi Ho Tae and Eun Soo Choi

Additional information is available at the end of the chapter

1. Introduction

Time-dependent behavior of polyester-based materials such as polymer mortar or polymer concrete is recognized as an important aspect of mix design. Typically, creep and shrinkage of polyester-based materials exhibit complex time-dependent behavior. Thus, such time-dependent behavior is considered an essential design factor of the safety and serviceability of precast structures, especially during construction and at early age of curing. In the construction industry, the early age properties of polyester-based materials have become increasingly important because the use of rapidly cured materials can accelerate a construction process or shorten the production cycle of precast members [2,3]. The early age material properties of polyester-based materials, as influenced by setting shrinkage, have been investigated theoretically and experimentally. In some cases, the extent of setting shrinkage can significantly change the early age material properties and their development over time [4]. Recently, many polymer-based materials have been developed for potential civil engineering applications. Polymer concrete is one of the most promising materials. Polymer concrete has various advantages over conventional Portland cement concrete. They have higher compressive, tensile and flexural strengths. They also have fast curing time , an important advantage in many construction applications; polymer concrete materials cure in several hours or less, whereas Portland cement based materials take 2~4 weeks [5,8]. Also, they have excellent resistance to impact, abrasion, weathering, chemicals, water and salt sprays. However, the early age properties of these polymer concrete with unsaturated polyester resin have not been carefully investigated. Unsaturated polyester resins are used extensively as matrix materials in resin mortar systems. Conventional unsaturated polyester resins characteristically shrink about 5 to 10% by volume during curing process [1].

This shrinkage is unfavorable for molding properties despite the many excellent characteristics of the resins. Because the resin mortars have high elastic modulus and their relatively low tensile strength owing to the high content of aggregates, the curing shrinkage is liable to induce cracks in molded parts [7,11]. It is well known that low shrinkage resin compounds consist of unsaturated polyester resins mixed with certain thermoplastics such as polystyrene, acrylic polymer, polyvinyl acetate, etc [13]. However, because low shrinkage behavior is complicated according to its dependent on many factors, such as components and curing conditions, several different hypotheses have been postulated [14].

In this study, shrinkage behaviors of recycled unsaturated polymer resin mortar were observed at the curing temperature, and creep tests on recycled unsaturated polymer resin concrete were carried out to achieve long term behaviors of creep. Results were extrapolated to develop a model that effectively creep characteristics from filler type, filler contents, and stress ratio and shrinkage mechanisms in recycled unsaturated polymer resin mortar were elucidate.

2. Experimental programs

2.1. Shrinkage experiment

2.1.1. Materials

The relevant details of the coarse and fine aggregates used are given in Table 1. The nature and properties of fillers are given in Table 2. The aggregate and fillers were dried in an oven at 110°C(230F) for 24 hours prior to their use. The properties of montmorillonite are given in Table 3.

	Size(mm)	Ratio of Abrasion(%)	Specific gravity	Bulk specific gravity
Coarse	≤ 13	10.7	2.60	2.61
Fine	≤ 6	-	2.63	2.60
	S ize(mm)	Unit weight (N/mm²)	Fineness modulus	Absorption (%)
Coarse	≤ 13	1,470,000	6.42	0.7
Fine	≤ 6	1,612,100	2.48	0.44

Table 1. Physical properties of aggregate

	Specific gravity	Fineness (cm²/g)	Moisture (%)	PH	Absorption (%)
Flyash	2.2	3,765	0.2	-	-
$CaCO_3$	2.7	2,500~3,000	0.3	8.8	0.1

Table 2. Physical properties of fly ash and $CaCO_3$

	Specific gravity	L.B.D (kg/L)	Moisture (%)	pH	Sieve (%)
Spec.	-	0.75~0.85	0.2	10~11.5	Max.20
Result	1.22	0.8	0.3	10.4	18.9
Test Method	KS M 1104	KSM 0009		-	KS A 0507
Components	Recycled P ET	Propylene glycol	Terephthalic acid	Styrene Monomer(SM)	OcCo
Percentage by whight	28.1	17.5	13.9	40	0.5

Table 3. Physical properties of Monmorillonite and polyester resin

2.1.2. Shrinkage measurement apparatus

A shrinkage measurement apparatus was specifically fabricated by modifying a commercially available length comparator (shrinkage measurement device conforming to BIS: 4031 – 1968), primarily used to measure the shrinkage of cement concrete and.. Shrinkage measurement devices were also devised earlier by Ohama, Stanley et al., and Dupont Company [9]. However, the modified length comparator was found to be more adaptable to the local laboratory conditions. A schematic diagram of the apparatus is shown in Fig. 1. It consists of two threaded circular rods, each of 25.4mm (1 in.) in diameter and 550mm (21.65 in.) in length. The threading was provided at both ends of each rod up to a length of 130mm (5.12in.). These rods were fixed to a channeled metallic frame (thickness 5.28mm [0.21 in.]) through bolts, and this frame was firmly fixed to a heavy table. Two cross plates, A and B, each of 0.002mm (78.7 micro in.) in sensitivity and 10mm (6.39 in.) in travel, were fixed at the middle of the two cross plates, A and B, respectively, such that the plungers of the dial gauges faced each other. A hole was provided at an appropriate place in the channeled frame so that the plunger of Dial Gauge B could protrude to come in contact with the test specimen. It was ensured that there was no hindrance of any kind to the movement of the plunger of Dial Gauge B when it passed through the hole in the frame. All the metallic parts of the device were made of mild steel. A preferred position could be to fix both cross plates on one side of the channeled frame, in which case no hole was needed in the frame. But the plates in this instance were placed on either side of the frame due to the limitations of rod lengths.

2.1.3. Shrinkage specimens

Polymer concrete specimens in this experiment were divided into three types. The first type consisted of 100mm square by 400mm bars for initial setting shrinkage, shown in Fig 2. The second type consisted of 25mm square by 285mm bars, whose mold is shown in Fig 3, tests setting shrinkage, coefficient of thermal expansion and strain by using the strain gauge inserted into above bars according to ASTM C531. And the last type is a φ75 ×

150mm cylinder, which was used to test the compressive strength of polymer concrete according to the content of montmorillonite. The proportions of binder and filler contents were varied, as shown in Table 4. The filler was mixed with the aggregate mixture and thoroughly dispersed. The binder, containing appropriate amounts of hardener and accelerator, was then added to the aggregate filler mixture, and a homogeneous polymer concrete mix was prepared. The dry coarse and fine aggregates were first added to the filler and montmorillonite in accordance with mix proportions and mixed for at least 2 minutes before adding unsaturated polyester resin. After mixing, Methy Ethyl Kepton Peroxide was slowly added in unsaturated polyester and was mixed for sufficiently long time. The mixture was poured slowly into the mixing equipment, which already contained the former mixture of aggregates, filler and montmorillonite. And final mixture was poured into respective appropriate molds to obtain a good mixture. The molds were then vibrated for 2 minutes to eliminate air voids in the mixture. The specimens were placed at room temperature for a day before demolding. The specimens demolded were placed at various temperatures and for various durations according to the experimental and material variables.

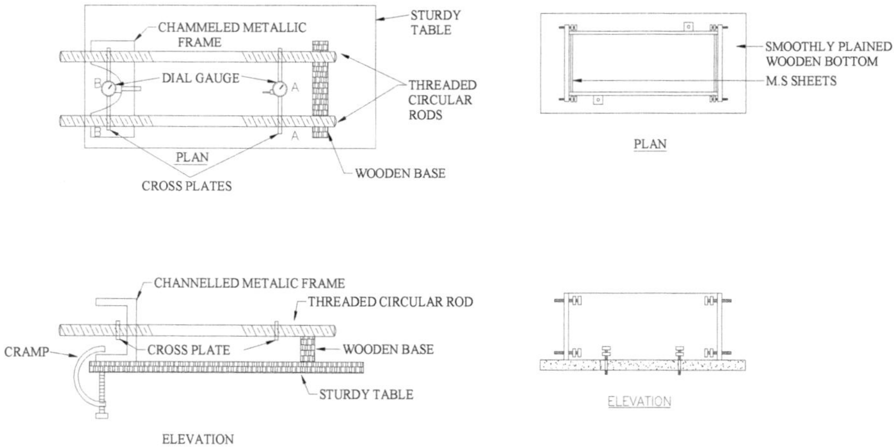

Figure 1. Shrinkage measuring device

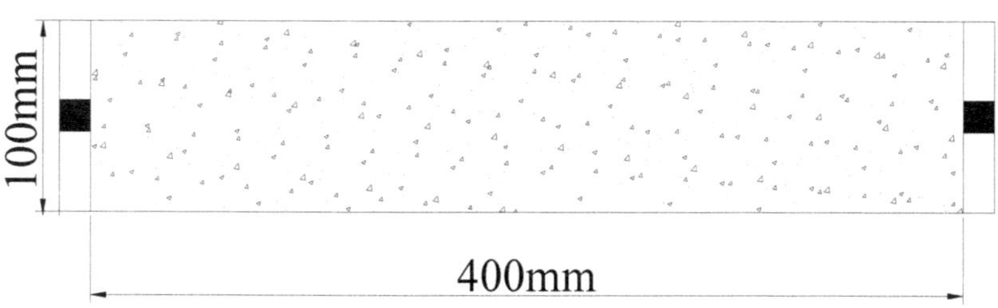

Figure 2. Sketch of mold of initial setting shrinkage

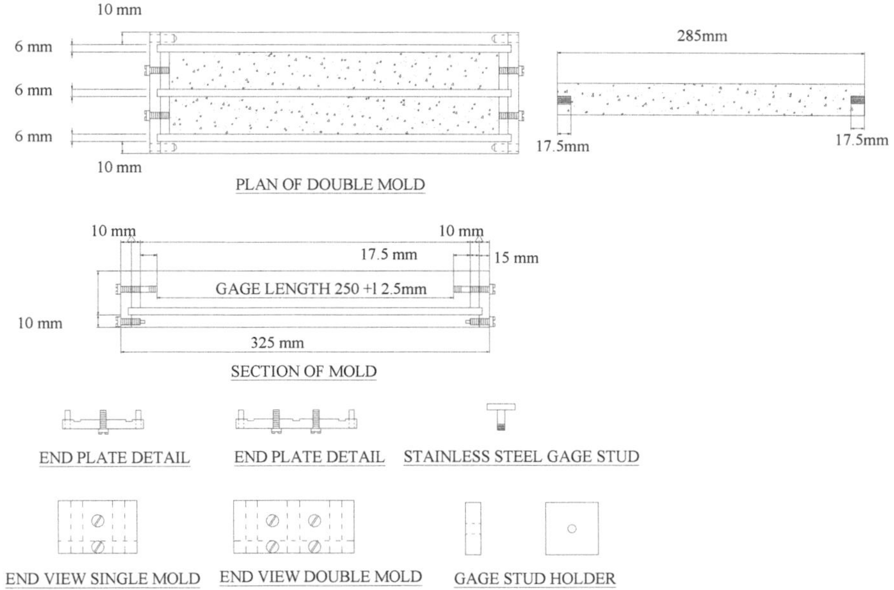

Figure 3. Sketch of mold of coefficient of thermal expansion

Specimens	Resin content (%)			Total Resin content(%)	Filler (%)	MMT (%)	Coarse Agg.	Fine Agg.
	UP	Mekpo	Total				<13mm	<6mm
PF					11	0		
PFM2					10	2		
PFM4					8	4		
PFM6					6	6		
PFM8	99	1	100	11	4	8	44	34
PFM10					2	10		
PFM12					0	12		
PC					11	0		
PCM5					6	5		

Table 4. Mix of Resin Paste

2.1.4. Curing method

30% methyl ethyl ketone peroxide (MEKPO) solution and 8% cobalt octoate (CoOc) solution were used as an initiator and an accelerator, respectively. The initiator and the accelerator were added in concentrations of 3.0 and 1.5 phr, respectively. N,N-dimethyl aniline (DMA) or acetyl acetone (AcAc) was employed to reduce curing time. These co-accelerator concentrations influenced the low shrinkage property of polyester resin. Furthermore, p-tertiary butyl catechol (PTBC) used as an inhibitor was kept at 20±1°C, the standard temperature before the addition of the initiator [6,10].

2.1.5. Measurements

The curing process of the resin mortar was continuously monitored by the strain gauge. The thermocouple embedded at the center of the test piece was removed so that the test piece could move freely. The curing process was monitored for over 24 hours. Measured values by the strain gauge were compared with the external dimension variations, as shown in Fig. 4. The exact relation between them was confirmed, though the measured values were 0.1% larger than the external dimension variations in either the expanded or shrunken state.

Figure 4. Strain gage setup for length change

The sides of the mold were opened immediately after the casing was completed. The specimen with the thin sheets at its ends was placed (along with wooden base) between the plungers of the dial gauges of the shrinkage- measurement device, and the initial readings were recorded. The entire casting operation took about 5 minutes.

The first readings in the dial gauges were treated as reference to calculate the shrinkage of polymer concrete. Subsequent readings on the dial gauges were taken after 20min, 30min, 1hr, 2hr, 3hr, 4hr, 5hr, 10hr, 15hr, 20hr, and 24hr of commencement of casting. During the period of the experiment, the laboratory temperature varied between 28 to 32C (82.4 to 89.6 F).

The change in length at one end of the specimen was calculated by multiplying the difference in dial gauge readings with the sensitivity of the dial gauge, i.e, 0.002mm. The total change in the length of the specimen would be the sum of the individual changes in length at both ends, calculated separately. The shrinkage of the specimen, in micro strains, was obtained by dividing the change in length of the specimen by its initial length [1,12].

2.2. Creep tests

The polymer concrete samples were mixed according to the polymer concrete test method 1.0 of the Society of Plastic Industry (referred to as SPI 1.0) [15]. The samples were mixed using a conventional concrete mixer for a period of about 3 min, poured into molds, vibrated, and cured at room temperature according to ASTM 1439 [16]. There are no standard tests that are directly applicable to polymer concrete specimens. Therefore, ASTM standards developed for cement were adopted as guidelines applicable. Polymer concrete cylinder specimens of 75mm diameter and 150mm height were tested in universal compression using hydraulic spring-loaded creep frames in the creep test room, which was equipped with temperature and humidity controls as shown in Fig.5. Electrical gages were bonded to the specimens at mid-height, and then connected to an automated data acquisition system. The short-term creep tests were carried out at 20°C, 30°C, and 40°C. Variables and specimen names are given in Table 5.

The temperature of the specimen was raised to the desired level and kept constant for 8h before the load was applied, in order to reach a stable temperature. The load was applied and kept constant for 24h. The strain was recorded before and after loading. At the end of the 24h the load was removed, the temperature was reduced to 20°C, and the specimens were allowed to recover for 16h from the creep strain before the temperature was raised to the next level (Fig. 6).

Fig.7 shows the stress level for the three short-term creep tests. Except for the stress ratio tests, the stress level f_0 was 20% of the ultimate compressive strength for short-term creep tests. The creep compliance curves of the 24h tests performed at elevated temperatures were used to predict the long-term creep compliance at 20°C. The creep strains were measured with concrete strain gages and recorded by an automated data acquisition system. The creep strains were recorded just after applied loading, every 10min for the first 1h, every 30min for the next 8h, and then every hour until the tests were finished.

Figure 5. Compressive creep test

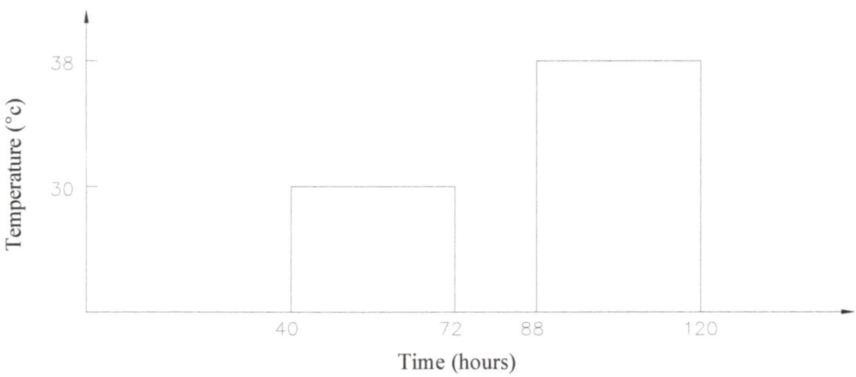

Figure 6. Temperature step for short-term creep test

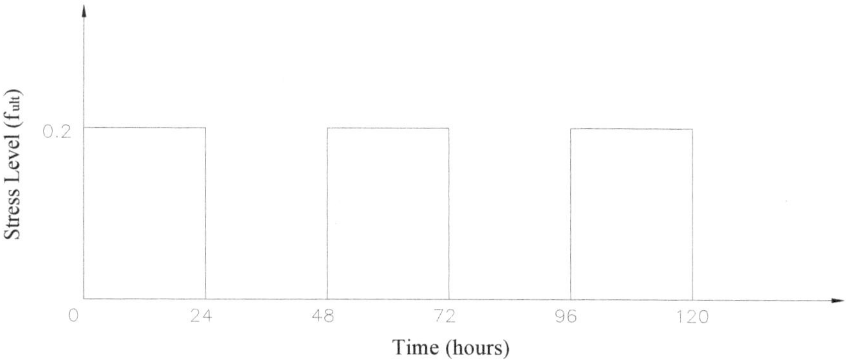

Figure 7. Stress level for short-term creep test

PC System	Classification	Change Factor
F-C-L20	Binder+ $CaCO_3$ 10%+Load 20%	Type of filler
F-Fa-L20	Binder+ Fly-ash 10%+Load 20%	
F-N-L20	Binder+ filler 0%+Load 20%	
F-C20-L20	Binder+ $CaCO_3$ 20%+Load 20%	Filler contents
F-C30-L30	Binder+ $CaCO_3$ 30%+Load 20%	
F-C-L30	Binder+ $CaCO_3$ 10%+Load 30%	Stress ratio
F-C-L40	Binder+ $CaCO_3$ 10%+Load 40%	

Table 5. Variables and names of specimens

2.3. Creep prediction method

A best way to achieve good long-term prediction is to conduct short-term creep tests on the given concrete and then extrapolate using Bayesian statistics [17]. A long-term creep compliance curve was constructed using the creep compliance curves of short-term creep tests performed at different temperatures. Based on the time-temperature analogy, a prediction was conducted using the results of short-term creep tests performed at 20°C, 30°C, and 40°C to predict the long-term creep compliance at 20°C [18,20].

This method takes the glassy creep compliance, measured at t=0 hours, and the 24-hour creep compliance values of each short-term creep test to develop a Prony series equation for the long-term creep compliance [22].

The creep compliance values at t=0 and t=24 are used to determine the coefficients of the following Prony series equation:

$$D(t) = D_g + D_1(1 - e^{-t/\tau_1}) \tag{1}$$

where D_g is the glassy creep compliance, which is measured at t=0. The retardation time τ_1 is set to the time coordinate t=24. The 24-hour creep compliance $D(t=24)$ is substituted in Eq. (1) and the coefficient D_1 is determined:

$$D_1 = 1.582\left[D(24) - D_g\right] \tag{2}$$

Eq. (2) is taken to determine the coefficient D_1 for each temperature. Coefficient D_1 is then substituted in Eq. (2) in order to develop a Prony series expression for each short-term creep compliance curve.

$$D^{20}(t) = D_g^{\ 20} + D_1^{\ 20}(1 - e^{-t/24}) \tag{3}$$

$$D^{30}(t) = D_g^{\ 30} + D_1^{\ 30}(1 - e^{-t/24}) \tag{4}$$

$$D^{40}(t) = D_g^{\ 40} + D_1^{\ 40}(1 - e^{-t/24}) \tag{5}$$

The superscripts 20, 30, and 40 indicate the temperature of the short-term creep compliance curve that each equation represents.

Eqs. (3), (4), and (5) are then used to develop the following expression for predicting the long-term creep compliance at 20°C:

$$D(t) = D^{20}(t) + D_{pred}(t) \tag{6}$$

where $D^{20}(t)$ is given by Eq. (3). The prediction term $D_{pred}(t)$ is given by two exponential terms:

$$D_{pred}(t) = D_2(1 - e^{-t/\tau_2}) + D_3(1 - e^{-t/\tau_3}) \tag{7}$$

where τ_2 and τ_3 are set at 1000 and 6000, respectively. The value of the exponential term $e^{-t/\tau}$ becomes zero when t increases above 4τ. The retardation constants τ_2 and τ_3 therefore have to be as large enough as that the exponential terms of Eq. (7) would have an effectively nonzero contribution to the long-term creep compliance predicted by Eq. (6).

It was observed that the coefficients of the exponential terms in Eqs. (3)–(5) increased with temperature. According to the time-temperature correspondence, increases in the value of the coefficients, i.e. an increase in the creep compliance, could have been caused either by a temperature or a time increase. The coefficients of the long-term 20°C curve, D_2 and D_3, are therefore expressed in terms of the coefficients of the 30°C curve and the 40°C curve, respectively, by the following expressions:

$$D_2 = rD_1^{\ 30} \tag{8}$$

$$D_3 = rD_1^{\ 40} \tag{9}$$

where r is the factor used to adjust the coefficients to minimize the error in the predicted creep compliance. The method of determining the parameter r is discussed below along with the presentation of the results [10].

Eqs. (3), (6), and (7) are substituted into Eq. (6) in order to develop the following equation that predicts the long-term creep compliance:

$$D(t) = D_g^{\ 20} + D_1^{\ 20}(1 - e^{-t/24}) + rD_1^{\ 25}(1 - e^{-t/1000}) + rD_1^{\ 30}(1 - e^{-t/6000}) \tag{10}$$

where the coefficients D_1^{20}, D_1^{30}, and D_1^{40} are determined by Eq. (4). Eq. (10) can predict the creep compliance values for times that do not exceed four times the larger retardation time, 6,000 hours. The predicted creep compliance value remains unchanged as time increases above 24,000 hours.

3. Results and discussion

To confirm the effectiveness of reinforcement, all results obtained from the specimens were compared with those from the control specimen. The following results are discussed: compressive and flexural strengths, thermal expansive coefficient, setting shrinkage for aging treatment, creep compliance, and predict model.

3.1. Compressive and flexural strengths

The effect of age on the compressive and flexural strength of recycled unsaturated polyester polymer concrete is shown in Fig. 8(a) and Fig. 8(b). Recycled unsaturated polyester resin-based polymer concrete achieves more than 80% of its 28-day strength in seven days. The high compressive strength of recycled unsaturated polyester resin-based polymer concrete allows the use of thinner sections in precast components, thus reducing dead loads in structures and minimizing transportation and erection costs. As a result, an increase in montmorillonite content from 1% to 5% increased the recycled unsaturated polyester polymer concrete compressive strength by about 12%. However, an increase in the montmorillonite content 8% decreased the strength of specimens.

The effect of temperature on the compressive and flexural strength of recycled unsaturated polyester polymer concrete is shown in Fig. 9. Fillers (fly ash, $CaCo_3$) were placed in an environmental chamber at the specified temperature 48 hr prior to mixing. After mixing, the specimens were again put in an environmental chamber at the designated temperature for a period of 28 days prior to testing. The selected temperatures were 20°C, 25°C, 30°C. Actual testing, performed at room temperature, was conducted immediately after removing the specimens from the environmental chamber. When the temperature increased, recycled unsaturated polyester polymer concrete lost strength because of the resulting loss in strength of the resin binder and the resulting decrease in bond strength between the inorganic aggregates and the resin binder. For example, Recycled unsaturated polyester resin in temperature from 20°C to 30°C is more temperature sensitive than the inorganic cement binder used in producing normal Portland cement concrete. However, despite this loss in strength at high temperatures, Recycled unsaturated polyester polymer concrete remains at least twice as strong in compression as regular Portland cement concrete.

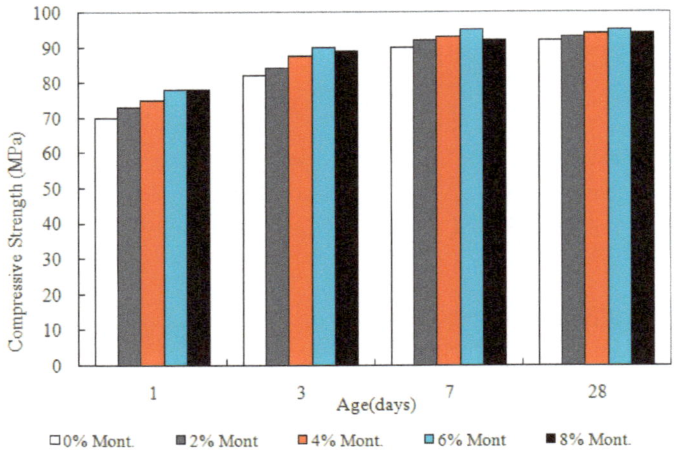

(a)

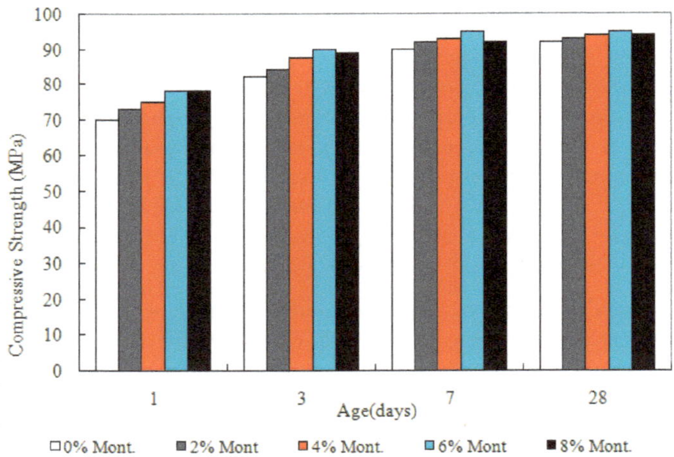

(b)

Figure 8. (a) Age effect on compressive strength, (b) Age effect on flexural strength

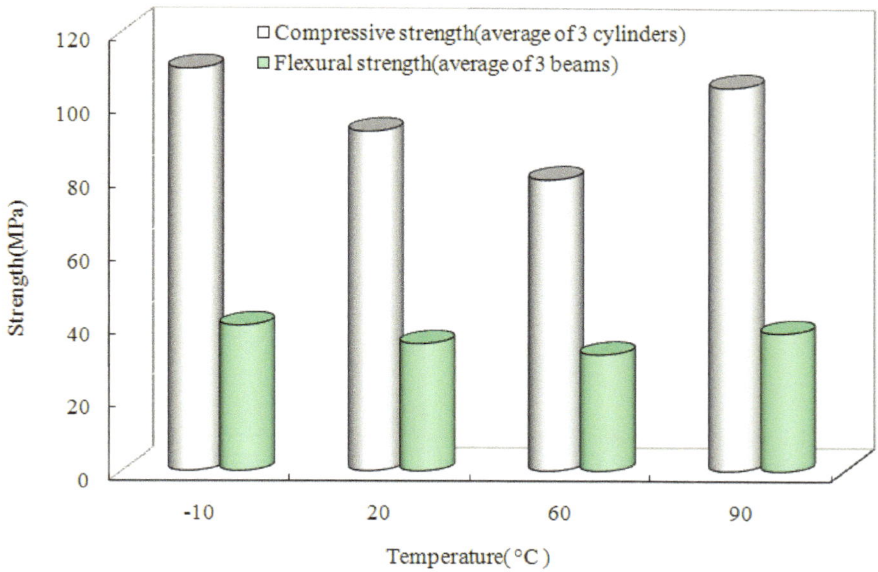

Figure 9. Temperature effect on compressive and flexural strength

3.2. Expansive coefficient

The thermal reaction of montmorillonite as the shrinkage reducing agent and fly ash and calcium carbonate as the filler is represented in Fig. 10.

The coefficient of thermal expansion decreases noticeably according to montmorillonite content. Moreover, specimens that used fly ash as the filler were less affected by temperature than those that used calcium carbonate among the specimens that contained 5% of montmorillonite evenly. The fact that fly ash is little sensitive to heat is useful when using polymer concrete made by recycled materials in construction materials. The coefficients of thermal expansion of polymer concrete produced by recycled PET maintained a good level than those of general polymer concrete ($2.5 \sim 3.5 \times 10^{-5}$) but showed a great difference from the coefficients of thermal expansion of general cement concrete ($0.7 \sim 1.2 \times 10^{-5}$).

The effect of curing temperature on specimens of montmorillonite 5% is shown in Fig. 11. The coefficient of thermal expansion decreased according to the increase of curing temperature because of the close up between polymer molecules and the complete hardening by high temperature during the initial curing process. Especially, the values from the specimens cured in water were relatively higher than those cured at high temperature.

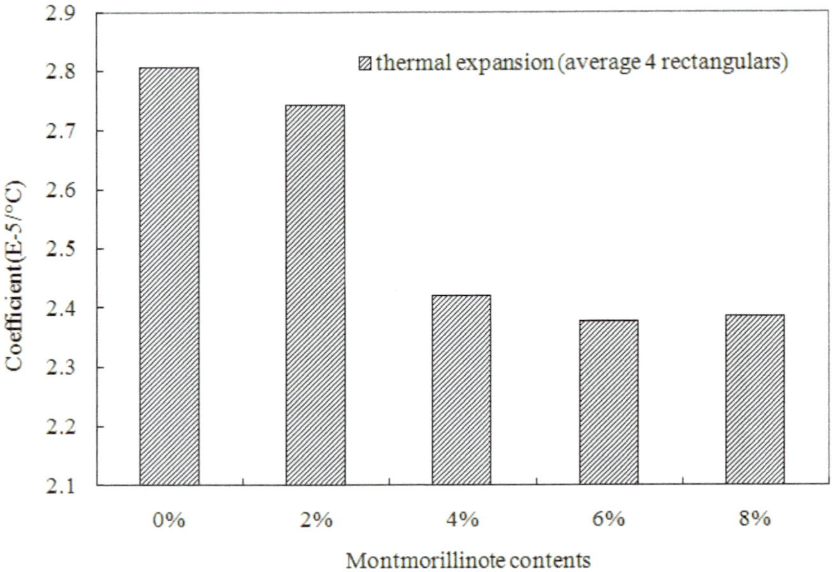

Figure 10. Coefficient of thermal expansion on montmorillonite contents

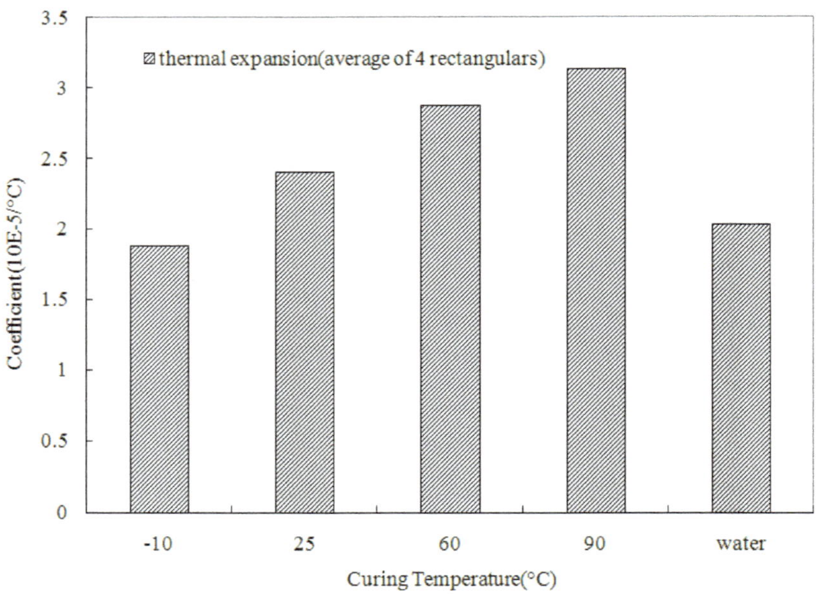

Figure 11. Coefficient of thermal expansion on curing temperature

3.3. Shrinkage behavior

The linear setting shrinkage values of different polymer concrete mixes investigated are presented in Table 6. The shrinkage of recycled unsaturated polyester polymer concrete was generally high, varying between 2400 and 3000 micro strains, as compared to that of Portland cement concrete whose (drying) shrinkage varied usually between 200 and 500 micro strains. However, the measured shrinkage values were in the range of 0.1 ~ 0.3 percent of the reported values were in the range of 0.2 ~ 0.3 percent. The reduced shrinkage values exhibited by the mixes in this study could be due to the use of shrinkage reducer such as montmorillonite.

Montmorillinote Content percent	Resin content percent		
	10	12	15
0	2401	2579	3117
2	2240	2305	2895
4	1938	2014	2405
6	1540	1655	2010
8	1220	1255	1520
10	1225	1250	1530
12	1240	1260	1535

Table 6. Shrinkage values of polymer concrete mixes (micro strains)

The development of shrinkage strains with time for recycled unsaturated polyester polymer concrete mixes containing different filler contents and montmorillonite contents are plotted in Figs. 12 and 13 for mixes made with resin contents 10, 12, 14 and 16 percent, respectively. Shrinkage developed at a rapid rate in the first few hours of casting (4 to 15hr), then slowed down and finally seemed to attain a constant value, with no further or very little increase thereafter, suggesting the near completion of the shrinkage process.

The variation of shrinkage (at 24 hr) with resin and filler contents is plotted in Figs. 14 and 15, respectively. The shrinkage increases linearly with resin content for all proportions of the filler content because the shrinkage of polymer concrete is principally due to its binder content and no shrinkage of the aggregate.

The shrinkage of polymer concrete increases with the filler content also, but the rate of increase seems to decrease at the higher filler content. A possible reason for these increases in shrinkage with filler content could be as below.

As the polymerizing resin shrinks, the aggregate offers a restraint to the shrinkage. Such a restraint is likely to be due to two mechanisms (:) a frictional component at the aggregate surface and the relative low compressibility of the aggregate. The filler, which is very fine and soft, when added, is likely to coat the surfaces of the aggregate particles and cause a decrease in the frictional component of the restraint. In such a case, restraint to shrinkage would decrease as the filler is added. However, when the filler is high (such as 5 percent), the influence of the filler content on polymer concrete is not proportional. This is possible since only a small amount of filler needed to coat the aggregate particles, and the filler content beyond this amount is unlikely to cause a further decrease of the frictional restraint. The low shrinkage of polymer concrete with low resin content and zero filler content has a practical significance in the repair and rehabilitation of deteriorated concrete structures.

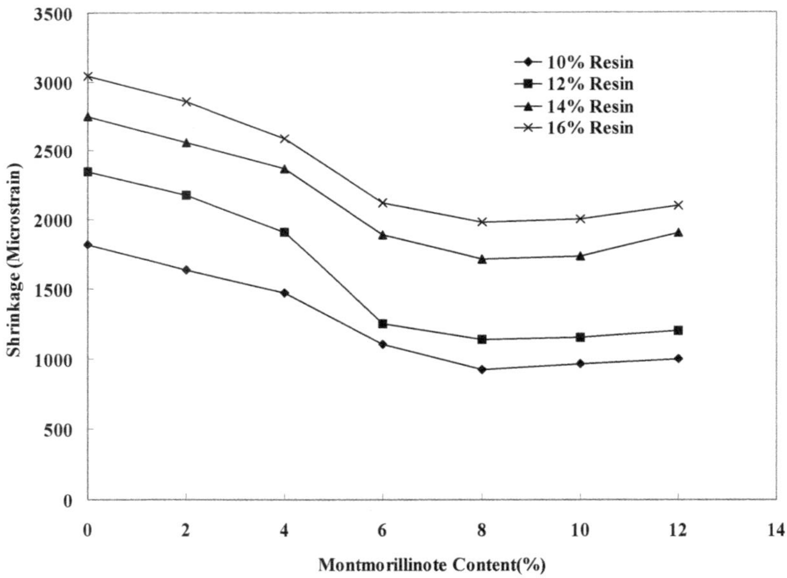

Figure 12. Shrinkage variation of recycled unsaturated polyester polymer concrete with montmorillonite content

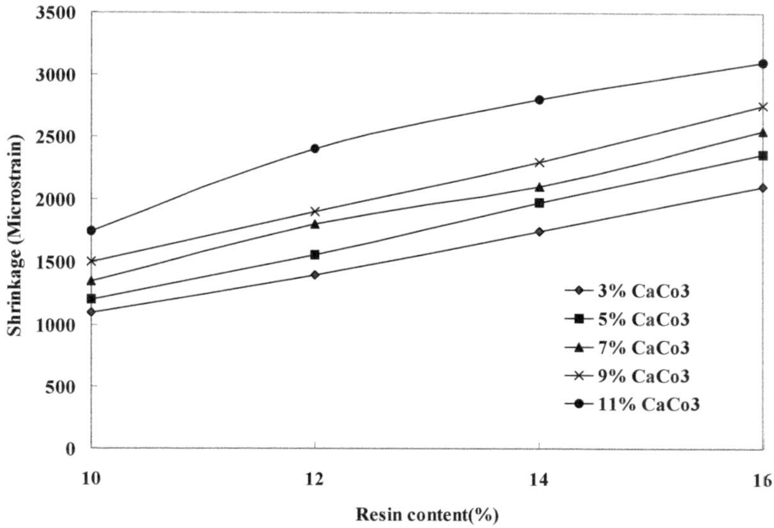

Figure 13. Shrinkage variation of recycled unsaturated polyester polymer concrete with resin content

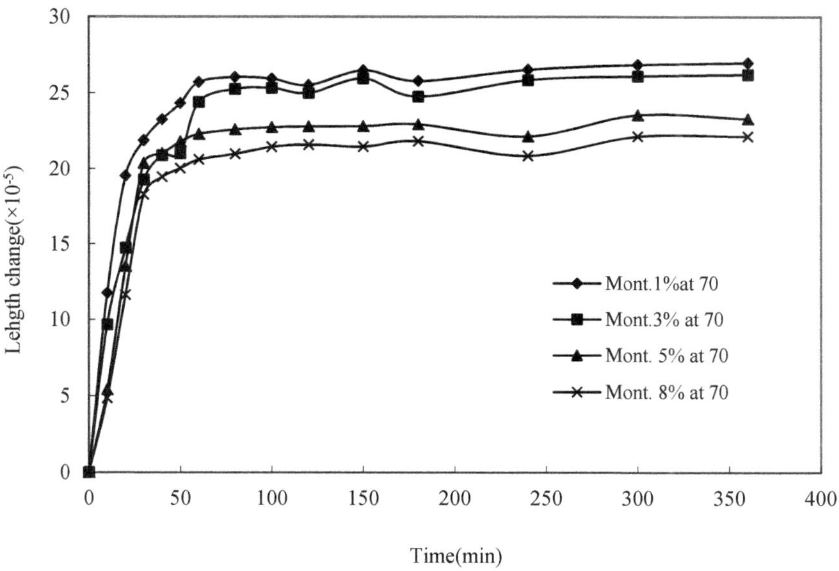

Figure 14. Shrinkage variation of recycled unsaturated polyester polymer concrete with time

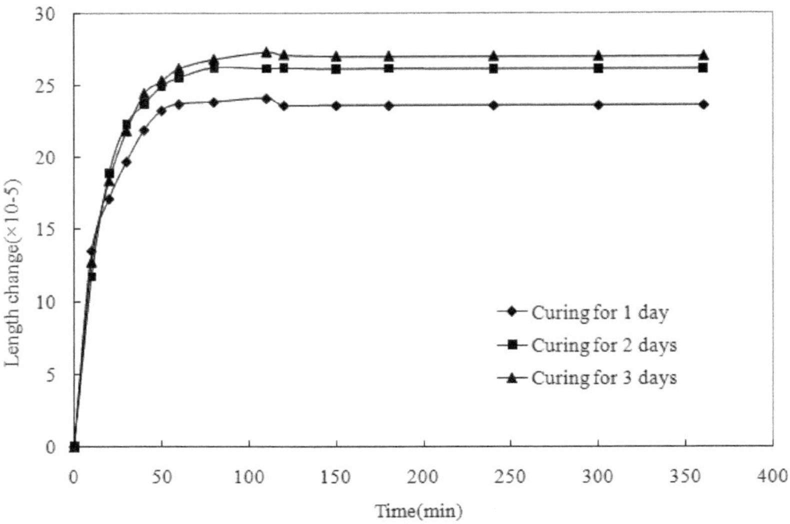

Figure 15. Length change of recycled unsaturated polyester polymer concrete with curing days

3.4. Prediction of long-term creep behavior

The short-term creep tests provided the 24 hour creep compliance values at the temperatures (20°C, 30°C and 40°C) to construct a Prony series equation that predicts the long-term creep compliance at the 20°C temperature.

The proposed method of short-term creep tests was used to predict the long-term creep compliance of recycled PET polymer concrete, incorporating types of filler, stress ratio and filler content. Each creep compliance curve is represented by the following Prony series equation:

$$D(t) = D_g + D_1(1 - e^{-t/24}) \tag{11}$$

where D_g, the glassy creep compliance, is the creep compliance at t=0 hours. The coefficient D_1, which is a function of the glassy compliance D_g, and the 24-hour creep compliance $D(24)$ are determined by Eq. (11). The values of the Prony series coefficients D_g, D_1 for compressive short-term creep tests performed on polymer concrete are shown in Tables 7 and 8. The 24-hour creep factor, $f(24)$, that is, the ratio of the specific creep $D(24) - D_g$ to the glassy compliance D_g, is also shown in Tables 7 and 8.

The coefficients of the Prony series that represents the short-term creep curves, shown in Tables 7 and 8, were used to calculate the coefficients of the following Prony series equation predicting the long-term compliance at the 20°C creep curve:

PC System	Temp. (°C)	$D(0)=D_g$ (μ/Mpa)	$D(24)$ (μ/Mpa)	D_1 (μ/Mpa)	$f(24)$
F-C-L20	20	204.60	223.91	30.55	0.094
	30	223.51	257.30	53.45	0.151
	40	260.00	332.60	114.85	0.279
F-Fa-L20	20	218.91	241.10	35.10	0.101
	30	240.98	275.85	55.16	0.145
	40	278.24	356.60	123.96	0.279
F-N-L20	20	328.79	369.67	64.67	0.124
	30	368.19	439.10	112.18	0.193
	40	437.36	592.70	245.74	0.355
F-C-L30	20	214.30	237.32	36.42	0.107
	30	237.89	276.86	61.65	0.164
	40	275.34	358.80	132.30	0.304
F-C-L40	20	247.34	277.32	47.43	0.121
	30	278.42	327.91	78.32	0.184
	40	322.30	431.91	173.40	0.340
F-C20-L20	20	200.11	219.30	30.37	0.096
	30	217.92	241.50	37.30	0.108
	40	239.67	280.20	64.12	0.169
F-C30-L20	20	195.58	211.50	25.19	0.081
	30	219.14	245.80	42.18	0.122
	40	243.14	284.80	65.91	0.171

Table 7. Results of short-term creep test

PC System	r	D_g^{20}	D_1^{20}	D_1^{30}	D_1^{40}
F-C-L20	0.862	204.60	30.55	46.07	99.00
F-Fa-L20	0.906	218.91	35.10	49.97	112.30
F-N-L20	0.811	328.79	64.67	90.98	199.30
F-C-L30	0.865	214.30	36.42	53.33	114.44
F-C-L40	0.842	247.34	47.43	65.95	146.00
F-C20-L20	1.272	200.11	30.37	47.45	81.56
F-C30-L20	1.110	195.58	25.46	46.82	73.16

Table 8. The coefficients of Prony series equation

$$D(t) = D_g^{20} + D_1^{20}(1 - e^{-t/24}) + rD_1^{30}(1 - e^{-t/1000}) + rD_1^{40}(1 - e^{-t/6000}) \quad (12)$$

where D_g^{20} and D_1^{20} are the coefficients of the Prony series equation that represents short-term creep curve at the 20°C temperature. The last two terms of Eq. (12), which constitute the prediction terms, are determined from the results of the short-term creep tests performed at elevated temperatures (30°C and 40°C).

Coefficients D_1^{30}, and D_1^{40} of Eq. (12) are the Prony series coefficients that correspond to short-term creep curves at the 30°C and 40°C temperature. The factor r is used to adjust the coefficient so that the error in the predicted creep compliance may be minimized. Different values of r were substituted in Eq. (12) in order to define the range of r for which the error is minimized. A single value of r does not minimize the error of the predicted creep compliance in all polymer concrete. We determined the value of r to minimize the error of the results of the short-term creep tests performed at the 30°C, 40°C temperature. It was observed that the values of r for which the error was minimized could be related to the ratio of the Prony series coefficients and the average creep factor of all temperatures f_{ave} by

$$r = \frac{D_1^{20}}{D_1^{30}} + \frac{D_1^{30}}{D_1^{40}} - f_{ave} \quad (13)$$

where D_1^{20}, D_1^{30}, D_1^{40}, and f_{ave} are listed in Table 8. The ratios of the coefficient in Eq. (13) represent the compliance curves. The ratio D_1^{20} to D_1^{30}, for example, is the slope ratio of the 20°C and 30°C creep curve at the 24 hour. This can be proved by using the first derivatives of Eq.3 and Eq.4, which represent the short-term creep compliance curves at the 20°C, 30°C temperature. The values of r shown in Table 8 were determined by Eq. (13).

The values of r, determined by Eq. (13), and the other coefficients of Eq. (12) display the coefficients of a different Prony series equation. The Prony series equation that predicts the compressive long-term creep compliance of the F-C-L20, for example, is constructed by using the values listed in the first row, as follows:

$$D(t) = 204.6 + 30.55(1 - e^{-t/24}) + 46.07(1 - e^{-t/1000}) + 99(1 - e^{-t/6000}) \quad (14)$$

3.5. Evaluation of the short-term creep test

The creep compliance curves that were predicted using the Prony equations, and experimental creep compliance curves, are shown in Figs. 16, 17, and 18. It shows the creep compliance curves of polymer concrete systems on the 20, 30, and 40% stress ratio namely F-C-L20, F-C-L30 and F-C-L40. The curves of the F-C-L30 and F-C-L40 are one year long.

The differences between the predicted and the experimental creep compliance curves that are shown in Figs. 16,17 and 18 were caused by: (1) the differences between the glassy modulus of the short-term creep test and the glassy modulus of the long-term creep test, and (2) the inaccuracy of the prediction method. The error that was caused by the inaccuracy of the prediction method was isolated by subtracting the glassy creep compliance from both the predicted and the experimental creep compliance values.

A more accurate evaluation is made by a quantitative comparison of the results. The predicted specific creep and creep compliance values are compared in Table 9 with the corresponding experimental values obtained from a long-term creep test. The difference between the predicted and the experimental values is expressed as a percentage of the experimental values.

The errors of the predicted creep compliance values, shown in Table 9, are less than 5 percent for all polymer concrete systems.

The difference between the two curves is due to differences between the glassy compliance of short-term creep test and the long-term creep test at the 20°C temperature. The difference in the glassy compliance was probably caused by a small difference in the elastic modulus of the polymer concrete that was not cast at the same time.

The minor difference between the predicted and the experimental creep strains indicates that the short-term creep test can be successfully used to predict the long-term creep deformation of the polymer concrete systems examined.

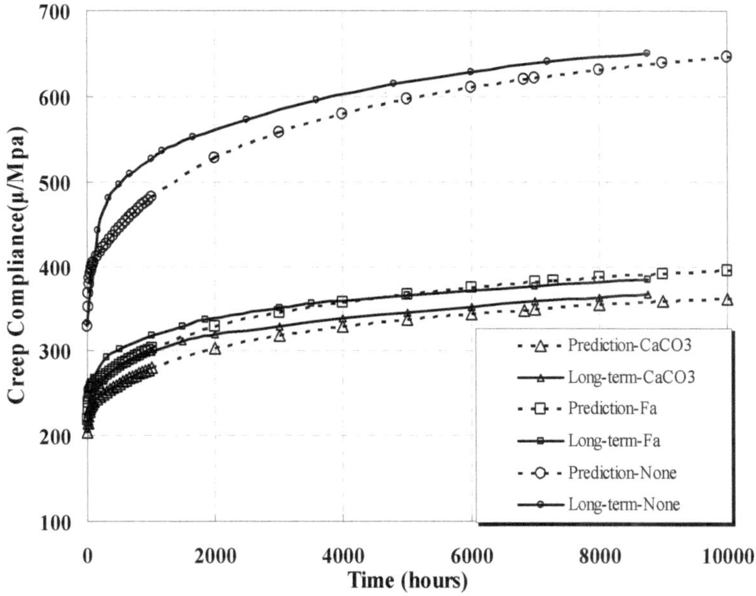

Figure 16. Creep compliance curves depending on filler types

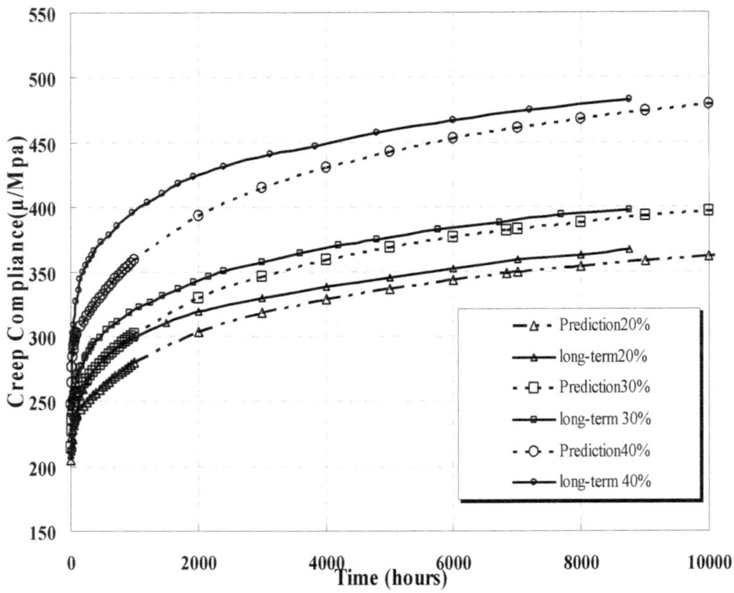

Figure 17. Creep compliance curves with stress ratios

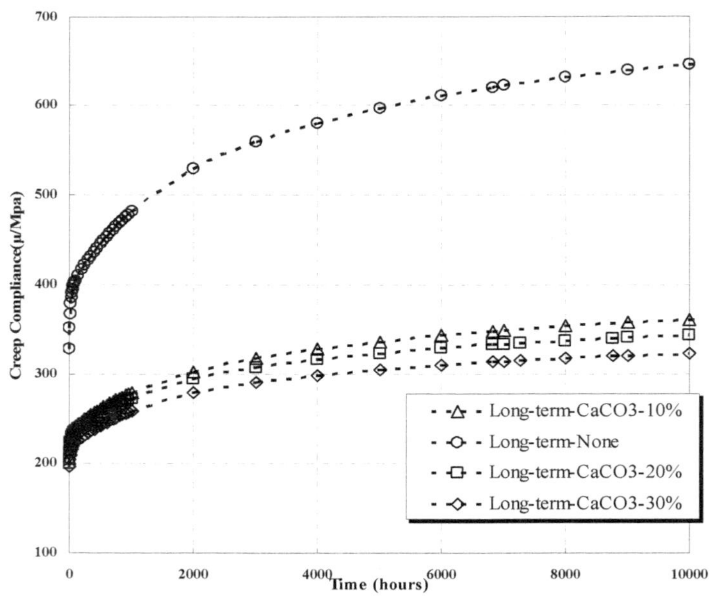

Figure 18. Creep compliance curves depending on filler contents

PC System	Creep Compliance			Specific Creep		
	Pred. (μ/Mpa)	Exp. (μ/Mpa)	Err. (%)	Pred (μ/Mpa)	Exp. (μ/Mpa)	Err. (%)
F-C-L20	358.1	366.9	2.5	152.6	157.9	3.5
F-Fa-L20	390.2	383.9	-1.6	171.3	169.5	-1.1
F-N-L20	637.4	650.0	2.1	308.7	316.4	2.5
F-C-L30	391.9	397.5	1.4	177.6	186.0	4.7
F-C-L40	472.8	482.7	2.1	225.5	232.7	3.2

Table 9. Comparison of the prediction and experimental creep

3.5.1. Effect of filler type

The filler plays an important role in restricting the deformation of polymer concrete. The filler decreases the quantity of resin per unit volume at mixing, and increases adhesion by increasing the viscosity [19].

In this study, resin and applied stress were fixed while different types of filler were used. Creep strain and specific creep prediction curves on the type of filler are shown in Figs. 19 and 20.

The creep strains were 394.87μ with the $CaCO_3$ and 433.9μ with the fly-ash, by prediction equation. The creep strain and specific creep were found to be smaller when $CaCO_3$ was used than when fly-ash was. The long-term creep strain of $CaCO_3$ was 408.52μ and that of fly-ash was 429.4μ.

The heavy calcium carbonate has a greater specific surface area than fly-ash. So, it has more bonding area with resin binder than the fly-ash does. Therefore, the adhesion between the resin binder and the aggregates is stronger, reducing creep strain. Fineness of fly-ash is smaller than that of $CaCO_3$, approximately 2,800 cm^2/g and 3,765cm^2/g respectively.

Very small fineness interrupts the cross-linking between the resin binder and the aggregates during polymerization. The adhesions decrease in the interface between the aggregates. Reduction of adhesion causes weak restriction of the resin, increasing creep deformation.

The creep strain of F-N-L20, which did not use the filler, is dramatically increased. This illustrates the importance of the filler in restricting the resin's deformation. The predicted creep strain and the specific creep (without filler) were 557.94μ and 308.65μ. The creep strain of the concrete without a filler thus approximately 30 percent to 40 percent higher than that with a filler. The specific creep was about two times higher than with a filler.

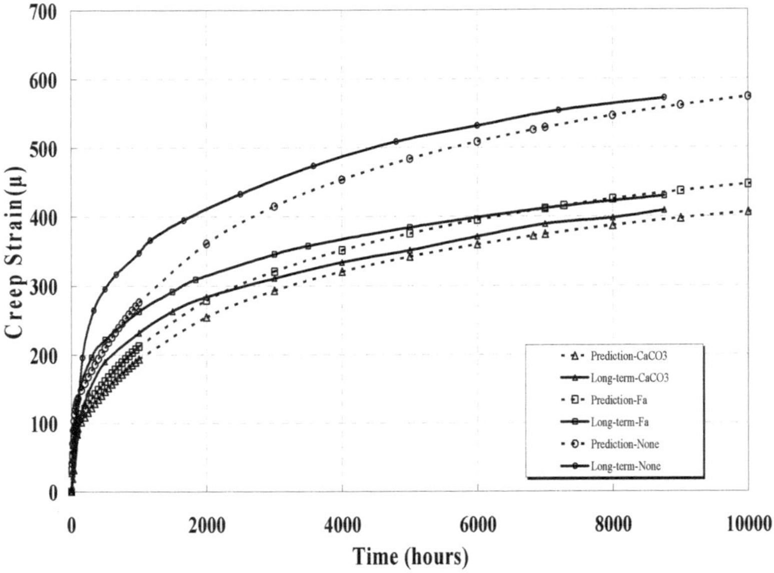

Figure 19. Creep strain curves depending on filler types

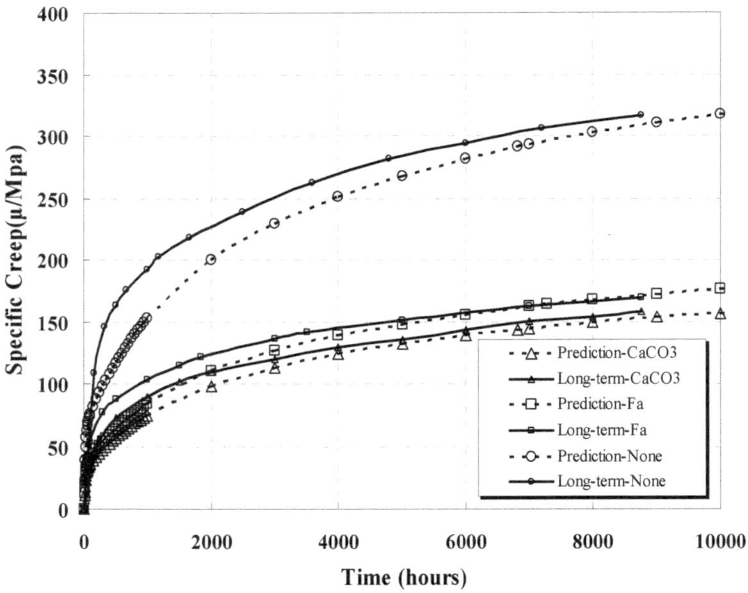

Figure 20. Specific creep curves depending on filler types

3.5.2. *Effect of the stress ratio*

The resin and filler contents were constant and the polymer concrete systems were subjected to progressively increasing stress ratios, ranging from 0.2 to 0.4 of the ultimate compressive strength. Creep strain and specific creep prediction curves on the stress ratio are shown in Figs. 21 and 22. In comparison with the predictions given by the Prony series equation, the creep strain and the specific creep curves given by the prediction curves are slightly lower than those of the long-term creep curves.

In the prediction creep strain, more than 20 percent of the final creep took place within the first two days and nearly 50 percent during the first 20 days, as shown in Fig. 21. In the long-term creep strain, however, 20 percent of the final creep was measured within first 1.5 days and 50 percent during the first 20 days. On the other hand, in the case of conventional concrete, 25 percent of the final creep takes place within the first month of loading and 50 percent within the first 3 months.

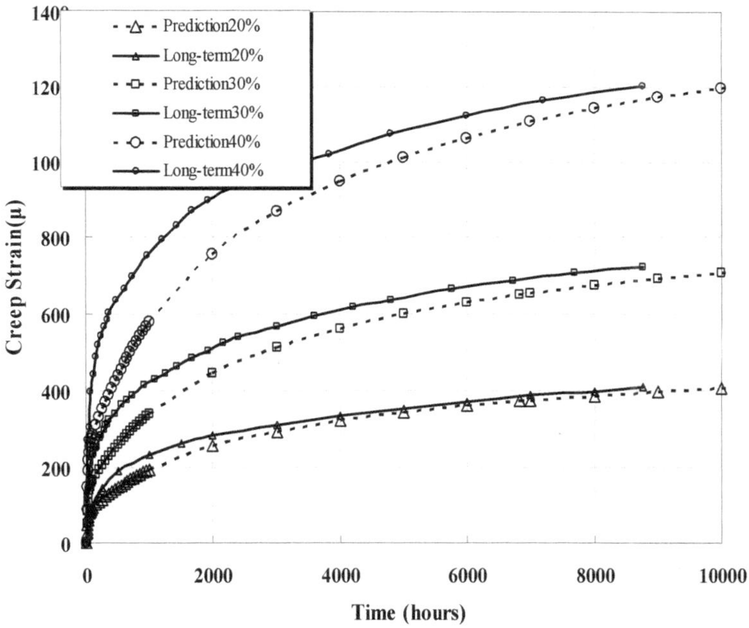

Figure 21. Creep strain curves with stress ratios

The prediction creep curves are smaller than long-term creep curves at an early stage, although the prediction creep strains are approximately equal to long-term creep at one year. After the rapid increase in strains at the early stages of loading, the deformation decreased and the creep strain curves tend to level off after approximately the first two to three months of applied load.

As expected, a higher stress ratio resulted in a larger creep strain. However, in comparison with the stress ratio results, the creep strain is not proportional to the rate of stress increase. During the first year, the prediction creep strain and long-term creep strain were 394.87μ and 408.53μ at 0.2, 689.35μ and 722μ, at 0.3, and 1166.65μ and 1204μ at 0.4 stress ratio. As the applied stress increased about 50 percent, the creep strain increased 75 percent. It increased 190 percent as the applied stress rose up to 100 percent. The specific creep, the creep strain per unit applied stress, is shown in Fig. 22.

The prediction and the long-term specific creep, at one year, are 152.62μ/Mpa and 157.9μ/Mpa at 0.2, 177.6μ/Mpa and 186μ/Mpa at 0.3, and 225.45μ/Mpa and 232.6μ/Mpa at 0.4 stress ratio. The trend of specific creep is approximately equal to the creep strain, regardless of the stress ratio. Creep behavior of polymer concrete using recycled polyester resin is viscoelastic behavior.

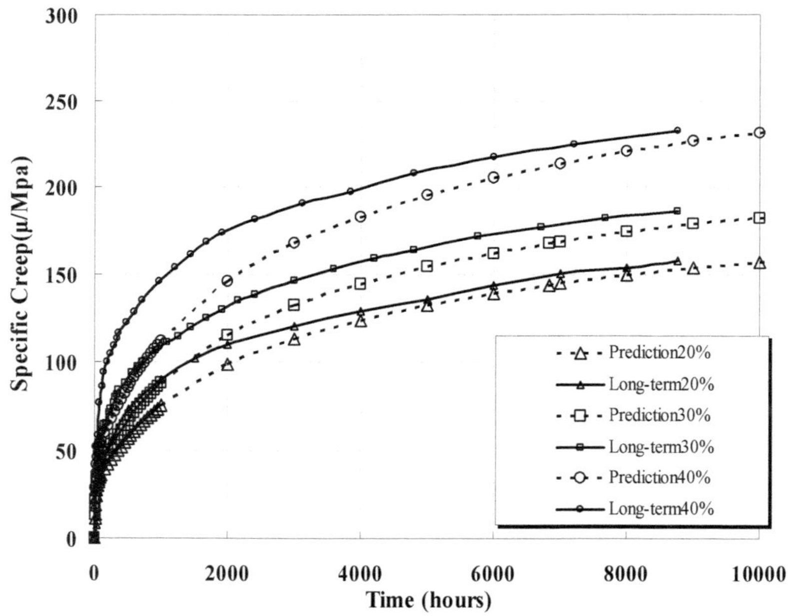

Figure 22. Specific creep curves with stress ratios

3.5.3. *Effect of filler contents*

In this study, the resin content and stress ratio were fixed and the filler contents ($CaCO_3$) varied from 0 to 30%. The creep strain and the specific creep prediction curves on the contents of filler are shown in Figs. 23 and 24. The predicted creep strains and specific creeps are 557.94μ and 308.65μ/Mpa without filler, 394.87μ and 152.62μ/Mpa with 10% filler, 343.99μ and 140.43μ/Mpa with 20%, and 341.27μ and 125.17μ/Mpa with 30% by prediction equations. In comparison with the results, it is clear that as the filler content increased, the creep strain and the specific creep decreased. As the filler content increased from 10 to 20 percent, the specific creep decreased about 8 percent. As it increased from 20 to 30 percent, the specific creep rose about 11 percent. As it increased from 0 to 10 percent, the specific creep decreased about 50 percent. This indicates that the use of filler had more of an effect than filler content.

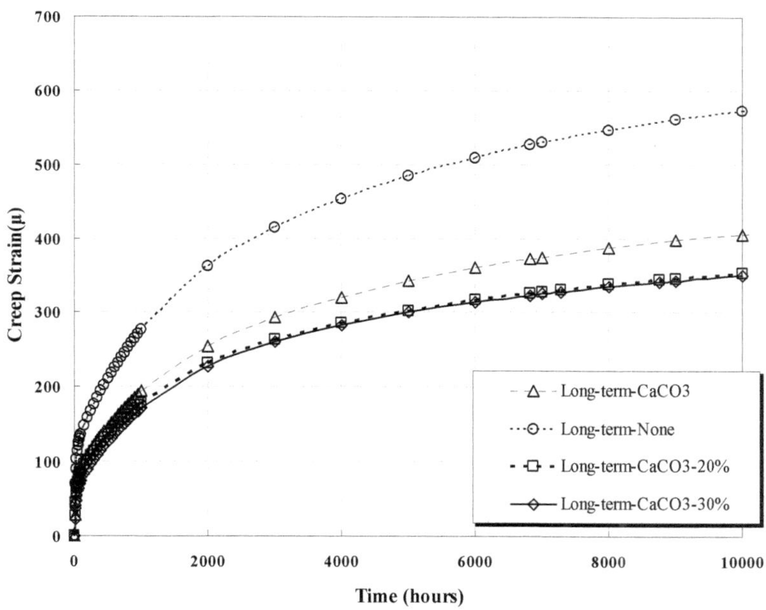

Figure 23. Creep strain curves depending on filler contents

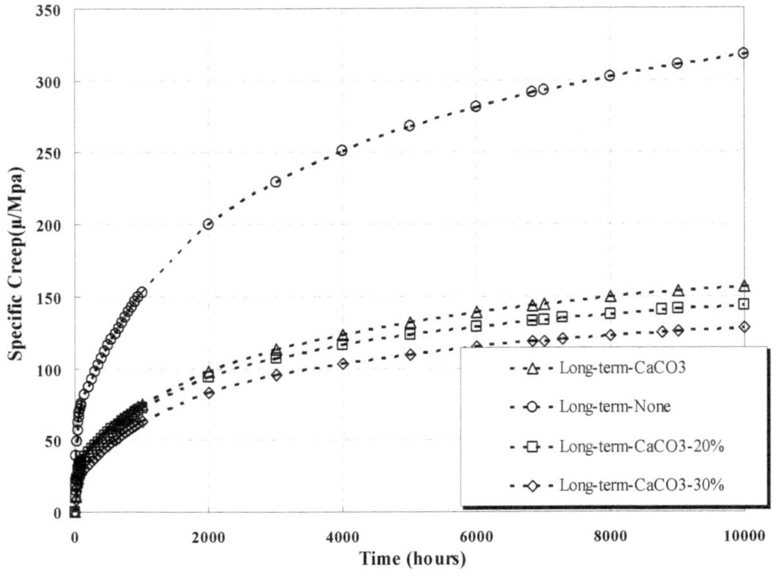

Figure 24. Specific creep curves depending on filler contents

4. Conclusion

Based on the results obtained in this investigation on time dependent behavior of polymer concrete, the following conclusions can be drawn:

1. Strength test results revealed that the material could achieve more than 85% of its final strength in one day. This result is an important advantage in many construction and structural applications. The material experiences a loss in strength at high temperatures; this result may be important if the polymer concrete is used in precast box culvert, for example. However, despite this loss in strength at high temperatures, the material remains strong compared to portland cement concrete.
2. Montmorillonite considerably affect various properties of polymer concrete. Also, the more the montmorillonite content increases, the more both the setting shrinkage and sensitivity to heat decrease. This result shows that montmorillonite without a special coupling agent could be used as one of the additives to enhance various properties of polyester resin.
3. The coefficient of thermal expansion of fly ash was relatively lower than that of calcium carbonate because similar chemical composition led to active reaction with montmorillonite.
4. With the addition of montmorillonite 5%, the compressive strength was found to increase by 12%. Improvements in the properties of recycled unsaturated polyester polymer concrete were true for the optimum values of 5% of montmorillonite content. If

the percentage of the montmorillonite exceeds 5%, recycled unsaturated polyester polymer concrete properties either remain constant or change in a negative way.

5. The long-term creep prediction models are formulated to facilitate an understanding of the behavior of polymer concrete structures through short-term creep tests. The error between the long-term creep prediction and experimental long-time creep compliance is so negligible as, less than about 4%.
6. The creep strains grow rather fast at early ages in comparison with ordinary concrete, because creep in polymer concrete results from molecular movement in the viscoelastic resin binder. The recycled-polymer concrete shows more than 20% of its long-term creep within the first two days, and about 50% during the first 20 days.
7. The creep strain of polymer concrete without a filler is much higher than that of polymer concrete with a filler, because the filler plays an important role in restricting the deformation of polymer concrete. Also, as far as the type of filler is concerned, the effect of $CaCO_3$ as a filler on polymer concrete is better than that of fly-ash in terms of the creep strain and specific creep. This is attributed to the larger surface area of $CaCO_3$ particles and the higher adhesion between the resin binder and the aggregates.

Author details

Ghi Ho Tae
Department of Structural Engineering, Seoul National University of Science & Technology, Korea

Eun Soo Choi
Department of Civil Engineering, Hongik University, Korea

5. References

[1] ACI Committee 548, "Guide for the use of Polymers in Concrete", *ACI Materials Journal*, V. 82, No. 5, September 1997, pp.11~13

[2] Vaidya, U.R.,and Nadkarni, V.M., "Unsaturated Polyester Resins from Poly(ethylene terephthalate) Waste", Industrial & Engineering Chemistry Research, Vol.26, 1987, pp.194-198

[3] Rebeiz, Karim Sami, "Structural use of Polymer Composites using Unsaturated Polymer Resins based on Recycled Poly(ethylene terephthalate)", Ph.D. at Austine in Texas, 1992, pp.59~62

[4] Edith A. Turi, "Thermal Characterization of Polymeric Materials", 2nd Edition Vol. 1, Academic Press, 1997, pp.227 pp.1145

[5] K.S.Rebeiz, "Precast Use of Polymer Concrete using Unsaturated Polyester Resin based on Recycled PET waste", *Construction and Building Materials* Vol.10 No.3, pp.216 ~ 218

[6] Kim ju hyeon, "A Study on the Strecture-Property Relationships of Unsaturated Polyester Resins Based on Recycled PET", Ph. M. in Chonbuk Uni, February 1995, pp.17-20, 49 ~ 50

[7] Peschke, H.J., "Stress and Strain Analysis Between Cementitious Concrete and Polymer Concrete", 3th ICPIC, pp.477 ~ 489

[8] Karim S.Rebeiz, "David W.Fowler and Donald R.Paul, Mechanical Properties of Polymer Concrete Systems Made with Recycled Plastic", *ACI Material Journal* V.91, No.1, Jan-Feb 1994, pp.40 ~ 45

[9] David Stanley, "Investigation of Dow Polyesteramide Resins for Use in Polymer Concrete", The University of Texas at Austin, 1984, pp.65 ~ 68

[10] C.E.Bream and P.R.Hornsby, "Structure Development in Thermoset Recycled-Filled Polypropylene Composites", Journal of Polymer Composites, June 2000, Vol.21, No.3, pp.418 ~ 423

[11] Dae-Keun Park and Jin-Hae Chang, "Nanocomposites Based on Montmorillonite and Thermotropic Liquid Crystalline Polyester", The Journal of Polymer, May 2000, Vol. 24, No.3, pp.339 ~ 406

[12] Scuk-Hyun Lee, "A Study on the Swelling Behavior of a Cured Unsaturated Polyester", *The Journal of Polymer*, December 1987, Vol. 11, No. 6, pp.557 ~ 562

[13] Alkonis, J. J., Maxknight, W. J., Shen, M., "Introduction to Polymer Viscoelasticity," Wiley-Interscience, John Wiley & Sons, INC., New York, 1972.

[14] E. Riande, R. Diaze-Calleja, Margarita G. Prolongo, Rosa M. Masegosa, Catalina Salom "Polymer Viscoelastisity Stress and Strain in Practice," Marcel Dekker, INC. 2000.

[15] Hsu, M., Fowler, D.W, "Creep and Fatigue of Polymer Concrete," Polymer Uses-Materials and Properties, ACI, SP 89-17, pp. 323~343, 1985.

[16] J.-P. Boehler, T. Dietl, A. Millien, "Creep Law for Polymer Concrete in the Hardening State"

[17] John D. Ferry, "Viscoelastic Properties of Polymers," third edition, John Wiley & Sons, INC, 1980.

[18] K. Aniskevich, J. Hristova, "Aging and Filler Effects on the Creep Model Parameters of Thermoset Composites," Composite Science and Technology, Vol.62, pp. 1097~1103, 2002.

[19] K. Aniskevich, J. Hristova, "Creep of Polyester Resin Filled with Minerals," Journal of Applied Polymer Science, Vol. 77, pp. 45~53, 2000.

[20] K. S. Rebeiz, "Time-Temperature Properties of Polymer Concrete Using Recycled PET," Cement&Concrete Composites Vol.17 pp.119~124, 1995.

[21] K. K. Charalambous, "Accelerated compression and flexural creep testing of polymer concrete", Ph.D. Dissertation, University of Texas at Austin, 1991.

Chapter 10

Characterizations of Polyester-Cement Composites Used for the Immobilization of Radioactive Wastes

Hosam El-Din Saleh, Talat Bayoumi and Samir Eskander

Additional information is available at the end of the chapter

1. Introduction

Radioactive waste is the ethical issue facing the development of peaceful applications of nuclear technology all over the world. It poses a serious threat to human health and dangerous impact to his environment. Hence, the problem needs early global cooperative solutions. Almost all the International Atomic Energy Agency's member states produce some types of radioactive wastes from their applications e.g. industrial, agricultural, medical, educational, nuclear power production plants, defenses aspects, accidents...

Solidification/stabilization (s/s) of radioactive waste is an attractive technology to reduce their risks and facilitate their handling prior to disposal. The transformation of those materials into monolithic homogeneous solid forms by immobilization in inert matrices; cement, polymer, polymer-cement composite, is essential to preserve man and his surrounding from the dispersion of these radioactive hazardous wastes (Tawfik et al., 2005). Owing to its advantage, cement matrices have been practiced for many years in different places for the immobilization of waste materials. Based on long experience on the suitability of cement for immobilization of numerous low and intermediate level waste residues, it is recommended to be used as an appropriate matrix for safe handling, transport, storage and disposal (Drace & Ojovan, 2009; Sobolev et al., 2006; Varlakova et al., 2009).

Cement and cement based materials have been successfully applied in multi barrier system of a repository. The first chemical and physical barrier is the cement matrix used for the immobilization of waste materials. The second important barrier is the cement container for packaging of waste forms. Among the advantages of the cement used, as engineering barrier, are low cost, simplicity of the process, abundance of raw materials, fire resistance, pronounced shielding effect, radiation and heat stability in addition to satisfactory mechanical properties (Burns, 1971; IAEA, 1983, 1993).

The high porosity of cement materials facilitates water percolation resulting in pronounced high leach rates of the radionuclides to the surrounding. To overcome this main disadvantage, polymer-cement composites have been developed as candidate matrices for radioactive waste incorporation.

The main component of liquid waste concentrates generated from the pressurized water reactor is boric acid (H_3BO_4). Such reactors generate about 50,000 gallons of the boric acid evaporator bottom yearly. Boric acid waste is mostly neutralized by using sodium hydroxide and then concentrated by evaporation. The resulted borate slurry should be immobilized usually in cement, polymer, bitumen or polymer-cement composites (Eskander et al., 2006; Greenhalgh, 1986; Huang & Yang, 1999). Direct solidification of borate waste with Portland cement is complicated through the retarding effect of both acidity of the waste (H_3BO_3) and specification of the borate ions ($B4O_7^{-2}$ and BO_3^{-3})(Tawfik et al., 2005).

2. Theoretical background

The polymer-cement composite is a combination between cement and polymers. Hence, it is essential to have a short theoretical background for those two key materials, i.e. cement and polymer.

Cement has been described as adhesive substance capable of uniting fragments of solid matter to compact solid blocks (Holcomb, 1978). There is a wide range in composition of cement materials used for the solidification of radioactive wastes and in the construction of various barriers in a repository. Their common feature is that the main constituents are lime (CaO), silica (SiO_2) and alumina (Al_2O_3). Portland cement is the most commonly used type in radioactive waste management applications (Peter C, 2003). It has lime as principal constituent which is obtained by mixing calcareous and argillaceous or other silica-alumina and iron oxide-bearing materials. The principal compositions and oxides percentages of Portland cement used in this study are represented in Table 1.

Chemical compositions, %		Compounds compositions, %	
SiO_2	19.84	**C_3S**	53.11
Al_2O_3	4.74		
Fe_2O_3	4.0	**C_2S**	16.89
CaO	61.01		
MgO	2.5	**C_3A**	5.81
K_2O	0.6		
SO_3	2.4	**C_4AF**	12.16
Insoluble residue	0.95		

-Loss on ignition = 3.96%.
-Lime saturation factor = 96%by wt.

Table 1. Chemical and mineral composition (%w/w) of Portland cement

Polyester is a category of polymers which contain ester functional group in their main chain (at least 85% by the weight). Although there are many types of polyester, the term "polyester" as a specific material most commonly refers to polyethylene terephthalate (PET). Polyesters include naturally occurring chemicals, such as in the cutin of plant cuticles, as well as synthetics through step-growth polymerization such as polycarbonate and polybutyrate. Natural polyesters and a few synthetic ones are biodegradable, but most synthetic polyesters are not. Depending on the chemical structure, polyester can be a thermoplastic or thermoset; however, the most common polyesters are thermoplastics (Rosato et al., 2004). Polyesters as thermoplastics may change shape after exposing to heat. While combustible at high temperatures, polyesters tend to shrink away from flames and self-extinguish upon ignition. Saturated polyesters refer to that family of polyesters in which the polyester backbones are saturated. Unsaturated polyesters refer to that family of polyesters in which the backbone consists of alkyl thermosetting resins characterized by vinyl unsaturation. On the other hand, unsaturated polyester resin (UPR) is thermosetting material. There are several reasons for the importance of polyester such as the relatively easy accessible raw materials, the very well understood and described simple chemical process of polyester synthesis, the low toxicity level of all raw materials and side products during polyester production and processing, the outstanding mechanical and chemical properties of polyester and the recyclability. Polyester is very durable: resistant to most chemicals, stretching and shrinking, wrinkle resistant, mildew and abrasion resistant. Polyester is hydrophobic in nature and quick drying.

This study is based on a large body of researches carried out at Inorganic and Applied Chemistry Unit, Radioisotope Department, Egyptian Atomic Energy Authority for nearly 5 decades. These published work, in addition, to that contained in literatures reviewed the development and applications of polyester-cement composite (PCC) in the field of solidification/stabilization of radioactive wastes. Throughout this review, it can be concluded that the proposed PCC, especially which based on the recycled PET, as a matrix for immobilizing the borate waste, is economic and safe.

3. Preparation of some polyester-cement composites

3.1. Water extended polyester (WEP)

The used water extended polyester is obtained either based on the virgin raw components or the recycled polyethylene terephthalate (PET).

3.1.1. Synthesis of water extended polyester from virgin raw components

Unsaturated polyester was prepared from high purity ingredients by adding 2 moles of maleic anhydride to 2.2 moles diethylent glycol in a round bottom three necks flask provided with a constant thermometer, a stirrer and a condenser. The reaction mixture was heated in an oil bath and maintained at 80°C for 1 hour, then to 190°C for 4 hours. The unreacted monomer and other products were removed from the reaction mixture under vacuum using oil pump and the acid number was adjusted at 50 mg of KOH/g. The temperature was then lowered and maintained at 170°C until an acid number of 50 or less was reached. Water extended polyester emulsions were prepared by dispersing water or waste solution samples containing 1.03% by weight triethanolamine as an emulsifier into the prepared unsaturated polyester containing 30% by weight styrene. The stable emulsion of polyester was cured by the addition of 2% by weight benzoyl peroxide (BP) as an initiator. More experimental details for the preparation of water extended polyester are given in previous publication (Saleh et al., 2005).

3.1.2. Synthesis of water extended polyester based on recycling of poly(ethylene terephthalate) (PET)

The overall annual consumption of PET, for packaging only, had increased by mid-2000 to about 30% compared to 1990. Over the last decades, PET consumption has increased tremendously due to its ever-increasing use in fibre manufacturing and in packaging. At present, the global yearly PET production is approximately 13 million of metric tons (Tawfik et al., 2005). PET is a synthetic semi-crystalline thermoplastic polymer, which has long life due to almost complete resistance to biodegradation and as a result large amounts of PET waste are accumulated. Physical, mechanical and chemical recycling processes have been developed even to industrial scales. Due to the well-known disadvantages of both mechanical and physical processes, chemical recycling has been the subject of interest all over the world. This technique is the most acceptable one, following the principles of sustainable development, since it leads to the formation of raw materials from polymer as well as to the secondary value-added products. PET is known to be attacked in chemical recycling by different alkalis, glycols, alcohols, ammonia and amines by making use of processes such as hydrolysis, glycolysis, alcoholysis, ammonolysis and aminolysis, respectively. The degradation of the main backbone of PET is supposed to take place through chain scission. The challenging disadvantage, accompanied with these processes, is the relatively high consumption of energy (Tawfik et al., 2005).

3.1.2.1. Glycolysis of PET waste

PET waste was depolymerized by diethelylene glycol (DEG)/propylene glycol (PG) mixture at a ratio of 1:0.5:0.5 by weight, in the presence of lead acetate as a catalyst (0.5% by weight of PET) at the reflux temperature of the glycols. The mixture was charged into a four-necked round bottom flask, which was fitted with a stirrer, reflux condenser, nitrogen inlet and thermometer. The flask was immersed in oil bath, and the temperature was brought up to 200°C for 5 hours. In order to check the extent of depolymerization, the glycolized product was analyzed for hydroxyl number value, and it was found to be 617 mg KOH/g.

3.1.2.2. Preparation of unsaturated polyester (UP)

Unsaturated polyester (UP) was prepared by the esterification of the glycolized PET oligomers with maleic anhydride (MA) and adipic acid (AA) at a molar ratio of 1.1:0.7:0.3 for hydroxyl to carbonyl groups, respectively. The polyesterification reaction was carried out in a four-necked round bottom flask, having a stirrer, reflux condenser, an inert gas inlet and thermometer. A Dean–Stark trap was used for measuring the water amount, which was removed throughout the reaction. The reactants were heated in oil bath, and its temperature raised to 150°C. The temperature was then increased to 210°C at a rate of 10°C/h until the acid number value reached nearly 45 ± 2 mg KOH/g.

3.1.2.3. Preparation of water extended polyester (WEP)

The obtained UP was dissolved in a styrene (S) monomer at a ratio of 60:40% by weight at room temperature (25 ± 5°C). WEP emulsion was prepared by dispersing with continuous stirring of 20% water by weight, containing 0.2 mmole DEA/g prepolymer as an emulsifier, into 10 g of the obtained styrenated polyester. The stable emulsion was cured by addition of 2% from the prepared WEP by weight benzoyl peroxide (Bz_2O_2) as a catalyst.

The obtained water extended polyester mixture was mixed with cement paste at various water/cement ratios to prepare either the final waste form by incorporating a predetermining weight of borate waste simulate or polyester-cement composite container.

3.1.3. Polyester-cement waste form

The final waste forms were prepared by mixing predetermined weight of powdered radioactive borate waste simulates with polymer-cement composite paste. This mixture was poured in a polyethylene moulds and allowed to set and harden at room temperature in its humid atmosphere for 28 days. At the end of curing time, the obtained blocks were demoulded giving solid cylindrical blocks of 60 ± 2 mm height and 31 ± 0.5 mm diameter. Various factors that may affect the physicochemical stability of the formed composite specimens were studied.

3.1.4. Polyester-cement container

Portland cement was thoroughly mixed with pure water at constant water/cement ratio of 1:2.8 by weight. Water extended polyester was mixed with cement paste to form a polymer-cement composite paste. The polymer-cement composite paste was poured in a special closed moulds and was left to solidify in a humid atmosphere for 28 days. The obtained solid samples were in the form of small containers having the dimensions of 7.6 ± 0.4 cm height, 5.92 ± 0.20 cm diameter and various wall thicknesses. A given weight of labeled powder simulating waste samples was introduced into these containers. The containers were closed with cover and then introduced to subsequent investigations.

4. Characterization of unsaturated water extended polyester (WEP)

4.1. Water extended polyester (WEP) based on virgin raw ingredients

Both IR and NMR analyses of the WEP prepared from its virgin raw materials illustrate the expected formula of the polyester given in the literature and determine the main characteristic groups of the prepared polyester used as an additive to the cement waste forms (Figs. 1,2)(Shatta, 1996).

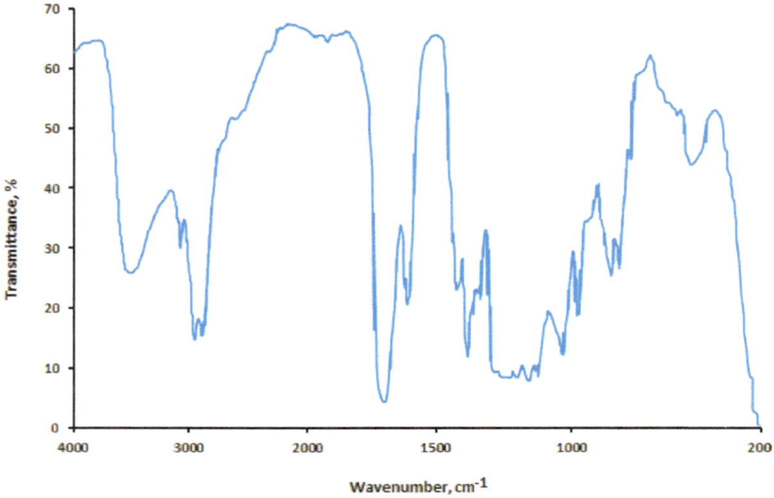

Figure 1. FTIR spectrum of the water extended polyester prepared from virgin raw ingredients

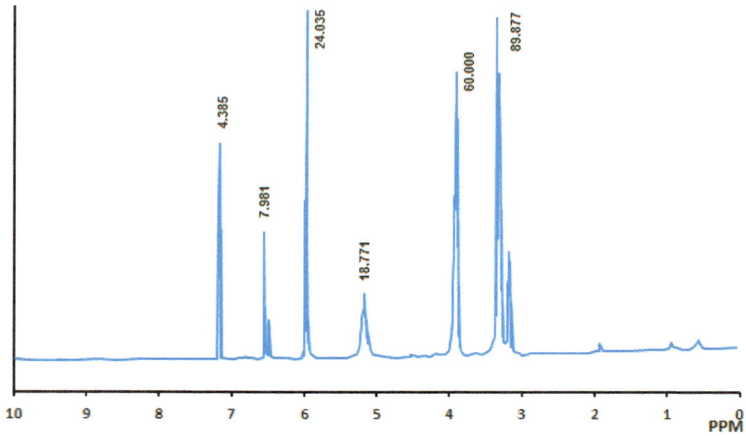

Figure 2. NMR spectrum of the water extended polyester prepared from virgin raw ingredients

4.2. Water extended polyester (WEP) based on recycled PET

WEP is used as an organic additive to improve the properties of cement waste forms. The resulting PCC products show a good mechanical integrity, enhanced physical properties, acceptable radiation stability, low leachability of radionuclide, and withstand relatively high temperatures. However, it could be stated that the characters of the prepared PCC depend mainly on the properties of UP used. Nuclear magnetic resonance (NMR, ppm) analysis for the UP based on PET was carried out using a Bruker AC-400 spectrometer and deuterated chloroform as a solvent. The spectrum represented in Fig. 3 shows the obtained signals for the UP analysis and their corresponding assignments represented the following groups; at 8.07–8.08 (4H, C_6H_4), at 6.85–6.89 (olefinic protons in the CO-CH=CH=CO chromophore of maleic and fumaric acid residues), at 5.2–5.6 (methine proton of PG residue), at 3.58–4.68 (-O-CH_2-CH_2-O and –O-C-CH_2-O), at 1.63–2.14 (adipic acid residue), and at 1.27–1.48 (methyl group in PG residue) (Tawfik et al., 2005).

The IR spectrum of unsaturated polyester (UP) obtained from recycled soft drink bottles of PET shows that the broad band that appeared around 3500 cm^{-1} is due to the stretching vibrations of the OH terminal in the hydroxyl group in the hard polymer. The band at 3020 cm^{-1} may be assigned to the aromatic CH while bands at 2950 cm^{-1} and 2880 cm^{-1} may refer to the stretching of CH in methylene and methyl olefinic groups, the stretching frequencies of the ester carbonyl group (C=O) in the polymer appeared near 1720 cm^{-1} and 1640 cm^{-1} while that for olefinic -C=C- appeared at 1580 cm^{-1}. The band in the region 1450–1380 cm^{-1} may be attributed to the bending vibration of CH_2 and CH_3 in propylene glycol. The curve shows also the bending in trans -CH=CH- groups, which appears at 980 cm^{-1}. On the other hand, bands at 880 cm^{-1}, 750 cm^{-1}, and 650 cm^{-1} may be attributed to cis olefinic and aromatic residues (Eskander et al., 2006). It could be clearly stated that there were non-significant differences between the WEP based on the recycled PET and that prepared from its virgin raw ingredients. However, it is worth mention that the economical and environmental aspects favored the preparation of WEP based on the recycled PET.

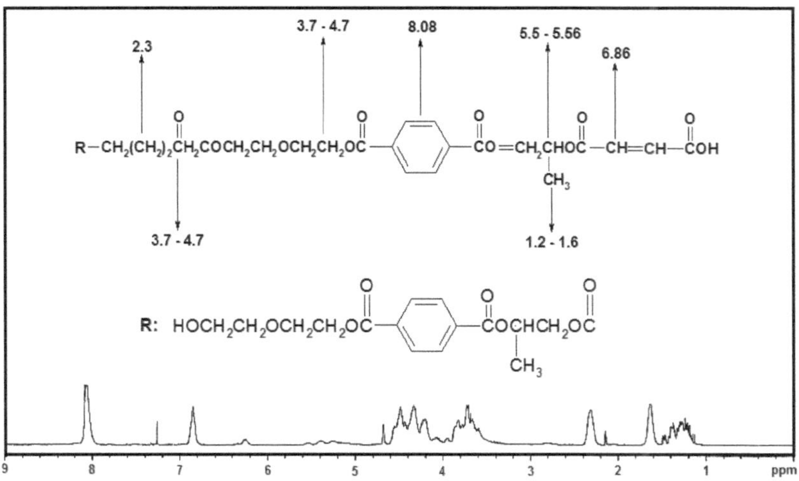

Figure 3. NMR spectrum of polyester based on glycolized PET waste

On the other hand, styrenated unsaturated polyester products based on both the recycled PET and poly(oxy-diethylene maleate) mixed with styrene monomer showed mechanical and physical characterizations closely similar to each other as reported in Table 2.

Composition	WEP based on recycled PET	WEP based on virgin raw ingredients
Styrene monomer	30 %	30 %
Specific gravity at 25°C	1.45	1.45
Appearance	Light yellow	Yellowish white
Water of emulsion	50 %	50 %
Initiator (Benzoyl peroxide)	2 %	2 %
Compressive strength, MPa	9.57	9.95

Table 2. Chemical, physical and mechanical characterizations of WEP

5. Evaluation of the final waste form

Transformation of borate wastes from pressurized water reactor into homogeneous monolithic solid waste forms was achieved using cement and polymer-modified cement as an inert matrices. The aim was to reduce the potential migration and dispersion of radionuclides from waste forms to the environment, the primary objectives were to provide possible options helping in choosing the proper immobilization media, to improve the quality of the final waste forms and process technology.

Compressive strength is generally given for the description of structural stability of the final waste forms. To evaluate the mechanical integrity of the solid PCC containing the borate waste simulate, compressive strength measurements were carried out for PCC solid blocks under variable factors that may affect e.g., polymer/cement ratio, water content, or borate waste simulate concentration. A remarkable change in compressive strength values for PCC samples containing an increasing ratio of WEP premixed with cement paste having different water/cement ratios were recorded, as shown in Fig. 4. Premixing cement with WEP affected the final properties of the final waste forms in two contradicting ways depending on the WEP/PC ratio used. In the case of low polymer content (up to 3% of hydrated cement), the compressive strength values were increased by increasing the WEP content, as shown in Fig. 4. On the other hand, increasing the polymer content, more than 3 weight percent, the mechanical integrity of the final waste forms diminished clearly. The improvement in mechanical integrity of the blocks containing low WEP concentration may be attributed to the positive effect of the organic WEP on the final waste forms. When premixing the polymer with the cement and water, it becomes uniformly dispersed within the fresh paste. On hardening, the polymer particles coalesce and gradually build up together to form a three-dimensional cross-linked network that contributes to the toughness and durability of the PCC-system (Tawfik et al., 2005). In other words, the WEP additives act as binding materials improving adhesion of the different heterogeneous cement components. Moreover, the WEP dispersion during hardening seals the pores of the cement matrix and, hence, diminishes the porosity of the products reflecting improvement of the mechanical integrity of the final waste forms. Also, the WEP may act as a filler material that has a plasticizing effect and results in a detectable increase in the durability of the final waste forms.

On the other hand, the high WEP contents negatively affected on the mechanical properties of the final solid blocks, as shown in Fig. 4. This may be due to the fact that the dispersion of the excess organic polymer materials in the paste restricts the access of water in cement grains by forming a thin coating skin over them and also enhanced the ettringitte formation. In addition, the hydration process of cement could also be negatively affected by the interaction of excess polymer with Ca^{++} in the cement materials. The carboxylate anions (COO^{-}) produced by the hydrolytic reaction of esters combine chemically with Ca^{++} cations released from cement grains during the hydration of cement material to form a molecular structure that prevents the scission of the polymer chain.

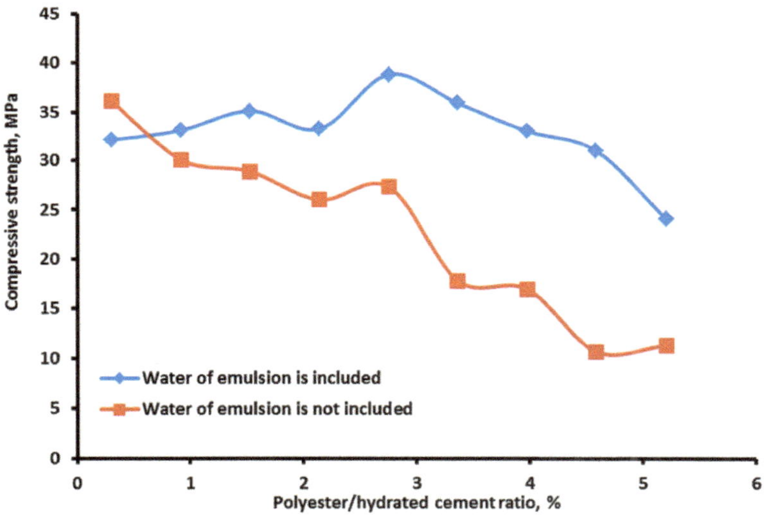

Figure 4. Mechanical integrity of the final polymer-cement composite waste form; Effect of the water content on the compressive strength of the final waste form

Basically, in the cementation process, the anhydrous tricalcium silicate (C_3S) and dicalcium silicate (C_2S), which contribute with the most abundant constituents of PC, commences the hydration reaction very rapidly. Water reacts chemically with the cement constituents to form hydrate silicate and aluminate compounds, which normally contribute to setting and hardening of cement (Tawfik et al., 2005). Fig.s 4 and 5 describe the effect of water content on the mechanical properties of PCC. Two sets of PCC solid blocks were prepared having different water contents. In the first set, the water of emulsion was included in the water of hydration, while in the second set the water of emulsion was not included in the water of hydration. It is clear from Fig. 4 that when the water of emulsion was not included in cement hydration water, solid blocks with higher compressive strength values were obtained, compared to that of the solid blocks in which the water of emulsion was included in the cement hydration water. This may be attributed to the fact that in the second set, part of the water was used for emulsion formation and at this low water content, the hydration reaction of cement may be stopped in the presence of significant amounts of non-reactive cement materials due to the lack of sufficient amounts of water needed for the hydration process. Therefore, it is recommended to calculate the cement hydration water separately from the polymer emulsion water. It is worth mentioning that the same trend was exhibited for different w/c ratio, i.e., 35 wt% or 45 wt% which was not including the water of emulsion. The ratio of 35 wt% of higher mechanical integrity was obtained compared to that at the ratio 45 wt%, but the workability at the w/c ratio 45 wt% was better than that at 35 wt% (Fig. 5) (Tawfik et al., 2005).

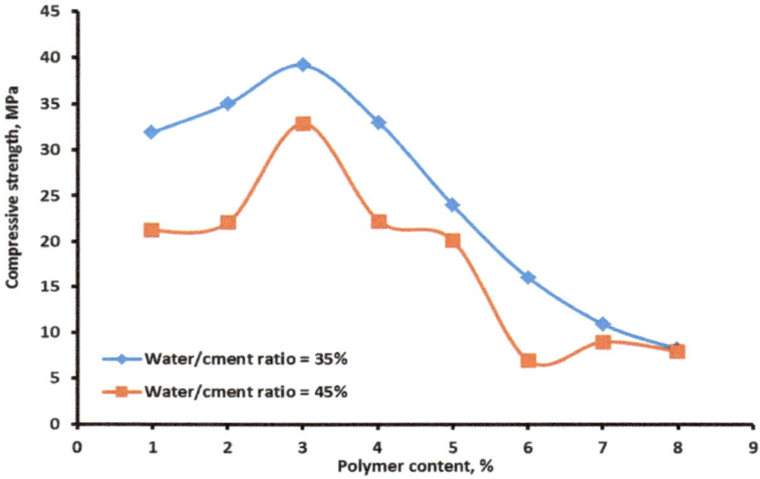

Figure 5. Mechanical integrity of the final polymer-cement composite waste form; effect of polymer content on the compressive strength of the final waste form

The effect of borate content on the mechanical integrity of PCC final waste forms were followed by increasing the waste simulate percentages that premixed with WEP/PC mixture containing 3% polymer (WEP) and at water/cement ratio 40%. After the curing period, the compressive strength measurements were performed for the solid blocks and the data obtained were represented in Table 3. Increasing the borate content premixed with the PCC caused a remarkable decrease in the mechanical integrity of the final waste forms.

Borate waste simulate mainly includes a large percentage of boric acid H_3BO_3 (i.e., 180 g/1) and some borates, e.g., BO_3^{-3} and $B_4O_7^{-2}$ are known as cementation inhibitors. Boric acid reacts with calcium hydroxide during cementation as follow:

$$2H_3BO_3 + Ca(OH)_2 \rightarrow Ca(BO_2)_2 . 2H_2O + 2H_2O$$

Calcium borate is sparingly soluble and is assumed to create a barrier at the cement particles and the solution interface. This barrier retards the diffusion of water and colloidal particles. Acceptable compressive strength values for solid blocks containing up to 3% borate waste simulate were achieved for PCC waste forms containing 3% WEP and w/c ratio 40%. At higher borate levels, a sharp deterioration in the mechanical integrity was detected (Tawfik et al., 2005).

Borate, wt.%	Compressive strength of blocks, MPa	
	Containing 3%, by wt WEP	Containing 6%, by wt WEP
0	33.6	32.74
1	30.06	28.74
2	19.0	0.38
3	20	brittle
4	2.5	brittle
5	brittle	–

Table 3. Compressive strength values of the final PCC waste forms containing, different percentages of borate waste simulates

6. Frost resistance of the polymer-cement composite immobilizing borate waste simulate

To evaluate the suitability of PCC to solidify and stabilize (S/S) radioactive borate waste resulting from pressurized water reactors (PWRs) under the aggressive frost conditions, experimental studies have been undertaken on a laboratory scale, aiming at the determination of the influence of freezing and thawing cycle (FTC) on the mechanical strength, structural deformation, and porosity of the candidate PCC after 115 cycles (Eskander et al., 2006).

The objective of immobilization of radioactive wastes is to convert the wastes into a form that is mechanically, physically, and chemically stable during handling, transport, and disposal. During the long term disposal, and at very low temperature, the frost resistance of the final solid waste form should be assessed. The adequacy of resistance of PCC, based on recycled poly (ethylene terephthalate) from postconsumed soft drink bottles as a candidate matrix for solidification of radioactive borate liquid wastes to frost attack can be determined by a freezing/thawing test. The repeated cycles of freezing and thawing have a cumulative effect rather than the single occurrence frost that causes damage (Eskander et al., 2006).

One of the classical criteria of frost resistance is the weight change of the casted specimens after a number of freezing and thawing cycles. Therefore, a simple formula was postulated to calculate the weight durability percent (Kw) for the candidate PCC subjected to 115 repeated FTC. Kw = Wn/Wo x 100 Where; Wo is the weight of solid specimen before FTC and Wn is the weight of solid specimen subjected to n FTC.

The data obtained for PCC and plain cement subjected to 115 freezing-thawing cycles were represented in Table 4. The Kw for the plain cement decreases by about 14% after 14 cycles and more than 17% after 115 cycles while the Kw for PCC decreases only 13.3% after 115 cycles (230 days) of freezing and thawing at -19°C and +60°C. It is also worth mentioning that the weight durability indices for the PCC exhibit constant value (87.5% ± 0.2) starting from the 28 cycles up to 115 cycles (i.e., end of the treatment). This indicates that the PCC shows acceptable frost resistance compared to the plain cement (Table 4).

FTC	Kw Durability index, %	
	PC	PCC
1	96.6	96.7
4	94.5	95.3
7	93.3	93.1
14	86.4	90.2
28	85.1	87.6
54	84.5	87.3
90	84.5	87.5
115	82.9	86.7

Table 4. The weight durability index (Kw) for plain cement and polymer-cement composite (PCC) subjected to 115 freezing-thawing cycles

Generally, the cement-water extended polyester system consists of three main interacting components: polymer, cement, and water. The polymer-cement modification is initiated in the presence of water to form the polymer phase and simultaneously the cement hydration occurs. Consequently, a co-matrix phase is formed with a network structure of interpenetrating polymer and cement hydration phase. As a result of this intrinsic structure, the mechanical strength and other properties of the modified polymer-cement matrix is improved (Eskander et al., 2006).

Table 5 represents the results of volume of open pores (cm^3), apparent porosity (%), and water absorption (%) for PC (w/c ratio = 40%) as well as PCC that are nominated for processing the borate waste (WEP/PC ratio = 3% and w/c ratio = 40%) and subjected to freezing/thawing treatment up to 115 cycles. It is clear that the PCC matrices have improved resistance to freezing and thawing, based on the improvement in the intrinsic properties, namely volume of open pores, water absorption, and apparent porosity compared to the conventional cement matrix Table 5. This may be attributed to filling the pores of cement matrix in PCC and air entrainment introduced by the WEP (Eskander et al., 2006). The apparent porosity percent for the PCC matrix decreased after the first cycle and then a very slight increase was detected (+4%) after 90 cycle, (Table 5). This may be explained on the basis that: the polyester-styrene binder is a thermoplastic resin and due to the heating effect during the thawing cycles (+60°C) enhancement in the polymerization may take place and, consequently, lower porosity is reached (Table 5). The slight increase in the apparent porosity after four FTC may be attributed to the deterioration that may start in the polymer-cement composite due to the repeated freezing and thawing effect.

The advantages of the addition of polymers to PC is reducing the water absorption, total porosity, and volume of open pores. Both water absorption and apparent porosity percentages were improved for PCC after 115 cycles compared to the plain cement. It is thought that this comes about by the effect of polymer on the size of the calcium hydroxide (CH) crystals formed during the hydration of C_3S, which in turns reduces this formation during the crystallization process, as well as the air entrainment and/or the filling effect introduced by the polymer and surfactants as previously stated (Eskander et al., 2006).

FTC	Composition	Exterior volume, (cm³)	Impervious Volume, (cm³)	Volume of open pores (cm³)	Apparent Porosity, %	Water absorption, (%)	Apparent specific gravity	Bulk density, (g/cm³)
0	PCC	26.85	18.05	8.8	32.77	20.28	2.4	1.62
	PC	26.40	16.55	9.18	35.67	21.8	2.54	1.47
1	PCC	25.55	18.74	6.81	26.67	15.08	2.41	1.77
	PC	25.44	16.56	8.89	34.95	21.17	2.53	1.65
4	PCC	25.7	18.28	7.42	28.25	16.68	2.43	1.73
	PC	25.86	16.85	9.00	34.82	20.99	2.54	1.66
7	PCC	24.94	17.51	7.43	29.81	16.61	2.56	1.8
	PC	26.04	16.96	9.08	34.87	20.8	2.57	1.66
14	PCC	26.02	18.34	7.57	29.2	16.43	2.46	1.74
	PC	26.09	17.28	8.82	33.8	20.17	2.6	1.68
28	PCC	25.32	17.69	7.63	30.13	17.3	2.49	1.74
	PC	25.65	16.19	9.46	36.87	20.4	2.6	1.65
54	PCC	25.09	17.34	7.65	30.52	17.41	2.52	1.75
	PC	25.54	16.55	8.98	35.78	20.96	2.59	1.68
90	PCC	25.54	17.17	7.36	30.76	18.71	2.61	1.75
	PC	25.84	16.52	9.33	36.09	21.69	2.64	1.67
115	PCC	24.89	17.83	7.05	30.3	18.36	2.40	1.82
	PC	24.71	16.72	9.98	36.31	21.04	2.51	1.7

Table 5. Some physical properties of PC and PCC subjected to freezing/thawing test

Certain minimum values of the mechanical integrity are required during long term stress of storage and disposal. An improvement in the compressive strength measurements was observed for polymer-modified cement candidate matrix up to 54 FTC, i.e., more than 35% compared to unmodified specimens (Table 6). On the other hand, after 115 FTC a slight decrease in the compressive strength values ≈12% of the candidate matrix was recorded, (Table 6). However, it should be mentioned that even after 115 cycles of freezing-thawing treatment at -19°C and +60°C, the compressive strength figures reached by the proposed matrix are still higher than that recommend by the Nuclear Regulatory Commission (NRC) for transport and disposal requirements (Eskander et al., 2006). On the other hand, obvious deterioration in the plain cement blocks was recorded due to the freezing/thawing treatment (Table 6). Similar trends were obtained previously in the literature. This improvement in mechanical integrity of the cementitious system is usually controlled by pore structure, especially porosity. The addition of polymer to the cement has the entrained air effect in the mixture that stabilized as small bubbles. The latter have a reproducible pore structure as well as good dispersion of air in the modified cementitious system and reflects a good frost resistance character. On the other hand, the increase in the compressive strength measurements of the candidate polymer-cement matrix up to 54 cycles (Table 6), can be attributed to the increase in the cross-linking of the polyester-styrene during the thawing cycles as previously stated (Eskander et al., 2006).

FTC	Compressive strength, MPa	
	PC	PCC
0	34.4	33.2
1	33.8	33.2
4	31.3	33.4
7	24.2	33.5
14	25.7	35.6
28	25.8	37.3
54	25.6	40.1
90	26.7	35.4
115	25.3	34.3

Table 6. Effect of freezing-thawing cycles on the compressive strength values of PC and PCC

7. Behavior of solid waste form during a scenario of flooding accident in the disposal site

The behavior of the PCC waste form during an accident of flooding disposal sites with any of three different types of water (tapwater, groundwater and seawater) was followed. Mechanical integrity, weight change, porosity, bulk density, water absorption, specific gravity, . . . were selected for laboratory testing under the immersion conditions in any of the three different types of water during a period of 260 days (Tawfik & Eskander, 2007).

Water is at the heart of most of the physical and chemical causes underlying the deteriorations of concrete structures. Cement-based solid products exposed to water undergo carbonation process that refers to the reaction of the carbon dioxide, in the surrounding, with water to form carbonic acid, which then combined with calcium ions leached from the hydrated cementitious materials to form insoluble calcium carbonate salts ($CaCO_3$). A white thin layer (≈ 0.1 mm thickness) started to appear very clearly on the surface of solid PCC blocks immersed statically in seawater after 30 days. On the other hand, by the visual examination of PCC blocks immersed in groundwater, this layer was less obviously seen, while for those immersed in tapwater this layer was hardly detected even after 260 days of immersion. This may be due to the high salt contents of both seawater and groundwater that affect highly the leaching of lime [$Ca(OH)_2$] from the surface of the PCC and its reaction with the carbon dioxide, of surrounding, forming the thin layer of calcite ($CaCO_3$). To confirm this finding, this thin layer was scratched from the surface of PCC blocks that immersed in seawater for 260 days, and subjected to XRD and the obtained diffractogram is represented in Fig. 6. Peaks at d = 1.63, 1.93, 2.63, 5.42 A° . . . are refer to the lime [$Ca(OH)_2$] leached from the PCC blocks, while that at d = 4.91 A° may attributed to the formed calcite ($CaCO_3$). The proposed reactions for the formation of this layer can be illustrated as follows. The slight acidity of naturally occurring water is generally due to the dissolved carbon dioxide (CO_2), which is found in significant concentration in mineral, seawater, and groundwater as previously stated (Tawfik & Eskander, 2007).

$$CO_2 + H_2O \rightarrow H_2CO_3$$

A cation exchange reaction between carbonic acid and calcium hydroxide $Ca(OH)_2$ leached from the hydrated PCC processed as follows:

$$Ca\left(OH\right)_2 + H_2CO_3 \rightarrow CaCO_3 + 2H_2O$$

This leads to the precipitation of the insoluble calcium carbonate salts ($CaCO_3$) and step further ingress of water into the interior of PCC products through reduction in permeability. This kind of protection would be available under static conditions. Nearly, similar behavior for the carbonation of concrete components was discussed in the literature (Tawfik & Eskander, 2007).

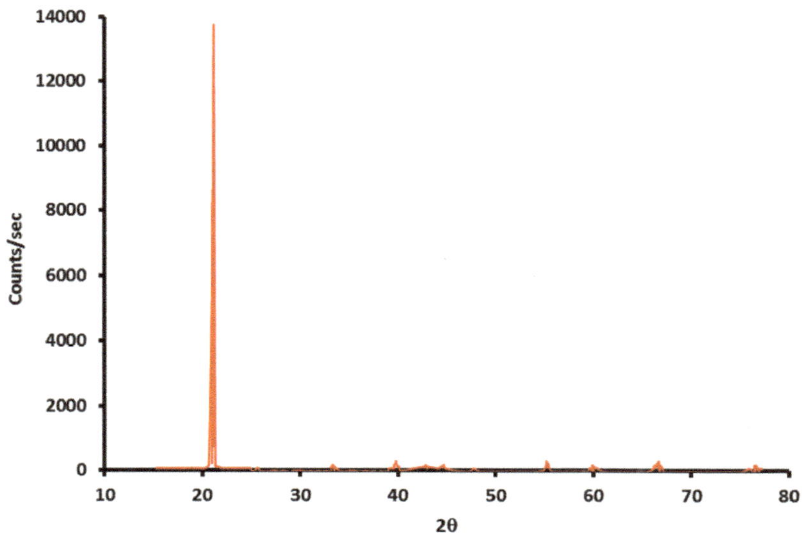

Figure 6. X-ray diffraction of the white thin layer scratched from the surface of polymer-cement composite (PCC) blocks immersed in seawater after 260 days

Porosity and other relating parameters for PCC blocks immersed in the three different types of water and for different periods were determined according to ASTM C20-74 based on BWS technique (Tawfik & Eskander, 2007) and the data obtained are represented in Table 7 for tapwater, seawater and ground water, respectively. Generally, it was observed that both porosity and water absorption percentages were increased for the PCC waste form immersed in tapwater to reach 39.55 and 24.44%, respectively, after 260 days of immersion at 30 ± 5°C compared to 32.77 and 20.28% for control non-immersed specimens (Table 7).

Immersion period, day	Immersion media	Exterior volume, (cm^3)	Impervious Volume, (cm^3)	Volume of open pores (cm^3)	Apparent Porosity, %	Water absorption, (%)	Apparent specific gravity	Bulk density, (g/cm^3)
0	----	26.85	18.05	8.8	32.77	20.28	2.4	1.62
7	Tapwater	27.32	18.12	9.20	33.67	20.99	2.42	1.60
	Seawater	26.32	16.68	9.68	36.78	22.54	2.58	1.63
	Groundwater	28.17	20.2	7.97	34.71	20.84	2.18	1.56
30	Tapwater	24.99	16.47	8.52	34.09	19.96	2.59	1.71
	Seawater	25.55	16.87	8.68	33.50	20.15	2.55	1.69
	Groundwater	23.7	15.54	8.16	34.43	19.24	2.73	1.79
45	Tapwater	25.16	16.40	8.99	35.41	20.58	2.66	1.72
	Seawater	25.53	17.66	7.87	30.83	17.62	2.53	1.75
	Groundwater	25.55	17.04	8.72	33.85	19.81	2.58	1.71
60	Tapwater	25.19	16.02	9.17	36.4	21.5	2.66	1.69
	Seawater	26.58	18.26	8.32	31.30	19.31	2.36	1.62
	Groundwater	26.64	17.92	8.72	32.73	19.82	2.45	1.65
260	Tapwater	25.26	15.59	10.2	39.55	24.44	2.68	1.62
	Seawater	26.63	19.66	6.97	26.17	15.36	2.31	1.7
	Groundwater	24.96	16.99	7.97	31.93	18.12	2.59	1.76

Table 7. Physical properties of polymer-cement composite (PCC) subjected to immersion test at different periods and in different leachants

On the other hand, very slight increase in the two parameters was detected for the PCC specimens immersed in groundwater at the same conditions of immersion up to 30 day. However, it should be noted that the values of porosity and water absorption percentages were declined again to reach 31.93 and 18.12%, respectively, after 260 days compared to 32.77 and 20.28% for non-immersed blocks. For the PCC blocks immersed in seawater, slightly different trend was observed where a detectable increase in porosity and water absorption percentages were occurred within the first 7 days, then a high decrease took place to reach a minimum value after 260 days (i.e. 26.17 and 15.36%, respectively) (Table 7). Generally speaking, the increase in the porosity values may refer to the leaching of the cementitious materials due to the diffusion of water into the pores of PCC specimens and removing lime and calcium sulfate components (Tawfik & Eskander, 2007). While the reduction in porosity values for the PCC blocks immersed in both groundwater and seawater afterwards can be attributed to the carbonization process where the calcite protective layers claim to be formed and fill the pores of immersed blocks as previously described. The pronounced reduction in the porosity and water absorption percentages for the PCC immersed in seawater may be due to the high salinity of the sea leachant solution compared to tapwater and groundwater. A direct relationship can describe the mechanical integrity of the concrete composite from the knowledge of its pore system characteristics (Tawfik & Eskander, 2007). The results obtained from the porosity and pore parameters measurements were confirmed by the compressive strength measurements for the PCC cured waste forms immersed in the three different leachants for different immersion periods and their strength were measured at the end of each period up to 260 days.

Fig. 7 represents the change in compressive strength of PCC blocks immersed in different types of water for different periods compared to non-immersed ones. It is clear that loss in compressive strength was observed for all PCC specimens immersed in the three leachants for 7 and 30 days. This result confirmed the result obtained from porosity measurements. Similar trend was reached for similar concrete products immersed in groundwater. The loss in compressive strength for blocks immersed in tapwater was continued while gain in compressive strength was obtained for blocks immersed in seawater and groundwater after 45 and 260 days, respectively (Fig. 7). The initial drop, as previously explained, may be due to the leaching of the cement constituents, which results in formation of microvoids, while the gain in compressive strength can refer to the deposition of the insoluble calcite ($CaCO_3$) salt inside the PCC pores leading to its filling. These results are in a satisfactory agreement with previously published work and are confirmed by porosity measurements (Tawfik & Eskander, 2007). However, it is worth mentioning that, in the case of PCC blocks immersed in tapwater even with the loss in compressive strength integrity, the PCC waste forms still had compressive strength values largely greater than that recommended for transportation, handling, and disposal of final radioactive solid waste form. Porosity and mechanical integrity measurements affirm that the PCC waste form made from PC and WEP based on recycled PET candidate for borate waste incorporation can withstand the different disposal environments even in case of flooding the disposal site for long periods.

Theoretically, any environment at pH less than 12.5 is aggressive to the cementitious products due to the reduction in the alkalinity of the pore fluids, which would eventually led to destabilization of the hydration products (Tawfik & Eskander, 2007). However, our candidate PCC waste form shows acceptable properties, visually and experimentally, in the three immersion solutions even at pH values less than 12.5 (Table 8).

The change in chemical constituents of the cementitious waste forms due to leaching has been intensively studied because it is assumed to affect the release of radioactive nuclides to surrounding environment in the repository. The leaching of soluble calcium from 28 days cured PCC blocks immersed in the three leachants for 260 days is shown in Fig. 8. It is clear from the Fig. 8 that the highest release of soluble calcium was detected for PCC blocks immersed in seawater and reached ≈360 ppm after 45 days. This value represents nearly 0.37% of the Ca added as CaO in PC. It is worth mentioning that the release of soluble calcium declines after 45 days to reach only 320 ppm at 60 days showed a tendency to coverage into a constant value for the next 260 days. The decrease in the Ca release may be due to the carbonation process that takes place for the PCC blocks during the immersion in the seawater. However, it could be predicted that, supposing the permanent flooding condition in the disposal site, more than 200 years are needed for the complete depilation of calcium from the PCC blocks. However, it should note that, although the PCC blocks immersed in groundwater and tapwater exhibited the same trend, the released Ca in both cases were less than that released from PCC blocks immersed in seawater (Fig. 8). These results lead to the suggestion that the proposed PCC made from PC at w/c ratio 40% and containing 3% WEP based on recycling PET waste could withstand the worst conditions of flooding in the disposal site.

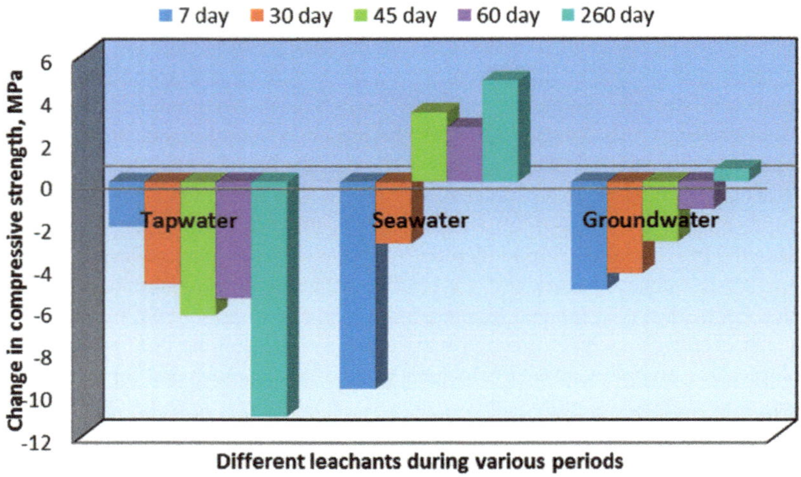

Figure 7. The variation in compressive strength integrity of polymer-cement composites (PCC) due to the immersion period in the different leachants

Immersion period, days	Tapwater	Seawater	Groundwater
0	6.72	8.02	7.2
7	11.1	10.7	10.9
30	11.1	10.9	10.9
45	11.1	11.0	11.0
60	11.1	11.0	11.0
260	11.1	11.0	11.0

Table 8. The pH of leachant solutions at the end of different immersion periods

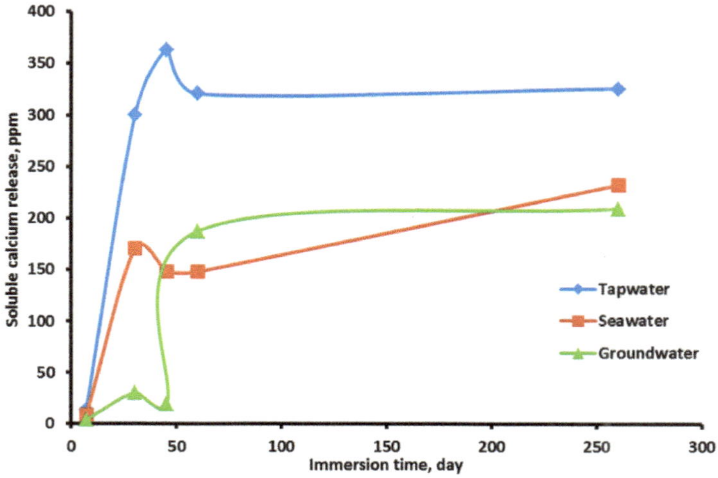

Figure 8. The release of calcium from polymer-cement composite blocks immersed in the leachant solutions versus the immersion periods

Fig. 9(a,b) shows the scanning electron micrographs (2500x magnification) of portion of a laboratory scale non-immersed plain cement, (w/c = 0.40) and polymer-cement composites (w/c = 0.40 and c/p = 3%) samples. Fig. 9(c–e) represents scanning electron micrograms of polymer-cement composite specimens, having the same composition but immersed in tapwater, groundwater and seawater, respectively for 260 days. C-S-H, which represents the main hydration product of the PC, has foil and plate like crystals (Fig. 9(a)). On the other hand, portlandite (CH), i.e. crystalline calcium hydroxide, which represents the second most abundant product in the hydrated cement paste, appears in the form of relatively large crystalline aggregate and supposed to be intimately mixed with the C-S-H. Their entanglement makes it hard to define their exact shape (Fig. 9(a)). It is clear from Fig. 9(b) that the surface of the non-immersed PCC based on PET waste is characterized by a continuous large polymer, mutually intergrown with considerable number of compact points and areas connecting the cement paste and forming the binding phase of the product (Tawfik & Eskander, 2007). Therefore, the hydration products of cement formed in the polymer layers cannot be distinguished out right by their appearance (Fig. 9(b)). The fracture of the polymer-cement composite samples after immersion in tapwater is characterized by porous surface (Fig. 9(c)), and its feature after that is comparable to the analogous sample before immersion (Fig. 9 (b)). On the other hand, the PCC specimens immersed for the same period in groundwater and seawater described by peculiar qualities to each other (Fig. 9(d,e)) but differ from these immersed in tapwater (Fig. 9(c)), in the predominant presentation of calcite crystals. However, it is worth mentioning that the calcium carbonate crystals in the micrograph of samples immersed in seawater (Fig. 9(e)) are abundant, widespread and lump. SEM imaging gives detailed information about the structure of the specimens of PCC before and after immersion in the three leachants, and consequently confirms the physicomechanical results obtained for the behavior of the PCC waste forms in the disposal environment.

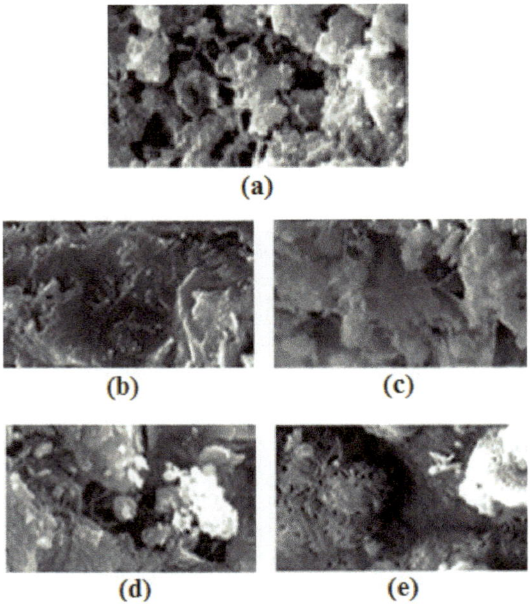

Figure 9. Scanning electron micrographs of Portland cement (PC), polymer-cement composite (PCC) before and after immersion in different leachants for 260 days. (a) PC non-immersed; (b) PCC non-immersed; (c) PCC in tapwater; (d) PCC in groundwater; (e) PCC in seawater

8. Performance of physical and chemical stability of aged polyester-cement composite containing borate waste simulate

8.1. Mechanical and physical characterizations

Much attention has been paid to study the long-term performance of polymer-cement composite from the point of view of durability related to mechanical characterizations. Fig. 10 represents the variation in compressive strength values of the aged final waste form specimens immersed in different water compositions. After 260 days of water exposure, the 7 years aged specimens exhibited a decrease in their mechanical durability through fluctuations in strength. The samples immersed in both seawater and groundwater showed greater fluctuations in strength compared to that dipped in tapwater. Loss of strength could be attributed to the formation of gypsum and ettringite in the pores of specimens (Thokchom et al., 2010). However, it is worth mentioning that the candidate polymer-cement composite immobilizing the borate waste simulate had compression strength greatly exceeds the value specified by the Nuclear Regulatory Commission (NRC) technical position paper for stabilized low-level radioactive waste forms even after 7 years aging and nearly 260 days of immersion in different compositions of water (Eskander et al., 2012).

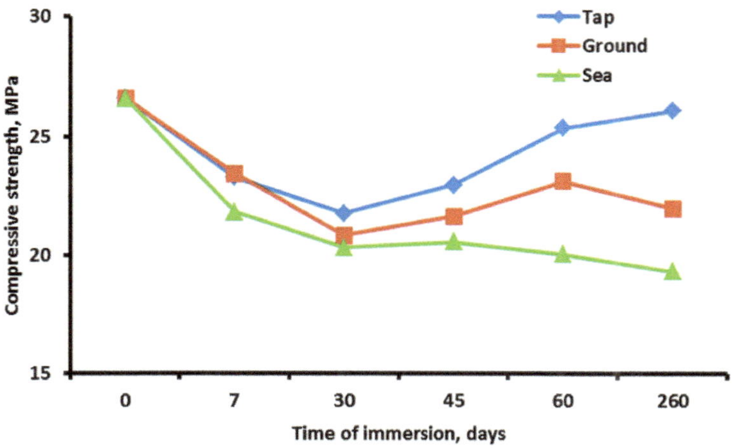

Figure 10. Compressive strength of aged PCC specimens immersed in different water compositions

The performance of plain polymer-cement composite [Portland cement (PC) and the recycled Poly(ethylene terephthalate)] incorporated borate waste simulates, which aged for 7 years immersed into the three types of water for 260 days was assessed. The internal microstructures and morphology of the aged polymer-cement composite incorporating borate waste simulate before and after immersions in various water types are represented in Fig.s 11, 12, respectively. Fig. 11 shows the plate like crystals of $Ca(OH)_2$, which found in parallel towards each other, and they are almost undistorted (Fig. 11a). The crystals are arranged in some sort of stack. The polymer films stretched and covered the C-S-H phases (Fig. 11b), giving them a more smooth texture.

Fig. 11c,d exhibits pronounced randomly oriented calcite crystals. At the end of the immersion period for the aged final waste form in various leachants, abundant C-S-H phases coated by thin polymer film were detected (Fig. 12a). The thin polymer film was also observed around the needle-like structure of ettringite that assumed to be recrystallized inside the final waste form holes due to the immersion in seawater and groundwater (Fig. 12b). Even after 260 days of immersion process, polymer was detected between $Ca(OH)_2$ crystals and may act as a gene bonding for the crystals together. In addition, a large amount of ettringite crystals was distributed in-between. It is worthwhile mention that, the solubility of the ettringite was much lower than that of portlandite and hence the former is predominant in the immersed samples (Fig. 12c).

Gypsum crystals could be distinguished in final waste forms immersed in seawater and groundwater due to their measurable contents of sulfate ions (Fig. 12d). Furthermore, this Fig. shows the gluing action of the polymer. Gypsum and ettringite are known by causing expansion in the polymer-cement composite immersed in groundwater or seawater and accordingly, some internal cracks could be present, (Fig. 12b). The compressive strength could be decreased due to the expansion cracks in the final waste forms. Similar trend was described by Thokchom et al. (Thokchom et al., 2010). The data recorded by SEM could confirm the adequate performance of the proposed polymer-cement composite based on the recycled PET waste to incorporate the borate waste up to 4% even for long time and under flooding condition for 260 days.

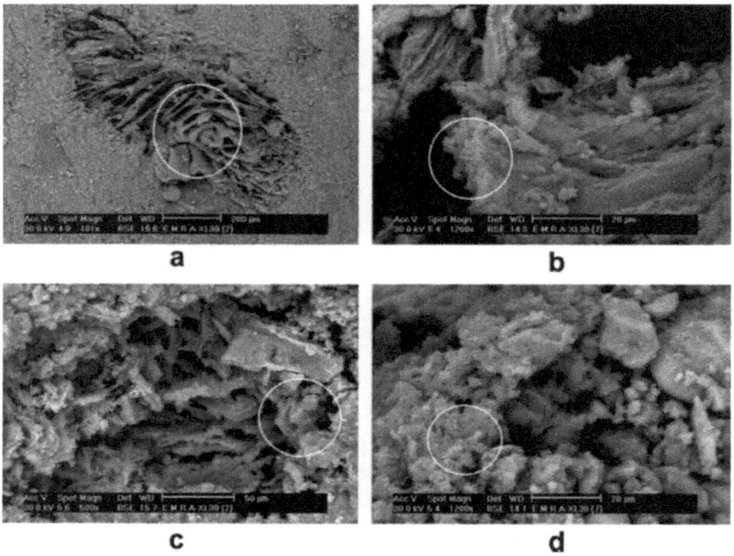

Figure 11. Scanning electron photomicrograph of the internal microstructure of 7 years aged polymer-cement composite incorporating borate waste simulate

Aged polymer-cement composite incorporating borate waste simulate did not show any changes in its shape and remained structurally intact without visible cracks even after immersion in the three types of leachants for 260 days. Specimen's surfaces received white deposits throughout the water exposure period while rare and minimum white deposits were appeared in case of tapwater. On the other hand, the maximum deposits were observed on the blocks dipped in seawater. These deposits were powdery and harden with time. The typical images of such structural formations are shown in Fig. 13, which revealed the photographs of the specimens after 260 days exposure to different water types tap (a), ground (b) and sea (c), respectively.

8.2. Chemical characterization

The long-term chemical stability of the polymer-cement composite waste (PCC) aged for 7 years under normal weather conditions and then exposed to different leachants was studied. Cumulative fraction leached of radiocontaminants and mass changes of the waste form during leaching periods were addressed (Saleh et al., 2011).

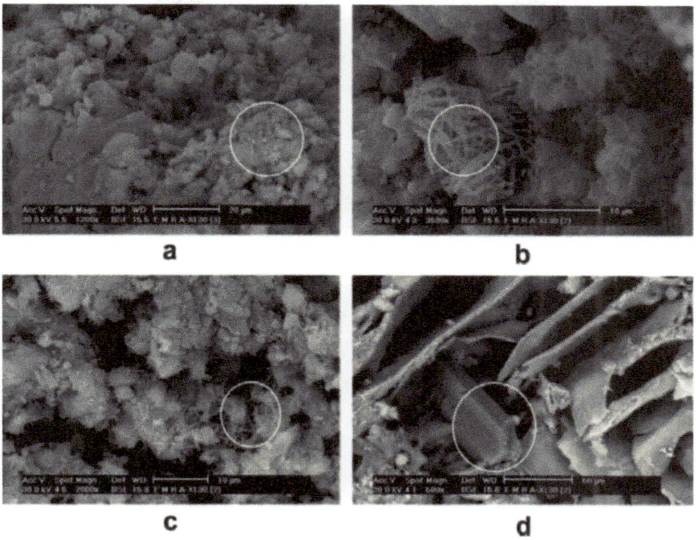

Figure 12. Scanning electron photomicrograph of the internal microstructure of 7 years aged polymer-cement composite incorporating borate waste simulate and immersed in different leachants for 260 days

(a) Immersion in tapwater (b) Immersion in groundwater (c) Immersion in seawater

Figure 13. The 7 years aged polymer-cement composite incorporating borate waste simulate after immersion in different leachants for 260 days

Development of leach resistant waste forms is a major area of research in the field of radioactive waste management in order to minimize the environmental impact through the back release of the hazardous materials. The leaching mechanisms of simulated low and intermediate level radioactive waste forms immobilized in polymer-cement composite are being determined as a support for the development of their chemical stability under different leaching parameters such as static and dynamic conditions, various dipping media, environment temperatures, etc. However, it should be noticed, and to the best of our knowledge, that no data have been published, which provides a basis and direct comparison of the leaching behavior of the candidate final waste form aged for 7 years before subjected to leachants flooding under previously stated conditions, which is supposed to play a role in the long-term durability of the prepared composite.

8.2.1. Static and dynamic leaching conditions

PCC samples spiked with ^{137}Cs and ^{60}Co were suspended in the center of a closed leaching jar containing groundwater (leachant volume of 150 ml) under static conditions (i.e. no agitation, no movement and the leachant was not replaced by fresh one during the whole test period). The whole volume of the leach solution was periodically withdrawn out of the leaching jar, (Fig. 14), subjected to counting and then returned back again. While in dynamic experiments, similar PCC blocks were treated in the same manner, except, the leaching solution was replaced periodically by fresh one at time intervals parallel to the sampling in a static test.

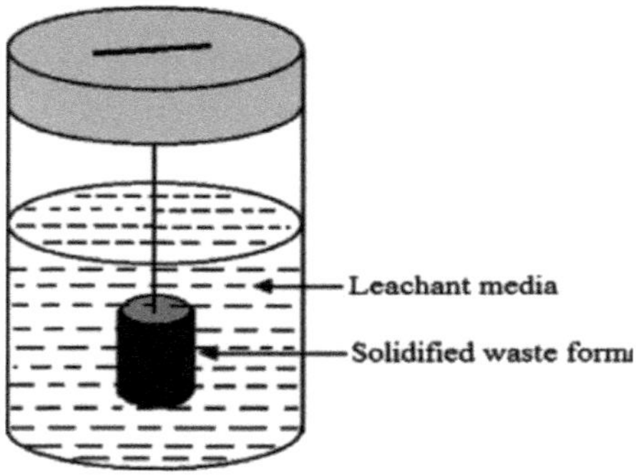

Figure 14. Schematic diagram of suspended solidified waste form during leaching test

The leachability, of ^{137}Cs and ^{60}Co radionuclides from the PCC monolith waste form, was calculated based on the upcoming relations.

Cumulative fraction leached (CFL) which is the sum of all fractions leached during the time intervals $\sum A_n/A_o$. S/V, was plotted versus the square root of time in days ($t^{1/2}$) where; A_n is the activity of radionuclides released during the particular leaching period with index n, A_o is the initial activity of radionuclides present in the borate waste immobilized in PCC and t_n is the elapsed time of the leaching period n in days. S and V are the surface area and volume of the tested block, respectively (Saleh et al., 2011). The counts of radionuclides were determined using multichannel analyzer, PCAP, USA. From the plot of the cumulative fraction leached as a function of the square root of time in the diffusion region of leaching, slopes were calculated and applied to determine the diffusion coefficient using the following formula: $De = \pi(mV{=}2S)^2$ where; De is the apparent diffusion coefficient (m^2/s), m is the slope of the curve demonstrating the $\sum A_n/A_o$ versus the square root of leaching time in seconds, V is the sample volume (cm^3) and S is the sample surface area (cm^2).

On the other hand, leach index (Lx) which denotes a direct measure of the extent of leaching of radionuclides from the candidate polymer-cement composite, was computed based on the relation, $Lx = \log(1/De)$.

For all leaching curves, it should be notified that the leaching behavior of radionuclides from the monolith waste forms revealed two distinguished performances. The first initial period characterized by a high leachability of radionuclides relative to the second final one. This may be explained by the fact that up to nearly 25 days from the starting time, the high release may be attributed to the washout of loose radioactivity from the surface of PCC waste form (Saleh et al., 2011). This can be due to the relatively rapid washing process occurred when the blocks were firstly exposed to the leaching solution. This was followed by nearly a steady state diffusion controlling periods that persist. Leaching solution velocity (m) is defined by the volume (v) of the leachant contacted with solid waste per unit of surface area (S) and unit of time (t) given as: $\nu = v/(S \cdot t)$.

At a low leaching solution velocities, i.e. static condition, where the leachant was not replaced by fresh one, the amount leached approaches the saturation limit and leaching solution allowed to equilibrate with waste forms. On the other hand, under dynamic conditions where the leachant was periodically replaced maximum saturation limits were not obtained (Saleh et al., 2011) and consequently, higher leach rate was reached as presented in Fig. 15a,b. In the forthcoming leaching experiments, the dynamic condition will be performed.

8.2.2. Chemical behavior of PCC in various leaching media

The long-term durability of the polymer-cement composite under the imposed disposal conditions is a key factor controlling whether the proposed PCC can be accepted as a solidification matrix for borate waste or not. One of the major concerns that meets the cementitious waste form products in the disposal site is the exposure to an aggressive flooding environment. The leaching test was used to predict the long-term durability of the waste forms when exposed to aqueous media.

Polymer-cement waste forms are alkali in nature, and they are vulnerable to acid attack either from acid rains, seawater or groundwater that may be reached the disposal environment.

Seven years aged PCC blocks, spiked with ^{137}Cs and ^{60}Co, were suspended in separate jars each of which contains one of the following leachants, 0.1 M H_2SO_4, groundwater, seawater or tapwater. Cumulative fractions leached of the two radionuclides in the different leaching solutions were plotted as a function of leaching time and the data obtained are shown in Fig. 16a,b. Both ^{137}Cs and ^{60}Co recorded the highest leachability in 0.1 M H_2SO_4 followed by seawater and then groundwater while the leachability in tapwater was the lowest.

In sulfuric acid medium, the dissolution of calcium hydroxide crystals and the highly decalcified calcium silicate hydrate (CSH) gels due to the acid attack result in increasing porosity and enlarged capillary pores in the leached specimens. This causes self-accelerating leaching (Saleh et al., 2011). Due to exposure to sulfuric acid for 260 days, the 7 years aged

PCC solidified waste forms showed some swelling and corrosion due to the formation of a gypsum layer on the surface. It is worth mentioning that the visual appearance of the blocks, shown in Fig. 17 is in consistence with the leaching rate results and confirms the data obtained for the high leachability of both radionuclides in sulfuric acid (Fig. 16a,b).

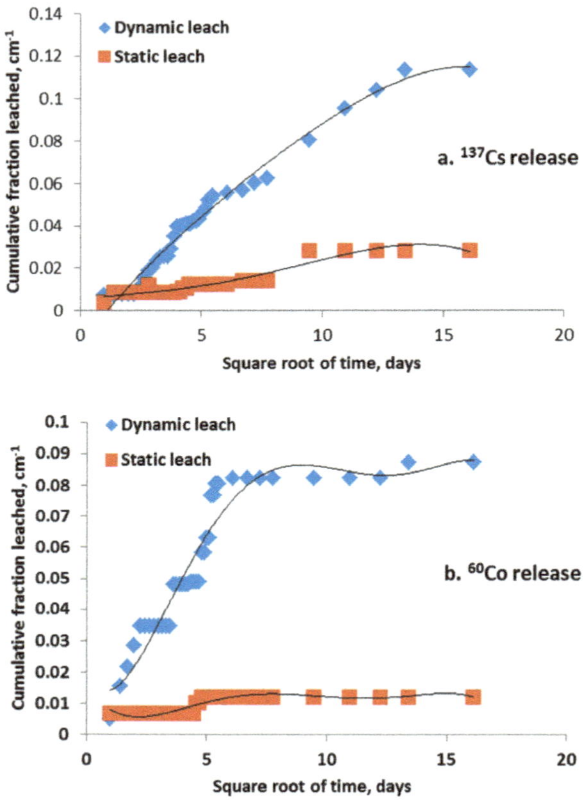

Figure 15. Cumulative fraction leached of ^{137}Cs and ^{60}Co released in groundwater under static and dynamic leach conditions at room temperature (25 ± 5°C)

The attack caused by solutions containing sulfate, e.g. seawater or groundwater can have the same diverse effects on the cementitious-base waste form. The effect of sulfate in both media may be associated mainly with the expansion of the gypsum and ettringite that can be formed as a result of contact with such leachants, as shown in the following formula:

$$\underset{\text{portlandite}}{Ca(OH)_2\,(s)} + H_2SO_4\,(a) \rightarrow \underset{\text{gypsum}}{CaSO_4{\cdot}2H_2O\,(s)}$$

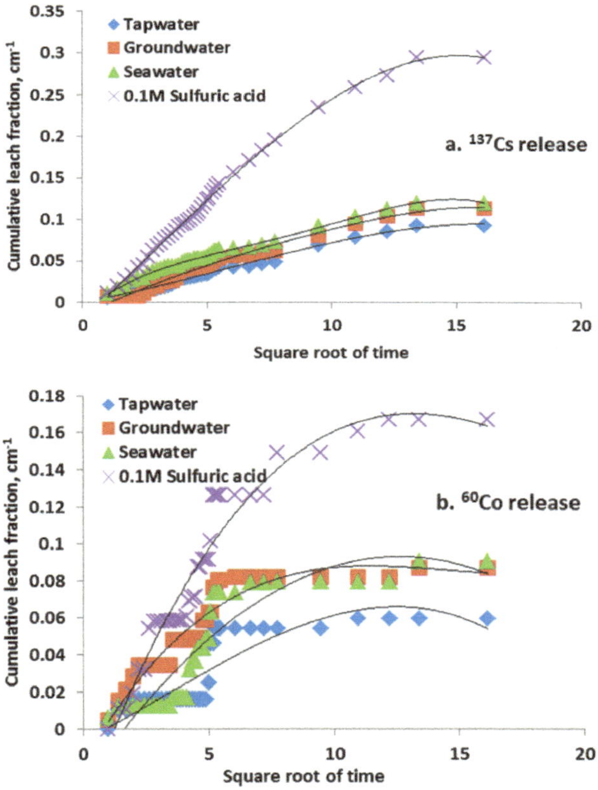

Figure 16. Cumulative fraction leached of ^{137}Cs and ^{60}Co released in different leachants at room temperature (25 ± 5°C)

Before leaching **After leaching**

Figure 17. Polymer-cement composite waste form (PCC) before and after immersion in 0.1 M H_2SO_4 for 260 days

$$3(CaSO_4{\cdot}2H_2O)\ (s) + 3CaO{\cdot}Al_2O_3{\cdot}12H_2O\ (s) + 14H_2O\ (a) \rightarrow 3CaO{\cdot}Al_2O_3{\cdot}3CaSO_4{\cdot}32H_2O\ (s)$$

gypsum hydrogarnet ettringite

The expansion of these two abrasive salts can cause cracking and consequently, increase in the exposed surface area of the PCC blocks to the leaching solution and accordingly results in higher leach rates (Saleh et al., 2011). Due to the absence of the most aggressive components (e.g. sulfate or chloride) in tapwater, the ^{137}Cs and ^{60}Co leachability represented the lowest rate even after more than 260 days (Fig. 16a, b). Hence, when PCC present in places flooded by tapwater or non-acidic rain water, reasonable leaching performance could be expected.

8.2.3. Effect of temperature on the leaching performance of the PCC

Waste forms may be subjected to severe changes in temperature during transport, storage or final disposal. The variation in temperature may affect the degradation of structural integrity of the final waste forms and consequently, their chemical stability. Fig. 18 describes the cumulative fraction leached, (cm^{-1}) of ^{137}Cs (a) and ^{60}Co (b) for PCC blocks aged for 7 years and then dipped separately in groundwater. The leaching process was performed at three different temperatures, namely, at $4 \pm 1^{o}C$ (in a refrigerator), room temperature ($25 \pm 5^{o}C$) and at $60 \pm 2^{o}C$ (in an oven). The obtained data showed that, as the temperature was increased from $4^{o}C$ to $60^{o}C$, the fractions leached for both radiocesium and radiocobalt were escalated. This effect of the temperature is in a good agreement with the results reported in previous reported work (Saleh et al., 2011). The high temperature may cause an increase in pore diameter dimension of the polymer-cement composite and consequently, raise their porosity that leads to elevation in the effective diffusion coefficient of both radionuclides from the final waste forms. On the other hand, when leaching was carried out at a lower temperature ($4^{o}C$), the water within the pores of the waste forms may be nearly frozen and consequently, the infiltration of groundwater leachant inside the PCC microstructures was highly slowed down and accordingly, the leachability of both radiocesium and radiocobalt was decreased (Fig. 18a,b).

8.2.4. Effect of volume of leachant to surface area of specimens on chemical durability of PCC

Leachant volume to surface area of PCC, (V/S), is an additional important factor that influences the leachability of radiocesium and radiocobalt from the final waste forms and consequently, their chemical long-term stability in the disposal site. Two of 7 years aged PCC similar specimens dopped with ^{137}Cs and ^{60}Co were hang in two leaching jars having various volumes of groundwater leachant. The duration of the leaching process was more than 260 days and the cumulative fractions leached for ^{137}Cs and ^{60}Co were calculated periodically and shown in Fig. 19a,b. The V/S ratio in repository environment will certainly affect the leaching behavior of radionuclides from waste form monolith. This was clearly illustrated by the data obtained where leaching of ^{137}Cs and ^{60}Co radionuclides was directly proportional to V/S values.

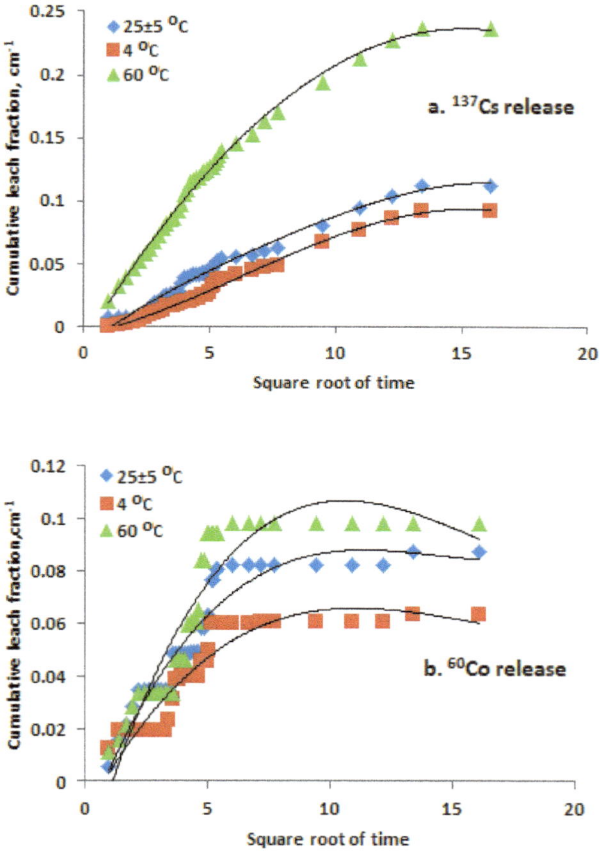

Figure 18. Cumulative fraction leached of ^{137}Cs and ^{60}Co released in groundwater as a function of leaching time at different temperatures

8.2.5. Mass changes and structural stability of PCC

The mass changes of the 7 years aged polymer-cement composite specimens during the immersion in the three leachant solutions are represented in Fig. 20. All the blocks recorded an increase in their masses over 260 days of exposure. The pattern of the mass gain was nearly identical to the three various immersion solutions. The highest gain in mass was recorded for blocks dipped in groundwater (10% after 260 days). While the lowest increase in mass was occurred for the specimens sank in tapwater. Insignificant difference in mass gain was observed for blocks immersed in seawater and those in groundwater (Fig. 20). Generally speaking, the increase in the mass of the specimens due to dipping in different leachants may be referred to the water absorbed by the slightly porous polymer-cement composite. On the other hand, the relatively high increase in mass of the PCC samples due to immersion in groundwater or seawater may be attributed to white salt deposits formed on the surface of the specimens. Similar trend was reported by Thokchom et al. (Thokchom et al., 2010). However, it is worth noting that even with this relatively high mass gain (about 10%) all the 7 years aged blocks sustained their formal structure under complete dipping in the aggressive leachant solutions where no microcrakings were observed for all tested blocks for more than 260 days of immersion. It is worth mentioning that, when the final waste form maintained their dimensional integrity, during leaching, it was indicated that internal bulk diffusion is the most likely rate determining mechanism during the phases of the leaching process (Saleh et al., 2011).

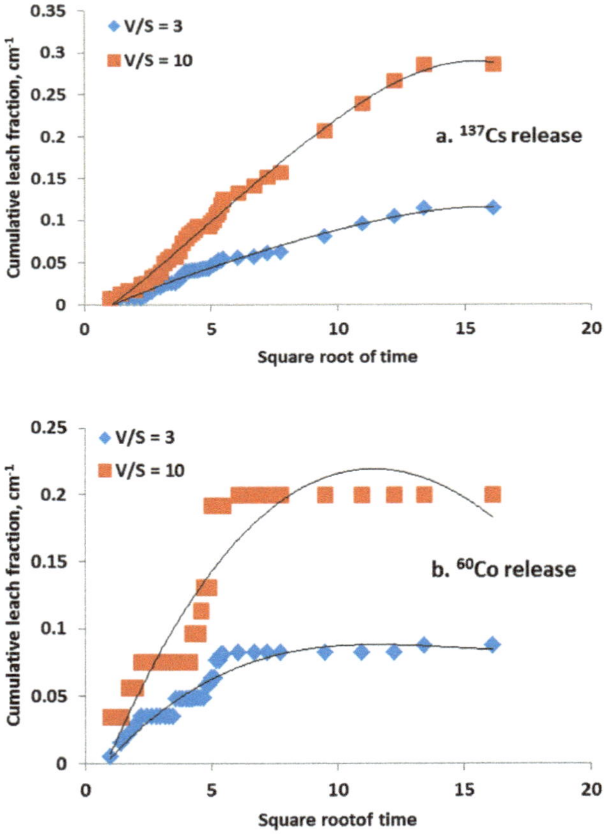

Figure 19. Cumulative fraction leached of ^{137}Cs and ^{60}Co released in groundwater at different ratios of leachant volume to surface area of waste form (V/S), under the room temperature (25 ± 5°C)

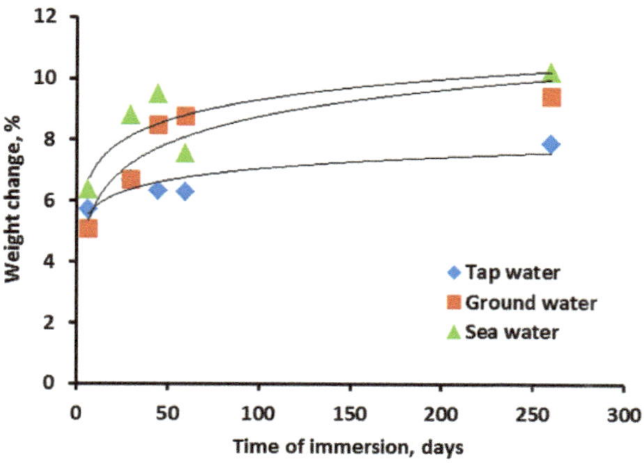

Figure 20. Mass change percent due to immersion period in various media

8.2.6. Diffusion coefficient and leach index

It is clear from all leaching curves that, after nearly 30 days a plateau of stabilization of the cumulative fraction leached is reached for the PCC blocks under all various studied conditions, which shows that the only leaching mechanism of the two radionuclides is controlled by diffusion mechanism.

Based on the diffusion coefficient (De), the leachability index (Lx) is calculated to characterize the efficiency of the candidate matrix to solidify the problematic borate waste concentrate. The value 6 of leach index is the threshold value to accept a given formulation as adequate for certain solidification/stabilization of nuclear wastes (Saleh et al., 2011). The higher values of the leachability index represent the lower release of the radionuclides from the monolith waste forms. The diffusion coefficients (De) in (m^2/s) together with leach indices data for the various leaching conditions are represented in Table 9.

Cesium, being the most mobile waste soluble species in the nuclear waste stream, is poorly adsorbed in the basic environment of cement (Saleh et al., 2011). On the other hand, cobalt may form insoluble hydroxide at this higher pH condition. Therefore, the leachability of ^{137}Cs was greater than that of ^{60}Co from the PCC-waste form immersed in the same leachants. The data in Table 9, indicated that all leach indices of either ^{137}Cs or ^{60}Co in various leachants and under different conditions were usually greater than the value of 6 (Saleh et al., 2011). The relative high Lx values may be attributed to the role played by polymers within the PCC in reducing the porosity of the final waste forms and hence retarding the release of radiocontaminants. Recently, it was reported that more than 60% of cesium activity was released from immersed PCC-waste form cured for 28 days only (Saleh et al., 2011). This value can be easily compared to 11.3% of radiocesium leached after 260 days from the aged PCC under investigation. This assures the task achieved by the polyester based on the recycled PET waste in increasing the chemical stability of the proposed PCC even after 7 years of aging and immersion for 260 days.

Based on all leach tests data obtained, it could be stated that the candidate polymer-cement composite is an acceptable solution for immobilizing the radioactive borate waste even after 7 years of aging and under a breakthrough of water into the disposal site.

Leaching Conditions	^{137}Cs		^{60}Co	
	D_e (m^2/s)	L_x	De (m^2/s)	L_x
0.1M acid*	1.02E-13	8.99	2.105E-14	9.67
Seawater*	1.29E-14	9.89	7.255E-15	10.14
Groundwater*	1.24E-14	9.91	2.786E-16	11.56
Tapwater*	9.49E-15	10	2.232E-16	11.65
4°C*	1.06E-14	9.98	6.509E-17	12.19
60°C*	3.48E-14	9.46	3.935E-17	12.41
Static condition*	1.03E-15	11	7.32E-15	10.14
V/S = 10	9.6E-14	9.02	3.327E-18	13.48

* leachant volume/surface area of waste form (V/S) =3

Table 9. Diffusion coefficients and leach indices at different leaching conditions

9. Conclusion

In conclusion, it could be stated that the polymer-cement composite prepared from PC premixed with WEP based on the recycled poly(ethylene terephthalate) wastes is a promising and economical matrix for incorporating borate waste solution originating from the primary coolant circuit of the pressurized water reactors. Also, the proposed treatment process serves in saving the environment from two types of wastes at the same time, namely the hazardous radioactive borate waste and PET solid waste.

Author details

Hosam El-Din Saleh, Talat Bayoumi and Samir Eskander

Radioisotope Department, Nuclear Research Center, Atomic Energy Authority, Egypt

10. References

Burns, R.H., (1971). Solidification of low and intermediate level waste. *Atomic Energy Review* 9, pp. 547-552.

Drace, Z. & Ojovan, M.I., 2009. The behaviours of cementitious materials in long term storage and disposal: an overview of results of the IAEA coordinated research programme. *Materials Research Society Symposium Proceedings* 1193, pp. 663-672.

Eskander, S.B., Bayoumi, T.A. & Saleh, H.M., (2012). Performance of aged cement–polymer composite immobilizing borate waste simulates during flooding scenarios. *Journal of Nuclear Materials* 420, pp. 175-181.

Eskander, S.B., Tawfik, M.E. & Bayoumi, T.A., (2006). Immobilization of Borate Waste Simulate in Cement-Water Extended Polyester Composite Based on Poly(Ethylene Terephthalate) Waste: 2-Frost Resistance of the Polymer Modified Cement Composite. *Polymer-Plastics Technology and Engineering* 45, pp. 939-945.

Greenhalgh, W.O., (1986). Process for Immobilizing Radioactive Boric Acid Liquid Wastes, US Patent 4595528, USA.

Holcomb, W.F., (1978). *A survey of the available methods of solidification for radioactive wastes.* University of Michigan Library, Washington.

Huang, C. & Yang, W., (1999). Method and agents for solidification of boric acid and/or borates solutions G21F 9/16 ed. Institute of Nuclear Energy Research (Tao Yuan, TW), US Patent 5998690.

IAEA, (1983). Conditioning of Low- and Intermediate-Level Radioactive Wastes. *Technical Reports Series* 222.

IAEA, (1993). Improved cement solidification of low and intermediate level radioactive wastes. *Technical Reports Series* 350.

Peter C, H., (2003). *Lea's Chemistry of Cement and Concrete* (Fourth Edition). Butterworth-Heinemann, Oxford.

Rosato, D.V., Rosato, D.V. & Rosato, M.V., (2004). *2 - Plastic Property, Plastic Product Material and Process Selection Handbook.* Elsevier, Oxford, pp. 40-129.

Saleh, H.M., Bayoumi, T.A. & Shatta, H.A., (2005). Mechanical and Chemical Characterizations of Polyester Modified Cement Waste Forms for Nuclear Wastes. *Isotopes and Radiation Research* 37, pp. 1587-1598.

Saleh, H.M., Tawfik, M.E. & Bayoumi, T.A., (2011). Chemical stability of seven years aged cement–PET composite waste form containing radioactive borate waste simulates. *Journal of Nuclear Materials* 411, pp. 185-192.

Shatta, H.A., (1996). Applications of polymer-impregnated cement in the field of radioactive wastes, Ph.D. Thesis, Science and Education, Chemistry Department. University of College of Women for Arts, Cairo, Egypt.

Sobolev, I.A., Dmitriev, S.A., Barinov, A.S., Varlakova, G.A., Golubeva, Z.I., Startceva, I.V. & Ojovan, M.I., (2006). 39-years performance of cemented radioactive waste in a mound type repository. *Materials Research Society Symposium Proceedings* 932, pp. 721-726.

Tawfik, M.E. & Eskander, S.B., (2007). Immobilization of borate waste simulate in cement–water extended polyester composite based on poly(ethylene terephthalate) waste, part 3: Behavior of solid waste form during a scenario of flooding accident in the disposal site. *Journal of Applied Polymer Science* 104, pp. 4113-4120.

Tawfik, M.E., Eskander, S.B. & Bayoumi, T.A., (2005). Immobilization of Borate Waste Simulate in Cement-Water Extended Polyester Composite Based on Polyethylene Terephthalate Waste 1-Mechanical Properties of the Final Waste Forms. *Polymer-Plastics Technology and Engineering* 44, 1355-1368.

Thokchom, S., Ghosh, P. & Ghosh, S., (2010). Performance of Fly ash Based Geopolymer Mortars in Sulphate Solution. *Journal of Engineering Science and Technology Review* 3, pp. 36-40.

Varlakova, G., Golubeva, Z., Barinov, A., Roshchagina, S., Dmitriev, S., Sobolev, I. & Ozhovan, M., (2009). Evaluation of the properties of cemented radioactive wastes with prolonged testing in mound type repository. *Atomic Energy* 107, pp. 32-38.

Section 3

Fabrics Manufacturing

Chapter 11

Giving Functional Properties to Fabrics Containing Polyester Fibres from Poly (Ethylene Terephthalate) with the Printing Method

Ewa Skrzetuska, Wiesława Urbaniak-Domagała,
Barbara Lipp-Symonowicz and Izabella Krucińska

Additional information is available at the end of the chapter

1. Introduction

Polyester fibres from polyethylene terephthalate have a leading position among synthetic fibres, it represent about 72%. According to Oerlikon Textile, the production and demand for polyester fibres were equal to 32 million tonnes in 2008. An increase in demand at the level of 3,2% was observed. The SRI (Socially Responsible Investment) Consulting analysts predict that till the year 2013 a further 4,2% increase in demand will occur. In 2008, there was an increase of about 1.6% in production of polyester filaments. The global production was equal to 17,3 million tonnes. The leader in the market of polyester fibres' producers is Asia including the first five companies of highest production capabilities: Reliance Industries (India), Sinopec (China), Formosa Plastics Group (Taiwan), Tuntex Group (Thailand and Taiwan) and Invista [1,2].

These fibres (PET) are used in different types of textiles and technical products due to very good dielectric, mechanic and strength properties, good resistance to aging and action of light, good thermal resistance, high chemical resistance to the action of diluted acids, alkalis, aliphatic and aromatic hydrocarbons, complete biological resistance, resistance to squishing and stability of dimensions [3,4].

According to the increasing demand for its included products, polyester fibers are applied widely by extended rate in technical and special products in various fields of life. In fabrics, the polyester fibers are often mixed with natural fibers such as cotton.

The polyester fibres are dyed almost entirely with suspension dyes using special dyeing methods. The positive results are also obtained during printing with those dyes.

The printing technology was successfully implemented in the textile industry for many years ago. For a very long time it fulfills the role of adornment thanks to placing decorative patterns. It can be noticed that printing on textiles is of increasingly broader use due to implementing/developing more and more modern printing technologies. It is an important fact that printing in the recent time, apart from the decorative role, fulfills also additional specific functions, such as giving antistatic properties [5,6], electroconductivity [5-8], bacteriostatic properties [9] and sensoric activity on the different exterior stimuli [10].

Printing is reckoned as a key and attractive technology in the range of electroconductive paths' design, leading to the creation of intelligent products. A constant progress in miniaturisation of microelectronics together with new technologies enables the integration of functionalities in clothing, enabling completely new applications. The vision of wearing intelligent clothes describes the future electronic systems as an integral part of casual clothes [11].

In the presented work carbon nanotubes were used for giving specified functionality to textile with the printing techniques. The characteristic properties of carbon nanotubes, such as: heat conductivity, electric conductivity, high elasticity module, high strength or resistance to chemicals induce its wide applications.

The carbon nanotubes, similarly as carbon fibres, are resistant to the action of chemicals. However, the physical properties differ greatly the nanotubes from fibres. The electric properties of nanotubes are typical for two-dimensional conductors, the electric conductivity changes, depending on the construction of nanotubes (one- and multi-wall) and on their structure (chirality); it is sensitive to the influence of external factors, such as electric and magnetic fields, mechanical factors, state of environment (temperature, content of vapours of specific chemical substances). The optical properties are characterised by the optical activity, susceptibility to selective absorption and photoluminescence in near infrared dependent on the electron structure of the nanotubes. The magnetic properties are characterised by high diamagnetic susceptibility, increasing with the increase of temperature, and after modification with iron the nanotubes show ferromagnetic susceptibility [10,12]. The mechanical properties are characterised by high mechanical strength, elasticity, susceptibility to deformations at bending and twisting, flexibility – e.g. during drawing the length of a nanotube can increase up to 40% without changing of its structure [12].

The rare properties of nanotubes nominate its application in many fields of nanotechnology, for example in creating nanocomposites with sharing nanotubes in the role of strengthening and functionalising material for the composite, in creating nanocontainers for storage of gasses, e.g. hydrogen [12], in effective removal of dioxins from the residues of burning medical and chemical wastes [12] (an important role of nanotubes in protection of the environment). However, the biggest hopes are seen in using the unique electrical properties of nanotubes, e.g. in microelectronics [13], and the advantages more than silicon technology.

In this work the electric sensorial properties of carbon nanotubes were used for functionalising the textile backing. The multi-wall nanotubes, combined with a binding substance in order to assure the durability of their connection with the surface, were used to creating printout on a textile surface. The printout is actually a composite of a binding agent, as a warp, and of nanotubes as a strengthening component that functionalises the surface of a fabric. In this solution the percolation threshold of electric conductivity was achieved at low share of carbon nanotubes in the printout.

The deposition of printouts was made based on two types of fabrics: polyester (100%) and cotton-polyester (70/30%). In the current work the electric, sensory and microbiologic properties of the printed textile backings, and their durability during multiple washings were examined.

2. Materials and methods

As a base for conductive printouts in the work, the water dispersion of carbon nanotubes with commercial name AquaCyl (AQ0101) from the Nanocyl company was used. The carbon nanotubes have a firm position in the group of nanomaterials used in sensorics. It is an open problematic to apply the nanotubes in such way, so that they are toxicologically safe in use, permanently connected with the substrate and so that they guarantee the highest sensitivity to the examined stimuli, with their minimum possible content in the sensoric element.

The electric properties of nanotubes were used by Kordas and his associates [14] to create conductive printouts on foils and on paper with the inkjet printing technique, where the ink was the water dispersion of nanotubes without binding agents. The authors have undertaken the work on creating water dispersions by sharing auxiliary agents with pastes used in film printing, in order to modify the fabrics containing polyester fibres from polyethylene terephthalate.

2.1. Materials used

The AquaCyl AQ0101 dispersion contains from 0,5 to 1,5% MWCNT of the Nanocyl®7000 series, which are characterised by purity of about 90%, average diameter of nanotubes is 9,5nm and average length up to about 1,5μm. They are also characterised by surface tension of about 57 mN/m, viscosity 36 cP and pH 7. These parameters were determined under the temperature of 25 °C. The dispersion contains additionally the dispersing agent of 0,1-3%.

For the modification of the commercial dispersions the following auxiliary substances were used: DBSA ($C_{12}H_{25}C_6H_4SO_3H$) solution 70 wt% in isopropanol (analytically Pure from Sigma Aldrich) and SLS ($CH_3(CH_2)_{11}OSO_3Na$, analytically Pure from Sigma Aldrich), Ebecryl 2002 (aliphatic urethane acrylate from Cytec, water compatible, UV curable system) and Esacure DP250 (water dispersion of photo initiators from Lamberti SPA).

The fabrics with different raw material content and different weaves were used in this work.

In current study, the polyester and cotton-polyester fabrics were used.

The cotton-polyester fabric with twill weave was purchased from a Dutch company Ten Cate Protect bv. The research of the fabric was made according to the 100 Oëko-Tex standard. Based on the results obtained it can be stated that the mentioned fabric meets the human-ecological standard required for products contact directly with skin.

The polyester fabric with plain weave is a product of a Polish company Miranda.

2.2. Characteristics of the paste and the used printing technique

Obtaining of an antibacterial and antistatic ink from AquaCyl is based on implementing into the 10% AquaCyl (in volume) of an addition of dodecylbenzenesulphonic acid (DBSA) or of sodium lauryl sulfate acid (SLS). These two substances were selected based on the antibacterial properties of sulphur. Such prepared ink is subject to filtration in order to remove the created agglomerates. The filtration is performed in a decreased pressure by using filter of pore sizes about 0.45 micrometer.

The obtained dispersions are used for printing fabrics containing polyester fibres with the use of a conventional technique (screen printing). The ink compositions obtained in this way were placed on fabrics in combination with the selected networking composition containing photo initiator (10% Esacure DP250 by volume) and the aliphatic urethane acrylate (0,7% Ebecryl 2002 by volume). The modified dispersion of carbon nanotubes after combination with the networking composition was mixed for 30 minutes with a magnetic mixer. Next it was placed on fabrics in A4 format with the use of a stencil. The whole surfaces of 0,05m^2 were printed and optionally the stripes with 1cm width and 20cm length (Fig. 1). Such prepared printouts, were subjected to the networking process under a radiator of UV lamp, length of 195mm (2100W), from Philips company.

UV-C 335 W of radiation dose equal to 3.5J/cm^2 was used [15].

Figure 1. Polyester fabric printed with a stripe pattern with 1cm width and 20cm length.

2.3. Investigation methods used

The investigation of physical properties of fabrics containing polyester fibres was characterised by examining the thickness using Arthur Meiber KG LTG thickness meter. The research was done based on the PN-EN ISO 5084:1999 standard. The value of pressure during the measurement was equal to 2 Pa, and the measuring surface was 1000 mm^2. Ten measurements were performed per 1 m^2 of the sample, and next the average thickness of the fabrics was determined with the accuracy of 0,01mm.

The measurements of linear mass according to the PN-EN 29073-1:1994 standard were also performed. From a unit of a product the sample was taken with full width, and length not smaller than 50000mm^2, in accordance with PN-EN ISO 186:2004. The sample was subject to the process of acclimatisation in 23^{0}C and 25%Rhw during 24 hours, and next its width, length and mass were measured. The mass of the fabric was determined with the accuracy of 0,1% of the weighted mass. Based on the obtained results the surface mass (m_ρ)was calculated, according to the equation (1):

$$m_p = \frac{m}{S \cdot L} \tag{1}$$

where:

m - average mass of the acclimatised samples, g
S - width of the acclimatised samples, m
L - length of the acclimatised samples, m

Based on the obtained results of research of thickness and surface mass of the sample, the apparent density was calculated. The apparent density is a quotient of the mass of a textile material and its volume, together with the volume of spaces between the elements of its structure, expressed in kg/m^3. The measurements were done in a room under normal climate, on acclimatised samples. The apparent density (ρ) of a sample was calculated with the accuracy up to 2 significant digits, according to the equation (2):

$$\rho = \frac{m}{S \cdot L \cdot G} \tag{2}$$

where:
m - mass of the sample, kg
L - average length of the sample, m
S - average width of the sample, m
G - average thickness of the sample, m

The electric conductivity of the printed fabrics was characterised by measuring the surface resistance according with the standard EN 1149-1:2008 – Protective clothing - Electrostatic properties - Part 1: Surface resistivity (Test methods and requirements)

The research were performed with a direct electromagnetic method using Keithley 610C electrometer. The source of electric voltage was the DC power supply 4218 (from RFT company) with the range of voltages 0-3000 V. The system of electrodes together with the investigated sample was placed on a Faraday screen. The processes of conditioning and investigating the samples were retained at a constant level: temperature 23^{0}C, RH=25%.

The electrostatic properties according to EN 1149-1:2008 standard are fulfilled by homogeneous materials, which show the surface resistivity below $2{,}5\times10^9\Omega$. In case of non-homogeneous, coated or laminated materials at least one of the surfaces should meet the requirements concerning the homogeneous materials.

The assessment of antibacterial activity of the printed fabrics was investigated on the plates with agar culture. The behaviour of bacteria was assessed in the zone of contact between the agar and the working sample, and the retardation zones around the sample were determined according to the EN ISO 20645:2006 standard. The assessment of the antibacterial activity was based on observation of the occurrence of the phenomenon of bacterial growth or its lack in the zone of contact between the agar and the working sample, and presumptive determination of the retardation zone around the working sample.

The width of the retardation zone (H), i.e. the zone without bacteria near the edge of the working sample was calculated from the equation (3):

$$H = \frac{D - d}{2} \tag{3}$$

where:
H – width of the retardation zone in mm,
D –diameter of the working sample together with the width of the retardation zones in mm,
d – diameter of the working sample in mm.

Then the working samples were placed under the microscope with 20x magnification and bottom lighting, and the growth of bacteria in the zone of contact on the bottom side of the sample was assessed. The assessment of the antibacterial effect of the investigated sample was in accordance with the data presented in Table 1.

Retardation zone, mm Average value	Growth of bacteria on the medium below the working sample	Description	Assessment
>1	None	The retardation zone over 1 mm, lack of growth	Good result
1-0	None	The retardation zone up to 1 mm, lack of growth	Good result
0	None	Lack of the retardation zone, lack of growth	Good result
0	Weak	Lack of the retardation zone, only some colonies were limited, growth almost totally stopped	Limited effectiveness
0	Medium	Lack of the retardation zone, growth lowered to half in comparison to the reference	Insufficient effect
0	Strong	Lack of the retardation zone, lack of lowering of growth in comparison to the reference or just a slight lowering of growth	Insufficient effect

Table 1. Assumptions for the assessment of the antibacterial effect of the investigated antibacterial sample [EN ISO 20645:2006].

The measurements of sensory properties to fluid vapours, temperature and deformations were performed with the use of a station constructed by the Department of Material and Commodity Science and Textile Metrology, equipped additionally in Optris Laser SIGHT pyrometer.

The sensitivity of the printed fabrics to changes of temperature was investigated by using a specially constructed station, which is presented in Fig. 2. The station consists of a Keithley multimeter (1) connected with measuring electrodes (2). In a distance of 50cm from the sample there was an Optris pyrometer placed (3). Due to connecting the pyrometer and multimeter with the computer (4) it was possible to simultaneously register the recorded results of temperature and resistance. Increasing the temperature was possible due to using the heat source (5). The sensory properties of the fabrics were investigated in the range of thermal changes of 20⁰C-70⁰C. For each sample there were performed minimum four cycles of heating and cooling. The investigated samples were of the dimensions 1,5cm x 4cm.

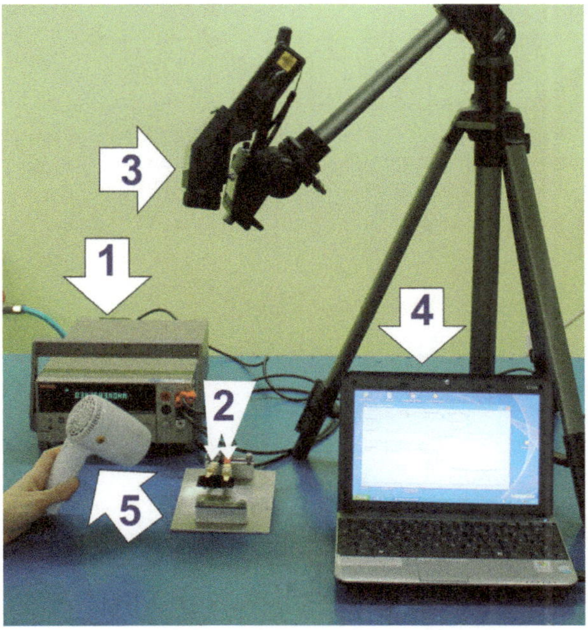

Figure 2. Measuring system for investigating the textile sensors to thermal stimulus: 1. Keithley multimeter, 2. measuring electrodes, 3. Optris pyrometer, 4. computer, 5. heat source.

The research of sensority to the presence of organic fluids was done in the same measuring system as the research of sensority to thermal stimuli. In case of the research of sensitivity to fluids the sample was cut in the shape of a letter U (Fig. 3). The bottom part of the sample was immersed in fluid up to 2/3 of the height from the base, so that the fluid would not get to the electrodes.

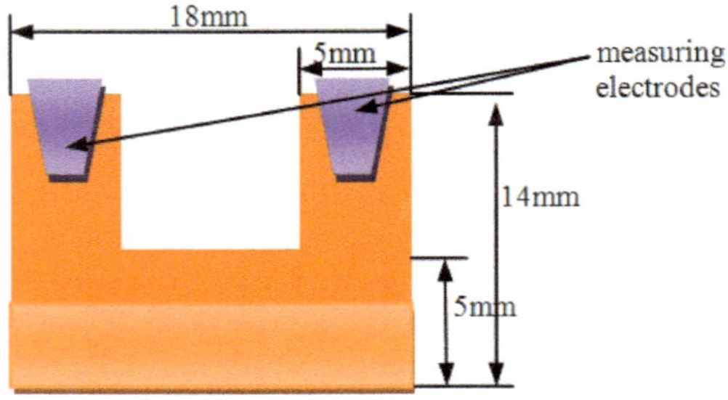

Figure 3. Scheme of preparing the sample to investigating the sensitivity to organic fluids.

The behaviour of sensority in the presence of solvents' vapours was investigated in a laboratory measurement system. The station allows to measure humidity and temperature of the system, and to create and implement into the measuring system of the fluids' vapours with a specified concentration. The sensory sensitivity of the printed fabrics was investigated with the help of a measurement station presented in Fig. 4, composed of an aquarium used as a gaseous chamber (1), pump for mixing the gasses' vapours (2), measuring chamber (3) containing measuring electrodes (4) connected to Keithley multimeter (5), coupled with a computer. The gaseous chamber is used for evaporating a proper amount of solvent. The amount of solvent which should be evaporated in the gaseous chamber in order to achieve a concentration of e.g. 100ppm is calculated according to the equation (4):

$$Y = \frac{X \times M}{24.45}; \qquad M = Y \times V \tag{4}$$

where:
Y - density mg/m^3,
X - parts per million,
M - molecular mass,
V - volume m^3.

The number 24.45 in the equations above is the volume (litres) of a mole (gram molecular weight) of a gas or vapour when the pressure is at 1 atmosphere (760 torr or 760 mm Hg) and at 25°C.

Inside the gaseous chamber there is a thermometer and humidity sensor, thanks to which the research is being conducted in equal conditions (at a level of 23^{0}C and 25%Rh). After evaporation of the solvent in the gaseous chamber, the vapours are transmitted with the use of a pump to the measurement chamber, in which the investigated sample, with the dimensions of 2cm x 4cm, is placed on the measuring electrodes. The sensory properties of the fabrics were investigated for the vapours of different solvents and the changes of resistance were registered. The measurements were done before and after implementation of the vapours of the investigated fluids.

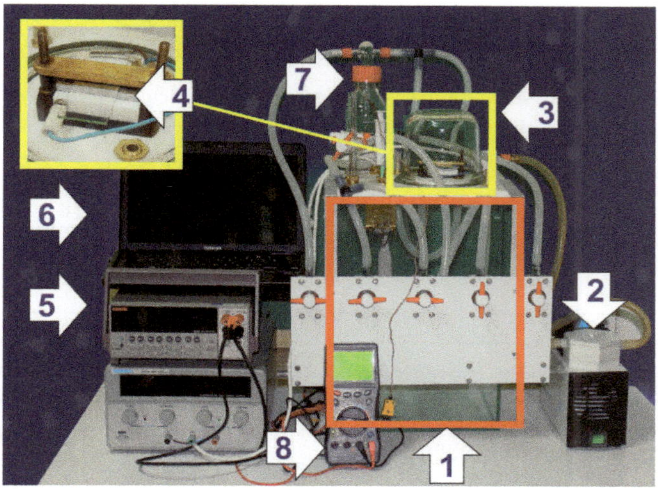

Figure 4. Measuring system for investigating the vapour textile sensors:
1. gaseous chamber with the volume of 0,024 m^3, 2. pump, 3. measurement chamber, 4. measuring electrodes, 5. Keithley multimeter, 6. computer, 7. system ensuring proper humidity of the environment, 8. thermometer.

3. Results and discussion

Physical properties of the functionalised fabrics containing polyester fibres were studied and represented in Table 2.

Material compositions	PET content [%]	Weave	Surface mass [g/m^2]	Thickness [mm]	Apparent density [kg/m^3]
PET	100	plain	214,9	0,53	421,3
PET/Cotton	30/70	twill	205,5	0,52	387,7

Table 2. Physical characterization of the fabrics used.

Fabrics with similar surface mass and thickness were selected for the research.

Table 3 presents the results of measuring the electric conductivity of the printed polyester and polyester-cotton fabrics.

By analysing the electric conductivity obtained from the examined fabrics, presented in Table 2, it can be noticed that for commercial form of AquaCyl water dispersion not containing the networking composition the conductivity is much worse after the process of multiple washings. The addition of the networking composition allows to obtain printouts resistance to the utilisation processes. It should also be noticed that printouts obtained on a polyester fabric with plain weave are characterised by a slightly worse conductive properties than printouts on a polyester-cotton fabric with twill weave, which can be connected with the increased adhesiveness of the printing paste to the cotton fibre.

Table 4 and Figure 5 are showing the assessment of the antibacterial activity of the obtained printouts on fabrics containing polyester fibres before and after the washing process.

Ink composition	Fabric type	**Surface electrical resistivity [Ω] (RH=25%, t=23^{0}C)**	
		Before the washing	After the washing (25 cycles)
AquaCyl	PET	538	22289
AquaCyl	PET/COTT	280	17103
Aquacyl+ networking composition	PET	1029	1601
Aquacyl+ networking composition	PET/COTT	800	856
Aquacyl+ networking composition +DBSA	PET	2285	2769
Aquacyl+ networking composition +DBSA	PET/COTT	1818	1890
Aquacyl+ networking composition +SLS	PET	498	617
Aquacyl+ networking composition +SLS	PET/COTT	181	321

Table 3. Results of electrical conductivity of the obtained printouts before and after the washing process.

Ink composition	Fabric type	Bacteria retardation zone [mm]			
		E.coli (gram-)		B.subtilis (gram+)	
		Before the washing	After the washing (25 cycles)	Before the washing	After the washing (25 cycles)
AquaCyl	PET	0,0	0,0	0,0	0,0
AquaCyl	PET/COTT	0,0	0,0	0,0	0,0
Aquacyl+networking composition	PET	0,0	0,0	0,0	0,0
Aquacyl+ networking composition	PET/COTT	0,0	0,0	0,0	0,0
Aquacyl+ networking composition +DBSA	PET	0,5	0,0	6,0	4,0
Aquacyl+ networking composition +DBSA	PET/COTT	0,5	0,0	6,5	4,5
Aquacyl+ networking composition +SLS	PET	0,0	0,0	3,5	2,0
Aquacyl+ networking composition +SLS	PET/COTT	0,0	0,0	4,0	2,0

Table 4. Results of microbiological activity for the obtained printouts.

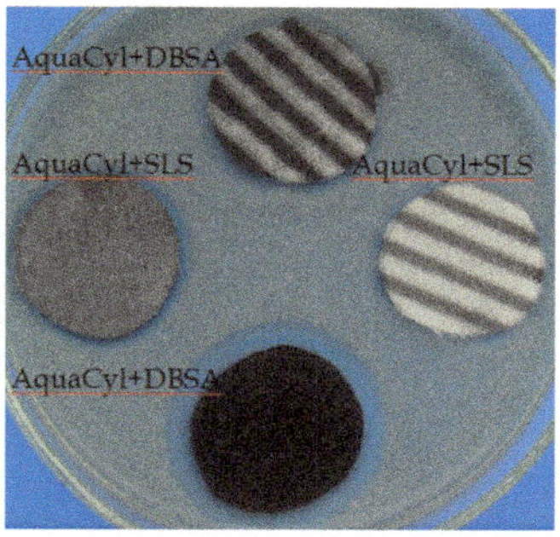

Figure 5. Bacillus bacteria growth retardation zone on polyester fabric.

By analysing the results obtained, it should be noticed that the commercial form of AquaCyl without additives used in this study does not show the antibacterial properties. The addition of dodecyl benzene sulfonic acid (DBSA) or sodium lauryl sulfate (SLS) causes the dispersion composition of ink with bi-functional properties, it is obtaining of carbon nanotubes' dispersion having at the same time electrostatic and antibacterial properties. It is a very important fact that after the process of multiple washings, the bacteriostatic properties are still retained. The obtained compositions show better bacteriostatic properties against gram-positive bacteria. In case of gram-negative bacteria there are no retardation zones around the sample, but below the sample there is not growth of bacteria. Such situation suggests that bacteriostatic properties are also retained for gram-negative bacteria.

In tables 5-8 there are presented the results of sensory measuring to fluids and their vapours. The sensority of the printed fabrics was investigated due to exposure of the selected polar and non-polar organic fluids, and the vapours of selected organic fluids:

- ethanol,
- acetone,
- toluene.

Below there are also presented the diagrams characterising the sensory behaviour of the ink compositions on textile backings containing PET fibres to the selected fluids.

As a quantitative coefficient of the sensory properties, the sensority coefficient (Ss) was taken, expressing the relative changes of electric resistance of the printed fabric's surface, caused by a chemical stimulus of a given type.

$$Ss = \left(\frac{\Delta R}{R_0}\right) \times 100\% \qquad (5)$$

where:
Ss - Sensority coefficient
ΔR – absolute change of electric resistanceof the fabric , $\Delta R = R - R_0$
R – resistance of the fabric under the influence of external stimulus
R_0 - resistance of the fabric before the action of stimulus

Type of chemical substance	Sensority coefficient [%]	Coefficient of variation [%]
Liquid vapours		
Ethanol	Ss = 42	3.96
Acetone	Ss = 37	4.72
Toluene	Ss = 27	6.57
Organic liquids		
Ethanol	Ss = 76	4.09
Acetone	Ss = 84	3.56
Toluene	Ss = 54	3.99

Table 5. Sensority coefficients for polyester-cotton fabric printed with AquaCyl dispersion modified with DBSA.

Type of chemical substance	Sensority coefficient [%]	Coefficient of variation [%]
Liquid vapours		
Ethanol	Ss = 47	3.69
Acetone	Ss = 39	4.50
Toluene	Ss = 28	6.59
Organic liquids		
Ethanol	Ss = 77	3.89
Acetone	Ss = 88	3.43
Toluene	Ss = 54	5.58

Table 6. Sensority coefficients for polyester fabric printed with AquaCyl dispersion modified with DBSA.

Type of chemical substance	Sensority coefficient [%]	Coefficient of variation [%]
Liquid vapours		
Ethanol	Ss = 31	4.47
Acetone	Ss = 25	8.44
Toluene	Ss = 17	13.44
Organic liquids		
Ethanol	Ss = 66	4.95
Acetone	Ss = 67	4.63
Toluene	Ss = 42	6.85

Table 7. Sensority coefficients for polyester-cotton fabric printed with AquaCyl dispersion modified with SLS.

Type of chemical substance	Sensority coefficient [%]	Coefficient of variation [%]
Liquid vapours		
Ethanol	Ss = 31	6.49
Acetone	Ss = 27	7.27
Toluene	Ss = 19	10.47
Organic liquids		
Ethanol	Ss = 67	4.48
Acetone	Ss = 72	4.31
Toluene	Ss = 44	7.30

Table 8. Sensority coefficients for polyester fabric printed with AquaCyl dispersion modified with SLS.

While analysing the results of research presented in tables 5-8 for the fluids' vapours, one can notice that they obtained printouts react the strongest to the vapours of polar fluids. The best sensory properties were observed for vapours of ethanol, at the level of S_s factor over 40% for printouts with DBSA and at the level of about 30% for printouts with SLS. In case of vapours of non-polar fluids the sensory reaction of the printed fabrics with the share of polyester fibres is much weaker – at the level of the S_s factor of about 30% for printouts with DBSA and about 20% for printouts with SLS.

The strongest sensory reaction of the printed backing in case of fluids was observed for polar fluids, e.g. acetone, at the level of the S_s factor of over 80% for printouts with DBSA and at the level of about 70% for printouts with SLS. In case of non-polar fluids the sensory reaction of the printed fabrics is much weaker – at the level of the S_s factor of about 50% for printouts with DBSA and about 40% for printouts with SLS.

It is a significant fact that the results are repeatable, which can be certified by a low coefficient of their changes in the presented tables.

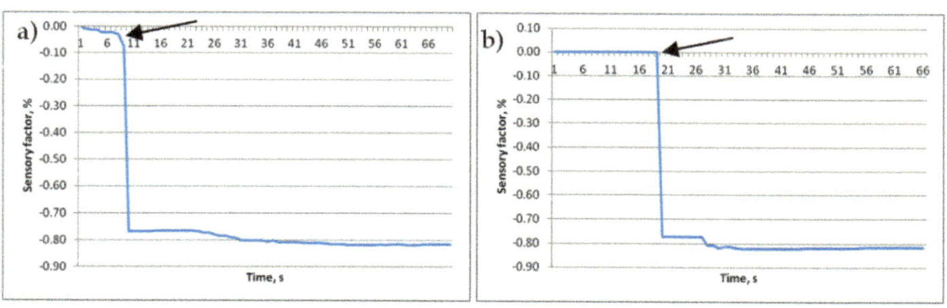

Figure 6. The sensory reaction curves for the polyester-cotton fabric printed with AquaCyl dispersion with the addition of DBSA for selected solvents: a) ethanol, b) acetone

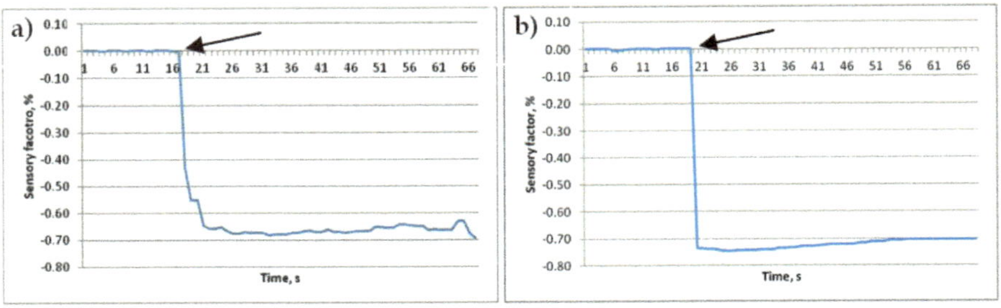

Figure 7. The sensory reaction curves of the polyester fabric printed with AquaCyl dispersion with the addition of SLS for selected solvents: a) ethanol, b) acetone

On the diagrams the arrow presents the moment of immersing of the samples in the selected solvents. The diagrams present the sensory reaction due to the given external stimulus.

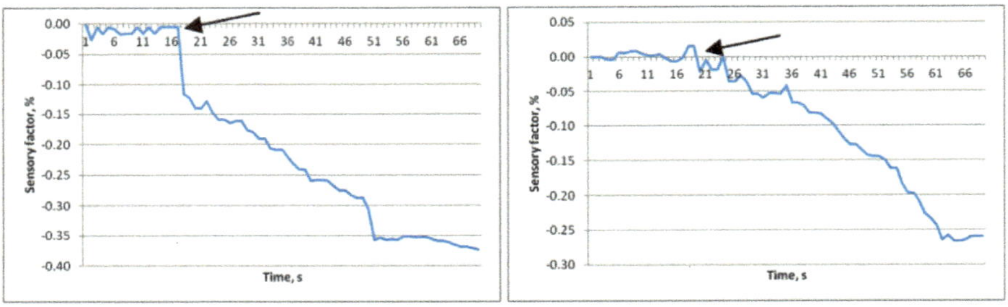

Figure 8. The sensory reaction curves of the fabric printed with the AquaCyl dispersion with the addition of DBSA for the vapours of the selected solvents: a) on polyester-cotton fabric – acetone, b) on polyester fabric – toluene

While analysing Figures 6-8 one can observe that the printed textile backings with the share of polyester fibres, subject to the influence of a fluid, show an immediate reaction, while in the case of fluids' vapours the reaction occurs after a few seconds.

Based on the presented diagrams the moment of implementing the vapours of selected solvents into the measurement chamber of the measuring system, in which the investigated sample is placed, is marked with an arrow. The diagrams show the sensory reaction to the given stimulus.

The results presented in Tables 5-8 and in Figures 6-8 show the sensority coefficient calculated on the basis of the absolute change of electric resistance. The investigated samples were subject to the action of fluids and their vapours. The effect of fluids' vapours on the sensority was investigated at the concentration of 100ppm. The research were repeated 10 times for each of the selected solvents and for each of the investigated samples. Each ink composition contained the networking composition, consisting of a photoinitiator and aliphatic urethane acrylate. The printouts were subjected to the networking process in order to fix them on the textile backing.

By analysing the results obtained it should be noticed that samples containing DBSA show better sensory properties than samples containing SLS. Based on the obtained results of research of the sensory properties there were no significant differences noticed between the printouts obtained on polyester and polyester-cotton fabrics.

A very important aspect of the performed process is the fact that the sensors can be used repeatedly, because the damages of the sensor under the influence of the used fluids were not noticed. The experiment was repeated 25 times for each of the used fluids.

Table 9 presents the results of the sensority coefficient to the thermal stimulus of the printed fabrics. The samples were subject to constant investigation of resistance and temperature. The measurements were repeated several times for the same sample.

Variant	Sensority coefficient Ss , %
Aquacyl+networking composition	47
Aquacyl+ networking composition +DBSA	53
Aquacyl+ networking composition+SLS	25

Table 9. Results of research of the sensority coefficient to the thermal stimulus of the printed polyester-cotton fabric.

The results of research of sensory properties to thermal stimulus, presented in Table 9, in the form of S_s factor, indicate that the textile backings printed with the ink composition with the addition of DBSA are at the level of 53% (high), while printouts with the addition of SLS are at a much lower level (around 25%). It means that the sodium lauryl sulfate limits the sensory properties of the investigated textiles.

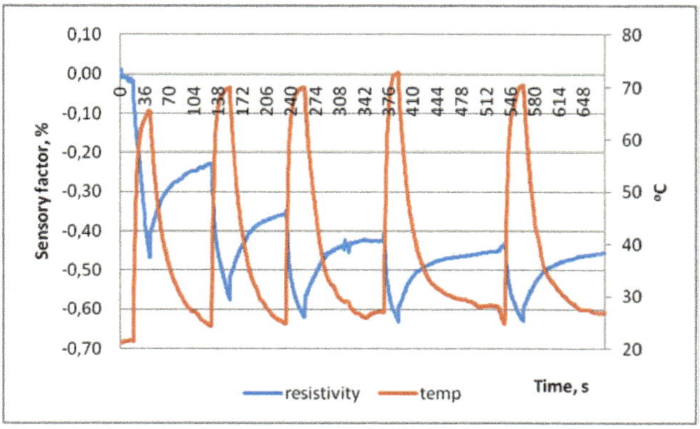

Figure 9. Dependence of the sensority coefficient on the temperature for the AquaCyl ink composition on cotton-polyester fabric

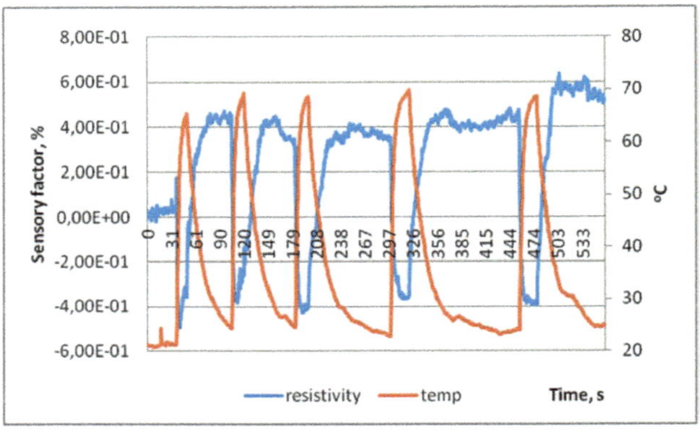

Figure 10. Dependence of the sensority coefficient on the temperature for the AquaCyl+DBSA ink composition on cotton-polyester fabric.

The presented graphs show the cyclical measurements of sensority for heated and cooled samples.

By computing the results obtained, it could be observed that the addition of SLS in a significant way deteriorates the sensory properties of the fabrics to thermal stimulus. It should also be emphasised that in the particular measuring cycles there are no changes observed in the character of the course of the curves, which indicates that the sample is not damaged during heating at a specified level of temperature.

The obtained results of this work show a potential possibility of utilising the proposed method of printing for manufacturing textile sensors used as elements of protective, technical or firearms clothing's and in the mining industry.

4. Conclusions

The performed research confirms the validity of using ink compositions with nanotubes, for modification of fabrics containing polyester fibres, as sensors to chemical and thermal stimuli. By using the screen printing technique for modification of polyester products' surface one can manufacture resistance sensors in a cheap and fast way.

The characteristics of resistance changes caused by thermal stimulus in the range of temperatures: 20°C-70°C are repeatable and reproducible. It was observed that the sensority coefficient was increased from 47% to 53% after adding the DBSA to the dispersion of carbon nanotubes.

In case of the characteristic of resistance changes caused by the chemical stimulus it was observed that the change of DBSA to SLS caused lowering the sensitivity of printed textile backings. For printouts containing SLS, made from polyester fabrics, the sensority coefficient for acetone was 72%, and for printouts containing DBSA was 88%. An addition of

DBSA causes deterioration of electric conductivity, but positively influences the improvement of sensory properties and achieving of bacteriostatic properties.

Using the UV networking composition composed of the aliphatic acrylic urethane and photoinitiator enabled to obtain printouts resistant to the utilisation processes. The electric resistance of a commercial dispersion of carbon nanotubes after 25 washing cycles was equal to $1,71*10^4\Omega$, and with the addition of UV networking mixture - UV was $8,56*10^2\Omega$.

The textile sensors presented in this article can find potential application in medical care, in areas where there is a risk of chemical substances' explosion, in military and sport applications. In order to apply the proposed solution at an industrial scale it is necessary to perform the integration of the manufactured sensor with the system of processing and sending the data to the computer.

Author details

Ewa Skrzetuska, Wiesława Urbaniak – Domagała,
Barbara Lipp-Symonowicz and Izabella Krucińska
Lodz University of Technology, Departament of Material and Commodity Sciences and Textile Metrology, Poland

5. References

[1] Oerlikon group (2007) The fiber year 2006/07, A world survey on textile and nonwovens industry, Issue 7.

[2] Sesto, B.; Yoneyama, M.; Xiaoxiong, O. (2010) Polyester Fibers, Chemical Economics Handbook, April 2010, Available from http://chemical.ihs.com/CEH/Public/Reports/541.9000/

[3] Kardas, I.; Lipp-Symonowicz, B.; Sztajnowski, S. (2011) The influence of enzymatic treatment on the surface modification of PET fibers, Journal of Applied Polymer Science, 119 (6): 3117-3126

[4] Podsiadła-Bulsa, Z.; Michalczewski, A.; Kałużka, J.; Wcisło, P. A. (2009) Poliestrowa włóknina do oczyszczania oleju silnikowego, Problemy eksploatacji, 1: 167-175

[5] Trans, S.J.; Verschueren, A.R.M.; Dekker, C. (1998) Room-temperature transistor based on a single nanotube, Nature, 393: 49-52

[6] Krucińska, I.; Skrzetuska, E.; Urbaniak-Domagała, W. (2012) Prototypes of Carbon Nanotube-Based Textile Sensors Manufactured by the Screen Printing Method. Fibres & Textiles in Eastern Europe, 91: 79-83

[7] Bachtold, A.; Hadley, P.; Nakanishi, T.; Dekker, C. (2001) Logic Circuits with Carbon Nanotube Transistors, Science, 294: 1317-1320

[8] Kong, J.; Franklin, N.R.; Zhou, C.W.; Chapline, M.G.; Peng, S.; Cho, K.J.; Dai, H.J. (2000) Nanotube Molecular Wires as Chemical Sensors , Science, 287: 622-625

[9] Wei, J.; Sun, J.; Zhu, J.; Wang, K.; Wang, Z.; Luo, J.; Wu, D.; Cao, A. (2006) Carbon Nanotube Macrobundles for Light Sensing, Small, 2: 988-993

[10] Przygocki W.; Włochowicz A. (2004) Fulereny i nanorurki. Właściwości i zastosowanie, WNT, Warsaw, Poland
[11] Krucińska, I.; Skrzetuska, E.; Urbaniak-Domagała, W. (2011) The use of carbon nanotubes in textile printing, Journal of Applied Polymer Science, 121: 483-490
[12] Huczko, A. (2004)Nanorurki Węglowe. Czarne diamenty XXI wieku, ISBN: 8388442864 WNT Warsaw, Poland
[13] Graham, A.G.; Duesberg; G.S.; Seidel, R.V.; Liebau, M.; Unger, E.; Pamler, W.; Kreupl, F.; Hoenlein, W. (2005) Carbon Nantubes for Microelectronics? Small, 1: 382-390
[14] Kordas, K.; Mustonen, T.; Toth, G.; Jantunen, H.; Lajunen, M.; Soldano, C.; Talapatra, S.; Kar, S.; Vajtai, R.; M.Ajayan, P. (2011) Inkjet Printing of Electrically Conductive Patterns of Carbon Nanotubes, Small,8-9: 1021-1025
[15] Stempień, Z.; Tokarska, M.; Gniotek, K. (2010) Laboratory Stand for the Optimisation of the UV Curing of Fluids Disposed on Textiles, Fibres & Textiles in Eastern Europe, 18: 65-69

Chapter 12

Hand Evaluation and Formability of Japanese Traditional 'Chirimen' Fabrics

Takako Inoue and Masako Niwa

Additional information is available at the end of the chapter

1. Introduction

Clothes, which are used in direct contact with the human body, are mostly made of fabrics of planar fiber construction, that is, they are manufactured for the most part from textiles. Needless to say, the quality of clothes directly affects both the human mind and body. For this reason, it is essential to have a system which allows us to accurately and thoroughly evaluate the qualities and use-value of textiles. Since Prof. Sueo Kawabata of Kyoto University developed the *KES* (Kawabata Evaluation System) in 1972 (Kawabata, 1972), research into fabric handle and quality based on physical properties has made remarkable progress, and objective evaluation of fabrics using the *KES* system is now common around the world. Evaluation formulas for fabric formability, tailoring appearance, and hand evaluation of the tailored-type fabrics represented by those used in tailored men's suits have been created, and allow us to objectively evaluate the fundamental performance capabilities of fabrics, unaffected by changing times and fashions (Kawabata, 1980). Furthermore, fabric formability, tailoring appearance, hand evaluation, and quality of tailored-type fabrics can now be influenced at every stage, even the very earliest: at the fiber-to-yarn stage, the yarn-to-fabric stage, and then at all the subsequent stages, up to the finishing of the material. This is invaluable for the fabric design process (Kawabata et al., 1992).

Each region of Japan has unique textile weaves and dyeing methods for traditional fabrics. Japanese traditional 'Chirimen' fabrics are used for making the traditional Japanese garments generically referred to as 'kimono', which have a fixed structure and are worn in very particular ways. These 'Chirimen' fabrics have also been used as dress fabrics in recent years, and polyester has come to be used in addition to the traditional silk in these fabrics or making kimonos for people of all ages, since it is easy to wash.

This chapter describes the mechanical properties of different types of 'Chirimen' fabric, including polyester 'Chirimen' fabric, by using an objective evaluation system for fabric. The differences in formation and performance among the various types are also examined here (Inoue & Niwa, 2010a). In our description of the mechanical properties, we look at the warp and weft direction of tensile properties and bending properties, because fabrics wrap around a cylindrical body beautifully in the direction of the weft direction and assume a form when worn that hangs down in the direction of the warp. Evaluation formulas derived to describe the specific handle (comprising *KOSHI* [stiffness] and *TEKASA* [hand quantity]) of 'Chirimen' fabric, and the mechanical properties which contribute to the handles are also explained and discussed. In addition, this chapter shows how to determine optimum silhouette design under the standard measurement conditions for ladies' thin dress fabrics (discussed in Inoue & Niwa, 2010b) rather than the high-sensitivity measurement conditions used hitherto (Niwa et al., 1998), and the criteria for the silhouettes using 'Chirimen' fabric are shown (Inoue & Niwa, 2011).

By examining 'Chirimen' fabric in the ways indicated above, the authors hope to clarify the importance of fabric mechanical properties contributing to fabric formability, hand evaluation, and silhouettes.

2. Japanese traditional 'Chirimen' fabrics

Each region of Japan has unique textile weaves and dyeing methods (Tomiyama & Ohno, 1967) for traditional fabrics. The textile weave and the dyeing method are two sides of the same coin, in the sense that either the weaving or the dyeing can be done first. Consequently, there are both fiber- or yarn-dyed fabrics and piece-dyed fabrics. Traditional fabrics called 'Chirimen' are produced in Tango, Nagahama, Hokuriku, and Gifu, as well as other districts, and are typical of woven piece-dyed fabrics produced in great amounts throughout Japan. 'Chirimen' is the generic name for silk fabric in which right-laid and left-laid hard-twist yarn is alternately woven to make weft yarn. There are crimps on the surface, and these crimps create a unique hanging down feeling and tinctorial effect. 'Chirimen' are considered very high-grade silk fabrics although artificial fiber has been used in them in recent years, and they are used to make formal 'kimono' (the famous traditional Japanese garment).

2.1. Construction of the traditional garment 'kimono' and how it is worn

The Japanese traditional garment called 'kimono' evolved over a long period into a garment well-suited to the climactic conditions peculiar to Japan. Stylistically, it is a formal garment similar to the Western formal wear and gowns worn on cermonial occasions, a result of its being constructed from one continuous piece of cloth (Yamamoto, 1960). The cloth is 36～38cm in width; using approximately 12m of cloth in total, the sleeves, main body parts (front and back), front panels, collar and collar cover are cut out and sewn together in a straight line. When it is put on, it is secured around the body with a waist cord, which is then covered with a decorative sash called an 'obi'. Fig. 1 shows the construction of a women's kimono and a photo of how it is worn.

2.2. Japanese traditional fabrics and varieties of 'Chirimen' fabrics

'Chirimen' fabrics were collected from the largest producing area, the Tango District near Kyoto, in 1981 (Komatsu & Niwa, 1981), and silk Chirimen and 33 polyester Chirimen samples were added to the collection of samples in recent years, resulting in a total of 304 samples. (The details are shown in Table 1.) Chirimen are classified by difference in yarn,

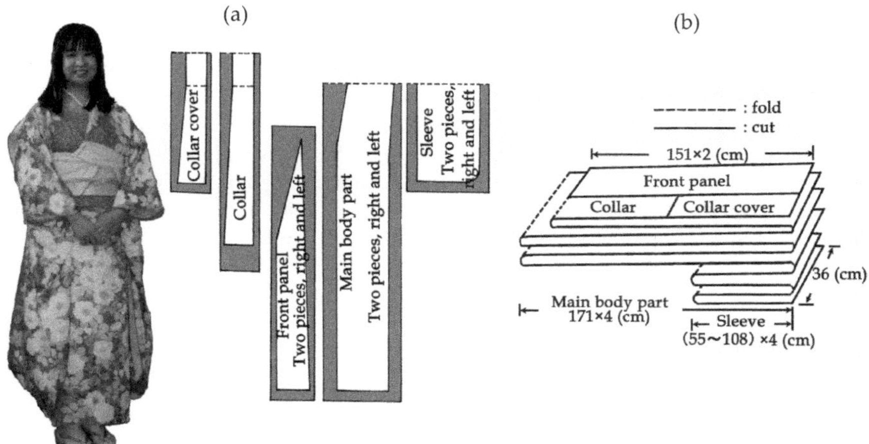

(a) The fabric sections used in a kimono (b) Example of how fabric is cut making kimono

Figure 1. Photo of women's kimono, showing its construction and how it is worn

Fiber	Kind of Chirimen		The number of Chirimen fabrics
Silk	Hitokoshi・Kodai Chirimen	Hitokoshi Chirimen	78
		Kodai Chirimen	26
	Kawari Muji Chirimen		14
	Mon Chirimen	Muji (Mon) Isho Chirimen	76
		Rinzu Chirimen	63
		Fuutsuu Chirimen	6
		Others	8
Polyester			33
		Total	304

Table 1. Chirimen fabrics

fabric construction, and the size of the rugged crimps on the surface (Nakae, 1993). Among the classifications of Chirimen fabrics examined are: 1) Hitokoshi Chirimen (crimp is minute, and filament yarn that is not twisted is used for the warp yarn; this is the typical Chirimen, with right-laid and left-laid hard twist yarn alternately woven to make weft yarn); 2)Kodai Chirimen (filament yarn that is not twisted is used for the warp yarn; thicker hard-twist yarn than that used in Hitokoshi Chirimen is alternately woven right-laid and left-laid to make weft yarn, namely "two jump" yarn, right/right-laid and left/ left-laid hard-twist yarn; there are larger crimps than in Hitokoshi Chirimen; also called "Futakoshi Chirimen", "Futakoshi" meaning "two jumps"); 3) Kawari Muji Chirimen (crimp is more minute than that of Hitokoshi Chirimen; hard twist yarn called "Chirimen Yoko" ["Yoko" meaning "weft"] is not used, rather, fancy twist yarn is used [i.e., yarn twisted differently from the usual way: perhaps one yarn made from non twisted yarn and right-laid yarn, another yarn made from non twisted yarn and right-laid yarn are twisted together to make one yarn], consequently, the degree of shrinkage is small; produced in greater amounts than Hitokoshi Chirimen in recent years); 4) Mon Chirimen (Chirimen which has warp yarn in plain weave fabric at the surface to highlight the woven pattern; comes in different varieties such as Mon Isho Chirimen [Muji Isho Chirimen] and Mon Rinzu Chirimen); 5) Fuutsuu Chirimen (woven pattern in the cloth is double-weave, using warp and weft yarn; the pattern is woven in high relief). In this chapter, each kind of Chirimen introduced here is examined.

3. Fabric handle and objective evaluation system

3.1. Measurement of fabric mechanical properties

Fabric mechanical properties were measured using the *KES-FB* system (Kawabata et al., 1996~1997) under the standard conditions (Kawabata, 1980) shown in Table 2.

The tensile properties, bending properties, shearing properties, surface properties, compression properties and the weight of the fabrics were measured.

3.2. The subjective hand-evaluation method

Sensory tests (Inoue et al., 2010) were performed for the subjective evaluation of the hand value of Chirimen. A standard sample was assumed to represent the standard feel of Chirimen; comparing Chirimen samples with the standard sample, we ranked the Chirimen samples by how strong they felt to the touch, using a scale of 0 to 10, with the standard sample being 5. This subjective hand-evaluation provided an indication of *KOSHI* (stiffness) and *TEKASA* (hand quantity), both of which influence the quality of Chirimen. (*TEKASA* is defined as a sense of the great bulk and rich feeling of cloth with substantial "give" to the touch, a feeling of thickness and warmth as well as elasticity under pressure [Kawabata, 1980].) Judges were four experienced technical engineers who were well acquainted with the silk 'Chirimen' fabrics of the special production site in the environs of the Kyoto Prefectural Institute of Northern Industry.

3.3. The mechanical parameters and three basic components of tailorability

The mechanical parameters and three basic components of tailorability were calculated to make a *TAV* (Total Appearance Value) prediction (Kawabata & Niwa, 1989). The mechanical parameters are as follows:

(Formability)
Weft-directional extensibility

$$\log_{10} EL2 = \log_{10} (EM2/LT2) \quad (1)$$

Effective bending stiffness in weft-bending mode

$$\log_{10} BS2 = \log_{10} [M2(1) + HB2],\ M2(1) = B2 \cdot K,\ K=1 \quad (2)$$

Effective shear stiffness

$$\log_{10} SS = \log_{10} [Fs(1) + HG5],\ Fs(1) = G \cdot \phi,\ \phi = 1^{\circ} \quad (3)$$

(Elastic Potential)

Bending elastic potential per unit area at K=2.5cm^{-1}

$$\log_{10} BP = \log_{10} [B(2.5 - HB/B)^2/2] \quad (4)$$

Shear elastic potential per unit area at ϕ=8°

$$\log_{10} SP = \log_{10} [G'(8 - HG/G')^2/2],\ G' = G + (2HG - 2HG5)/5 \quad (5)$$

(Drape)

Bending stiffness relating to drape

$$\sqrt[3]{BS/W} \quad (6)$$

Shear stiffness relating to drape

$$\sqrt[3]{SS/W} \quad (7)$$

Where, K : Bending curvature (cm^{-1}), ϕ: Shear angle (degree), M : Elastic bending moment (gf · cm · cm^{-1}), Fs : Elastic shear force (gf · cm^{-1}).

3.4. Determination of optimum silhouette design

Ladies' garments come in a wide variety of designs, and make use of fabrics with greatly varying mechanical properties, making a considerable range of silhouettes possible (Inoue & Niwa, 2003, 2009). Ladies' garment fabrics are divided into three categories, based on the silhouette types which they can yield: 1. Tailored Type, which results in the formation of a beautiful shape covering the female body; 2. Drape Type, which emphasizes a beautiful drape silhouette; and 3. *Hari* Type, or anti-drape silhouette which spreads out horizontally away from the surface of the human body.

The discriminant equations for ladies' garment fabrics related to silhouette design which divides the fabrics into three optimum silhouette types have been derived from the mechanical properties of fabrics under both high sensitivity conditions (Niwa et al., 1998), and standard conditions (Inoue & Niwa, 2010b). The tensile properties of these samples are measured under standard conditions up to a maximum load of 500gf/cm. In this chapter, three optimum silhouettet types are derived using the first canonical variate Z_1 and the second Z_2 obtained under standard conditions, as follows:

Symbols	Characteristic value	Unit	Measuring conditions Standard (Kawabata, 1980)
EM	Tensile strain at max. load	%	Strip biaxial deformation.
LT	Linearity	-	Upper limit tensile force (max. load) : 500 gf/cm
WT	Tensile energy	gf·cm/cm^2	
RT	Resilience	%	
B	Bending rigidity	gf·cm^2/cm	Pure bending.
2HB	Hysteresis	gf·cm/cm	Max. curvature, K = ±2.5cm^{-1}
MIU	Coefficient of friction	-	Contactor for friction measurement : Ten parallel steel-piano -wires with 0.5mm dia. and 6mm length simulating finger
MMD	Mean deviation of *MIU*	-	skin geometry. Contact force; 50gf.
SMD	Geometrical roughness	micron	Contactor for geometrical roughness : A steel piano wire, with 0.5mm dia. and 5mm length. Contact force ; 10gf.
G	Shear stiffness	gf/cm·degree	Shear deformation under constant tension of 10gf/cm.
2HG	Hysteresis at ϕ =0.5°	gf/cm	Max. shear angle, ϕ =±8°
2HG5	Hysteresis at ϕ =5°	gf/cm	
LC	Linearity	-	Upper limit pressure : 50 gf/cm^2
WC	Compressional energy	gf·cm/cm^2	
RC	Resilience	%	
T	Thickness at 0.5gf/cm^2	mm	Thickness at 0.5gf/cm^2 pressure.
W	Weight per unit area	mg/cm^2	Weight of specimen per unit area.

Table 2. Characteristic values of basic mechanical properties and measuring conditions for *KESF* measurements.

$$Z_1 = \sum_{i=1}^{8} C_{1i} U_i \tag{8}$$

$$Z_2 = \sum_{i=1}^{8} C_{2i} U_i \tag{9}$$

Where, $U_i = (x_i - m_i)/\sigma_i$: normalized mechanical data, and x_i : mechanical data from a sample. The coefficients C_{1i}, C_{2i}, the mean m_i and standard deviation σ_i are listed in Table 3.

4. Characteristics and performance of 'Chirimen' fabrics

4.1. Characteristics of mechanical properties

In order to clarify the mechanical properties of Chirimen fabrics, a data chart was created using the mean value and the standard deviation for 271 silk Chirimen fabrics (Fig. 2). We added warp and weft direction to tensile properties and bending properties in Fig. 2 because kimono fabrics wrap around a cylindrical body beautifully in the direction of the weft and assume a form when worn that hangs down in the direction of the warp. The mean value and the standard deviation for men's suiting, Women's Suiting, dress-shirt fabrics, and polyester Chirimen were used to represent the features of Chirimen. From the chart, it is clear that men's suiting and dress shirts fabrics have narrower ranges of mechanical properties than women's suiting.

i	Parameter	Description	Unit	m_i	σ_i	C_{1i}	C_{2i}
1	LT	Linearity of tensile curve	none	0.6094	0.1614	-0.0477	0.0892
2	logEM	Elongation at Max. load	%	0.9179	0.2545	0.0786	-0.3391
3	log$B$	Bending rigidity	gf・cm²/cm	-1.1222	0.4658	0.6852	-0.0333
4	log$2HB$	Hysteresis at K=1cm^{-1}	gf・cm/cm	-1.4469	0.5826	0.1658	-0.4825
5	log$G$	Shear stiffness	gf/cm・deg	-0.3607	0.2204	0.4434	0.1804
6	log$2HG$	Hysteresis at ϕ=0.5°	gf/cm	-0.3748	0.4967	-0.1794	2.4200
7	log$2HG5$	Hysteresis at ϕ=5°	gf/cm	-0.0645	0.4443	0.0053	-3.0857
8	log$W$	Fabric weight	mg/cm²	1.1782	0.2406	0.1649	1.2555

Table 3. Values of C_{1i}, C_{2i}, m_i and σ_i to calculate Z_1, Z_2

The distinctive features of silk Chirimen are its low values of bending properties, shearing properties, and thickness & weight, but the values for the weft direction of tensile properties *EM2*, *LT* and the bending properties *B2* and *2HB2* have the same ranges for men's and women's suit-fabric characters, and the ranges are wide.

The key characteristics of the polyester Chirimen indicated by the dashed line in Fig. 2 are: 1. the values for bending rigidity and the hysteresis of warp direction of polyester Chirimen are low and it bends softly, 2. the values of hysteresis at shear angle ϕ=0.5°of polyester Chirimen are higher than those for silk Chirimen. Further, 3. the values of the surface properties of polyester Chirimen are higher than those for silk Chirimen, men's suiting, women's suiting, and dress shirt fabrics. In addition, 4. the thickness and the weight of the fabric is higher than that of silk Chirimen: When we wear it, we feel the weight. From this list, it should be clear that there are differences as well as similarities between polyester Chirimen and silk Chirimen. It is understood that the feature of polyester Chirimen distinguishing it from silk Chirimen is its ability to create a silhouette as clothing when used in clothing.

The mean value and the standard deviation of each Chirimen group are plotted in Fig. 3 and Fig. 4. The characteristic ranges of mechanical properties are shown for each group. The values of tensile properties *EM1*, *EM2*, *WT* of Hitokoshi Chirimen and Kodai Chirimen are high, as are those of the surface properties and the thickness and weight ; this is most likely due to the fact that there are crepes in the surface. Tensile resilience *RT* is low, the reason for these Chirimen shrinking easily. The values of the surface properties of Rinzu Chirimen are the lowest in the Chirimen groups, and its surfaces are the smoothest. On the other hand, Fuutsuu Chirimen is made from double-weave cloth, so the thickness values of the fabrics are high, and with the woven patterns brought into high relief, the coefficient of friction

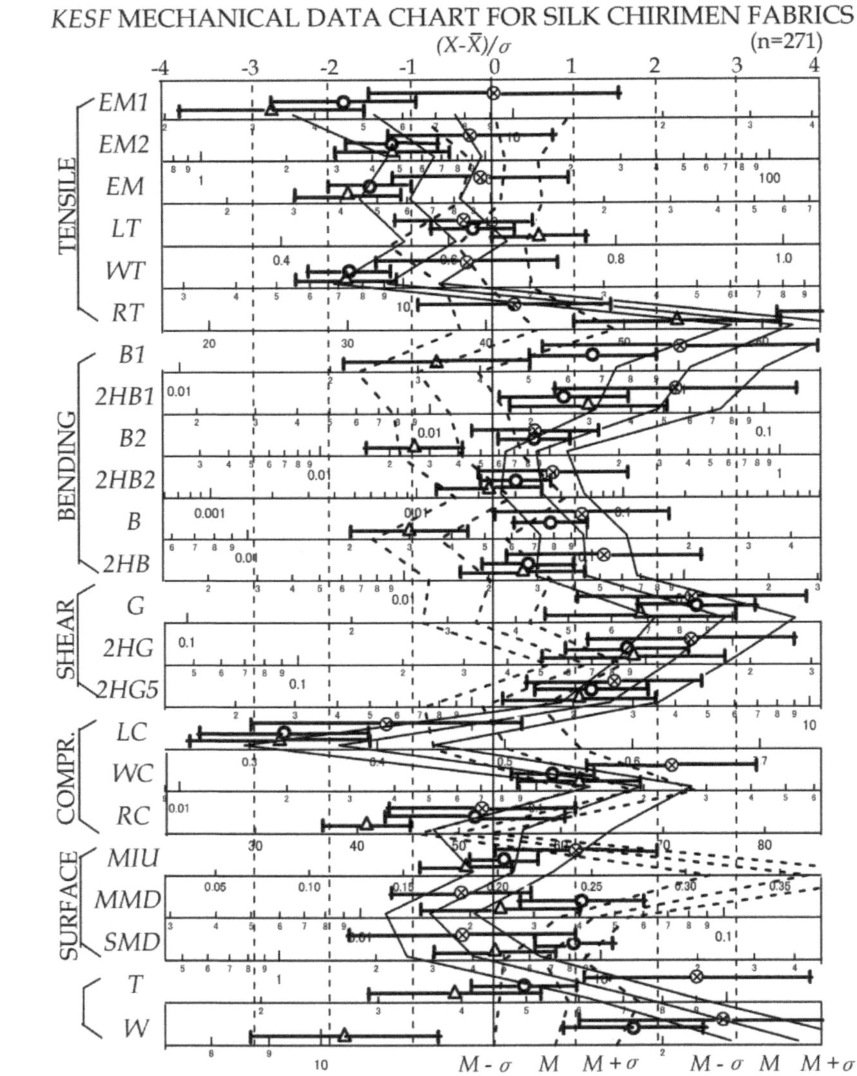

Suffix *1*: warp direction, suffix *2*: weft direction,

├Δ┤: dress shirt (n=116), ├⊗┤: women's suiting (n=220), ├O┤: men's summer suiting (n=156), ├──┤: mean value and ± standard deviation.

Men's winter suiting (n=214) is indicated by the solid line.

Polyester Chirimen (n=33) is indicated by the broken line.

Figure 2. The Mechanical properties of fabrics used in clothing in the west and polyester Chirimen fabrics

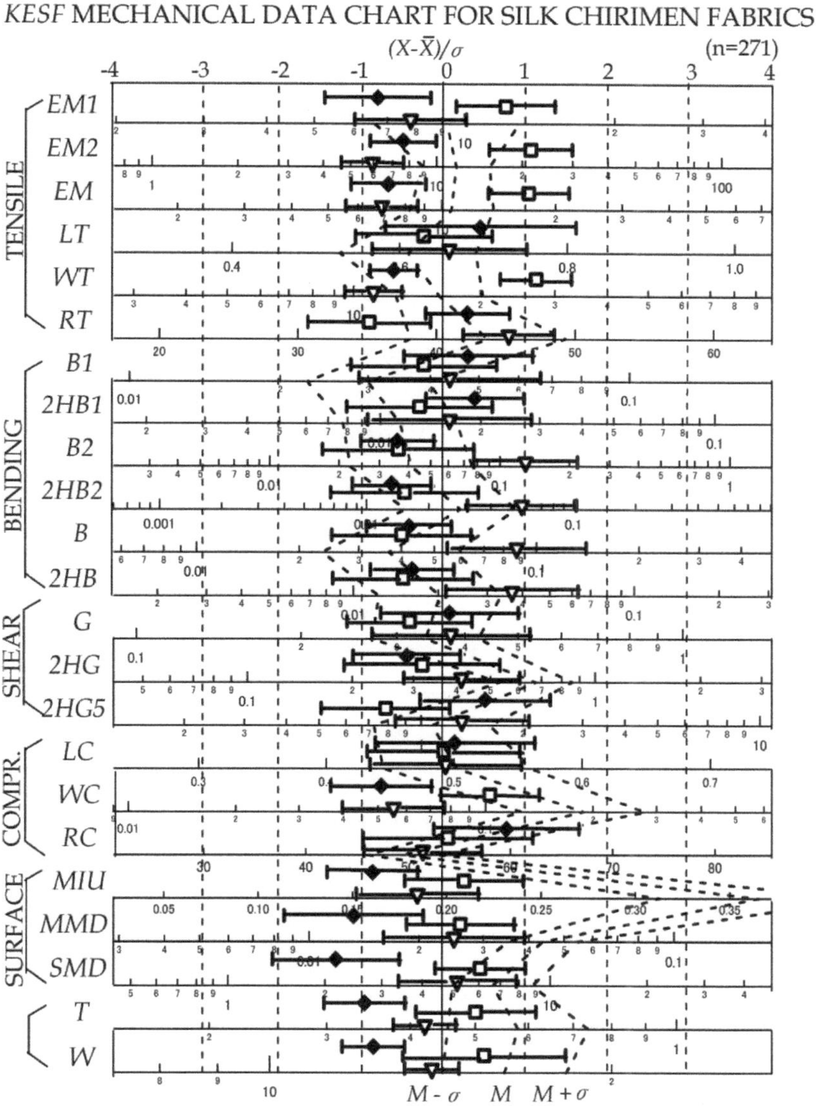

Suffix *1*: warp direction, suffix *2*: weft direction, ├─□─┤ : Hitokoshi Chirimen (n=78), ├─▽─┤ : Muji Isho Chirimen (n=76), ├─◆─┤ : Rinzu Chirimen (n=63), ├────┤ : mean value and ± standard deviation.
Polyester Chirimen (n=33) is indicated by the broken line.

Figure 3. The mechanical properties of Chirimen fabrics

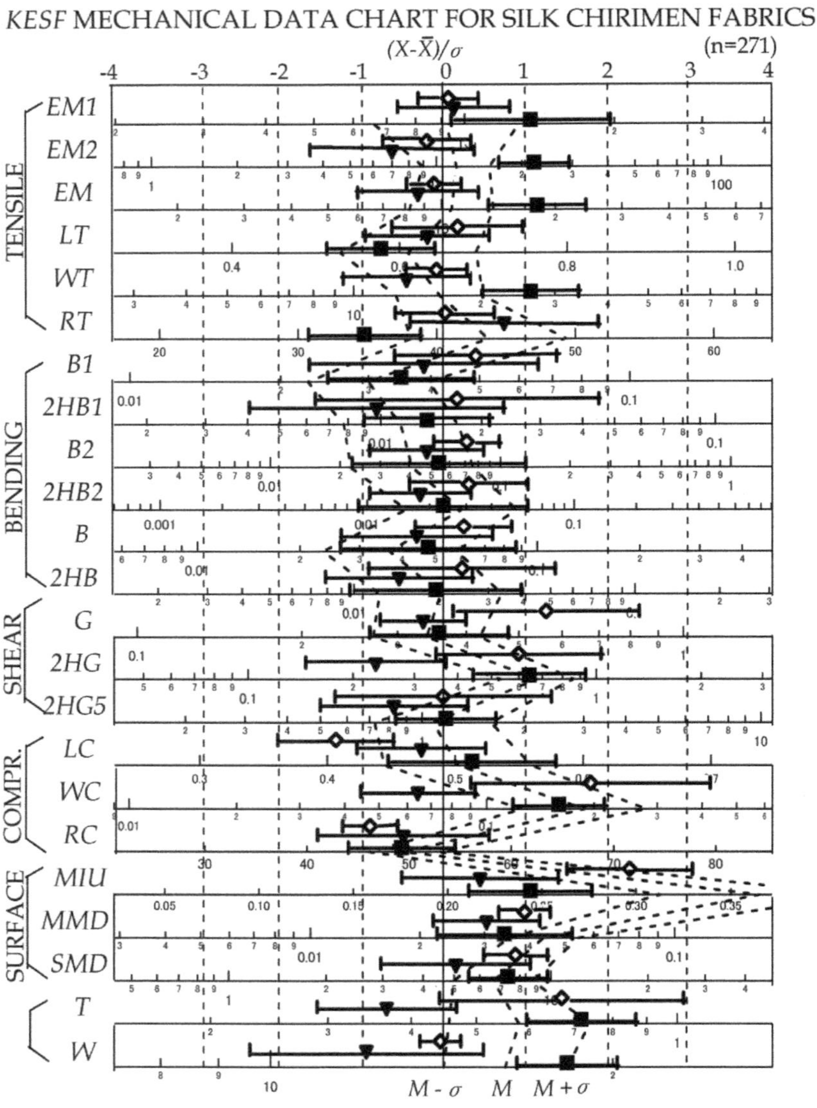

Suffix *1*: warp direction, suffix *2*: weft direction, ├▼┤ : Kawari Muji Chirimen (n=14), ├■┤ : Kodai Chirimen (n=26), ├◇┤ : Fuutsuu Chirimen (n=6), ├──┤ : mean value and ± standard deviation.
Polyester Chirimen (n=33) is indicated by the broken line.

Figure 4. The mechanical properties of Chirimen fabrics

MIU is also high. Regarding Kawari Muji Chirimen, its thickness and weight values are low and it has the distinctive feature of lower-value tensile properties than those of Hitokoshi Chirimen and Kodai Chirimen. The values for the bending properties of weft direction of all Chirimen groups are at the same level, and they have a range corresponding to those of men's suit fabrics, women's suit fabrics and dress shirt fabrics; the values of tensile properties are at the same level as those for women's suit fabrics. This can be said to be a distinctive characteristic of Chirimen fabrics.

4.2. Hand value (*HV*) and formability

The criteria chart for ideal men's autumn/winter suiting (Kawabata et al., 2002) was normalized again using the mean value and the standard deviation for women's suit fabrics; the mean value and standard deviation for hand value and the mechanical parameters of suit appearance for all Chirimen groups were calculated using the *KN*201 equation (Kawabata, 1980) and are shown in Fig. 5. The values for *KOSHI*, *NUMERI* (smoothness) and *FUKURAMI* (fullness and softness) of silk Chirimen are at the same level as those of dress shirt fabrics, but the value for *SOFUTOSA* (soft feeling) is high, at the same level as the values for men's suit fabrics and women's suit fabrics. A distinctive characteristic here of Chirimen is that the values for mechanical parameters for suit appearance are within the range of those for ideal men's suiting, excluding *SP* (equation (5)). The values for the basic components of tailorability are located almost in the middle of the range for men's suit fabrics and dress shirt fabrics.

The mean value and the standard deviation for each Chirimen group are plotted in Fig. 6 and Fig. 7. Values for *KOSHI* and *SOFUTOSA* in each group are different, but the values for the mechanical parameters in each group are located within the range of those for ideal men's suiting (excluding *SP* and *BS2*). Values for the basic components of tailorability are slightly different from group to group: the drapability values for Muji Isho Chirimen are high, so it is possible that this type can create a beautiful silhouette; the formability and elastic potential values for Muji Isho Chirimen are higher than for other Chirimen groups, which accounts for the fact that Muji Isho Chirimen is used to make Tomesode (married women's formal kimono decorated with five crests and a pattern around the skirt, the highest-grade kimono and the most formal ceremonial dress) and Houmongi (kimono for formal visiting, quasi-ceremonial dress decorated with a full-body pattern).

4.3. Analysis of mechanical properties affecting hand evaluation

How the mechanical properties of Chirimen affect the criteria for subjective evaluation of hand evaluation was examined. First, 139 samples of Chirimen were chosen without disparity among groups from a total of 271 samples of silk Chirimen. 47 Hitokoshi Chirimen, 7 Kodai Chirimen, 7 Kawari Muji Chirimen, 40 Muji Isho Chirimen, and 38 Rinzu Chirimen were chosen.

The stepwise-block regression method was applied, using 19 characteristic values of six blocks of the mechanical properties—tensile, bending, surface, shearing, compression, and thickness & weight—and the mean value of each hand evaluation value was used as the subjective value (*KOSHI, TEKASA*). The regression equation is shown below.

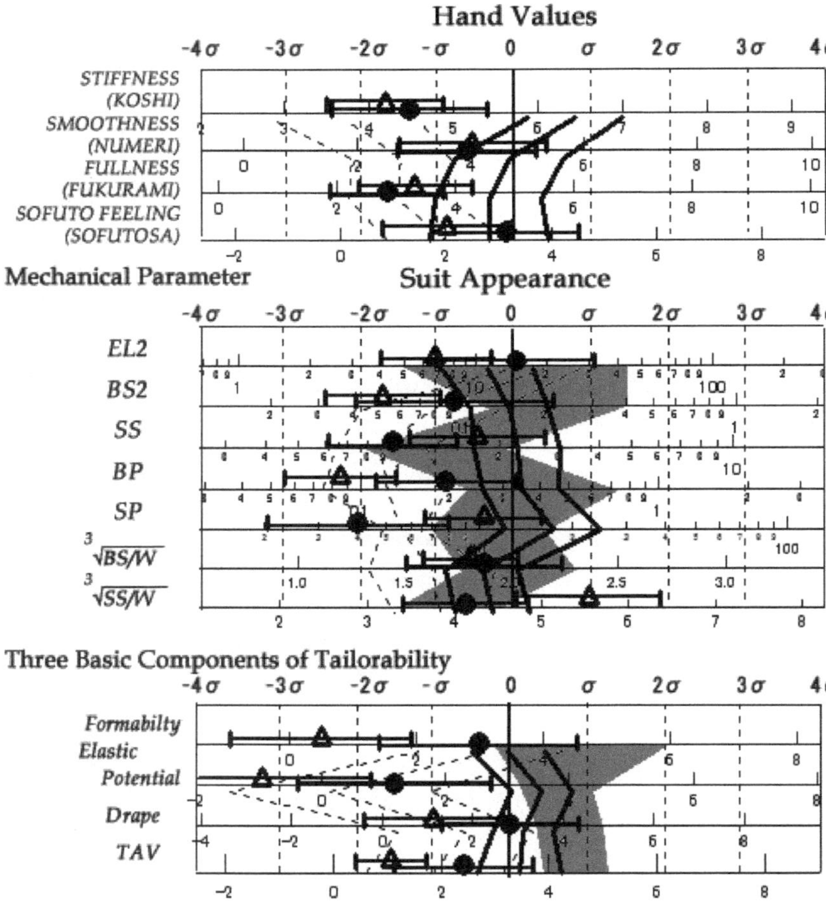

Suffix *1*: warp direction, suffix *2*: weft direction, |—Δ—|: dress shirt (n=116), |—●—|: silk Chirimen (n=271), |——|: mean value and ± standard deviation, men's winter suiting (n=214) is indicated by the solid line.
Polyester Chirimen (n=33) is indicated by the broken line.
Perfect property zone of men's suiting is indicated by the shaded zone.

Figure 5. Hand value and the mechanical parameters of Chirimen

$$HV = C_0 + \sum_{i=1}^{19} \left(C_{i1} \frac{X_i - M_{i1}}{\sigma_{i1}} + C_{i2} \frac{X_{i2} - M_{i2}}{\sigma_{i2}} \right) \tag{10}$$

Where C_0, C_{i1}, C_{i2} = constant coefficients of the ith variable terms; X_i = mechanical property of the ith variable term M_{i1} ; σ_{i1} = the population mean and standard deviation M_{i2} ; σ_{i2} = the square mean and standard deviation (listed in Table 4). From this analysis, the order of the contribution to HV can be clarified, and we can discover the relationship between the criteria of subjective evaluation and the mechanical properties.

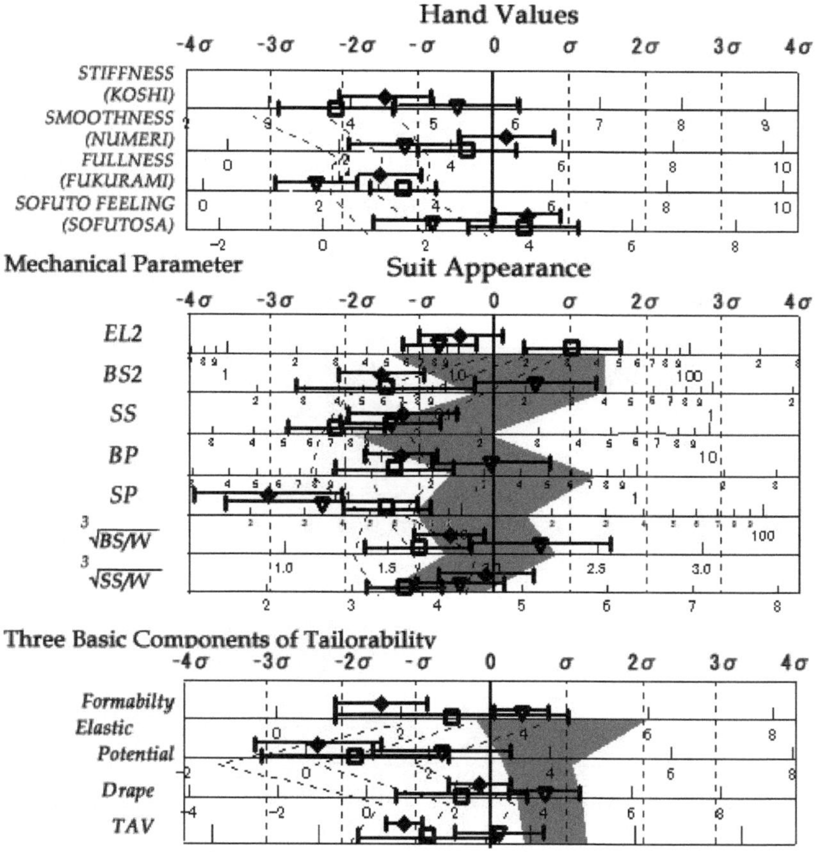

Suffix *1*: warp direction, suffix 2: weft direction, |—□—| : Hitokoshi Chirimen (n=78), |—∇—| : Muji Isho Chirimen (n=76), |—◆—| : Rinzu Chirimen (n=63), |——| : mean value and ± standard deviation.
Polyester Chirimen (n=33) is indicated by the broken line.
Perfect property zone of men's suiting is indicated by the shaded zone.

Figure 6. Hand value and the mechanical parameters of Chirimen

The contributions of each characteristic block to hand value were examined; the contributions to *KOSHI* are shown in Fig. 8 and the contributions to *TEKASA* are shown in Fig. 9. The results were as follows: 1) the contributions of bending properties to *KOSHI* are significant, the multiple correlation coefficient being 0.742; 2) the bending properties of weft direction contribute more to *KOSHI* than those of warp direction, judging from the figures indicating contributions; 3) thickness & weight contribute to *KOSHI*, the close positive relationship between weight and *KOSHI* being particularly clear; 4) thickness & weight contribute significantly to *TEKASA*, while the other mechanical properties hardly contribute at all: *TEKASA* will be high when the thickness & weight values are high; 5) the subjective evaluation of *KOSHI* and *TEKASA* are especially affected by the thickness & weight of fabrics.

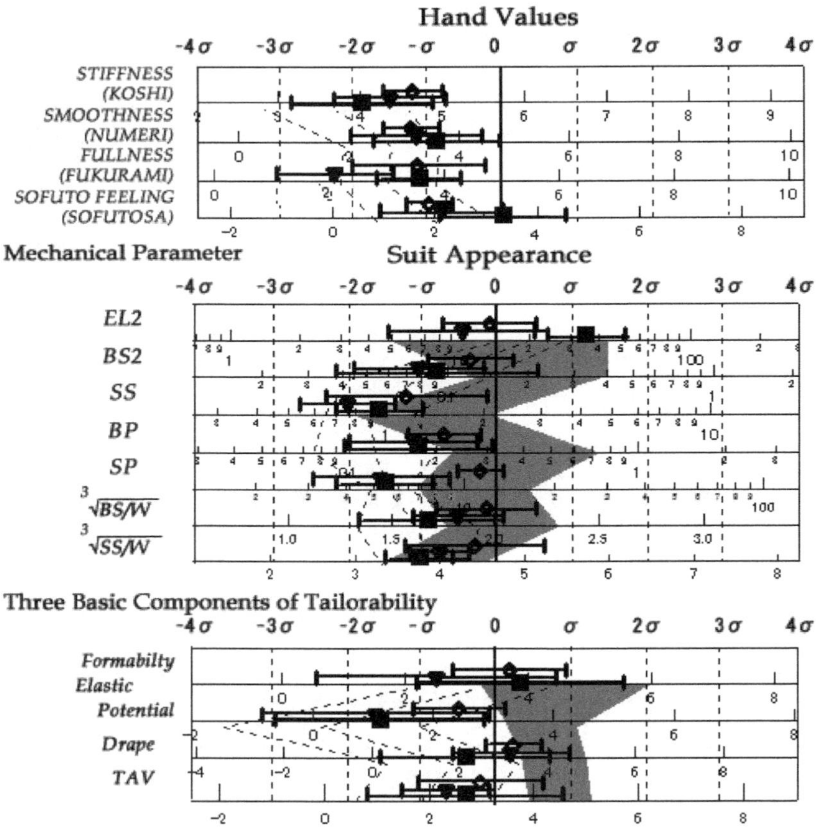

Suffix *1*: warp direction, suffix *2*: weft direction, ├▼┤ : Kawari Muji Chirimen (n=14), ├■┤ : Kodai Chirimen (n=26), ├◇┤ : Fuutsuu Chirimen (n=6), ├──┤ : mean value and ± standard deviation.
Polyester Chirimen (n=33) is indicated by the broken line.
Perfect property zone of men's suiting is indicated by the shaded zone.

Figure 7. Hand value and the mechanical parameters of Chirimen

(a) C_{1i}, C_{2i} table

KOSHI			TEKASA		
Importance order Xi	C_o: 3.7821 C_{i1}	C_{i2}	Importance order Xi	C_o: 3.7755 C_{i1}	C_{i2}
1. Bending			1. Construction		
logB1	6.6610	6.4144	*logT*	1.2341	0.3498
logB2	7.7449	6.3543	*logW*	-7.0396	8.3058
log2HB1	-7.0768	-6.7456	2. Tensile		
log2HB2	-6.7346	-6.7956	*LT*	0.2771	-0.1679
2. Construction			*logEM1*	6.1277	-5.9271
logT	0.3426	0.8121	*logEM2*	-4.0092	3.6738
logW	-0.2476	1.4344	*RT*	0.0651	-0.4849
3. Tensile			3. Bending		
LT	3.8422	-3.9285	*logB1*	3.2375	3.1516
logEM1	0.7865	-0.9586	*logB2*	3.2438	2.3411
logEM2	2.4756	-3.0446	*log2HB1*	-5.8773	-5.8259
RT	-2.6962	1.8785	*log2HB2*	-5.1234	-4.0671
4. Surface			4. Surface		
MIU	0.1663	-0.2863	*MIU*	1.4497	-1.5179
logMMD	0.3402	0.2634	*logMMD*	1.1149	1.1132
logSMD	-0.6171	0.7915	*logSMD*	-0.1162	0.2109
5. Shear			5. Shear		
logG	0.7408	0.7425	*logG*	-0.3474	-0.2097
log2HG	0.4447	0.6033	*log2HG*	-0.3245	-0.2806
log2HG5	0.1643	-0.1160	*log2HG5*	0.0785	0.1211
6. Compression			6. Compression		
LC	0.4654	-0.4600	*LC*	0.3061	-0.3627
logWC	0.9128	0.8409	*logWC*	1.7455	1.7135
RC	0.5144	-0.5053	*RC*	0.0034	0.0438

(b) M_{i1} M_{i2}, σ_{i1} and σ_{i2} table

Mechanical properties Xi	Chirimen fabrics (n=139) M_{i1}	M_{i2}	σ_{i1}	σ_{i2}
Tensile				
LT	0.7103	0.5112	0.0819	0.1140
logEM1	1.9060	3.6570	0.1536	0.5955
logEM2	1.9920	4.0520	0.2920	1.2029
RT	40.500	1672.0	5.7041	449.39
Bending				
logB1	-1.2750	1.6360	0.1042	0.2735
logB2	-1.1510	1.4460	0.3483	0.7825
log2HB1	-1.7040	2.9230	0.1433	0.4997
log2HB2	-1.5250	2.4630	0.3714	1.0944
Shear				
logG	-0.4257	0.1900	0.0935	0.0803
log2HG	-0.4486	0.2343	0.1819	0.2156
log2HG5	0.1141	0.0665	0.2313	0.0714
Compression				
LC	0.4772	0.2300	0.0480	0.0470
logWC	-1.2010	1.4760	0.1851	0.4252
RC	57.550	3380.0	8.2580	1008.4
Surface				
MIU	0.1765	0.0324	0.0349	0.0137
logMMD	-1.7330	3.0570	0.2299	0.7874
logSMD	0.5596	0.3851	0.2683	0.2791
Construction				
logT	-0.3769	0.1512	0.0959	0.0692
logW	1.1510	1.3280	0.0607	0.1404

Table 4. Parameters for translating mechanical values into hand values for Chirimen fabrics

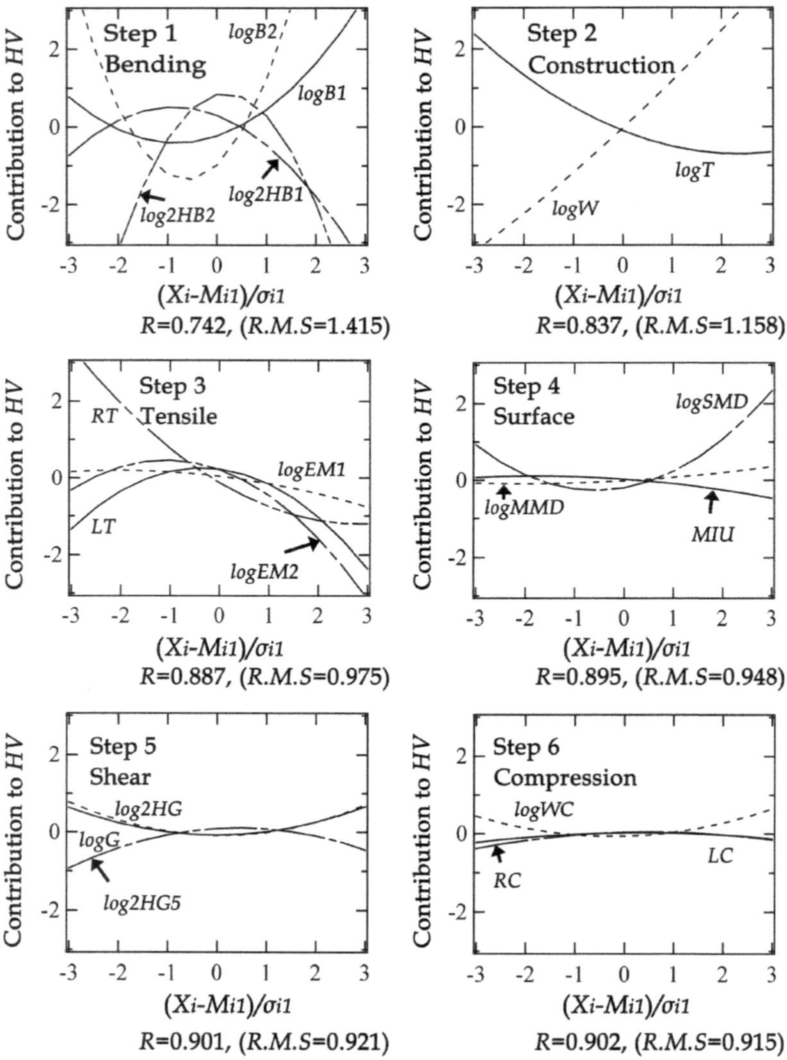

(Number of samples =139, number of subjects =4, suffix *1* in the figure of tensile and bending properties: warp direction, suffix 2 in the figure of tensile and bending properties: weft direction.)

Figure 8. The contribution of each mechanical property to the hand value *KOSHI* of Chirimen

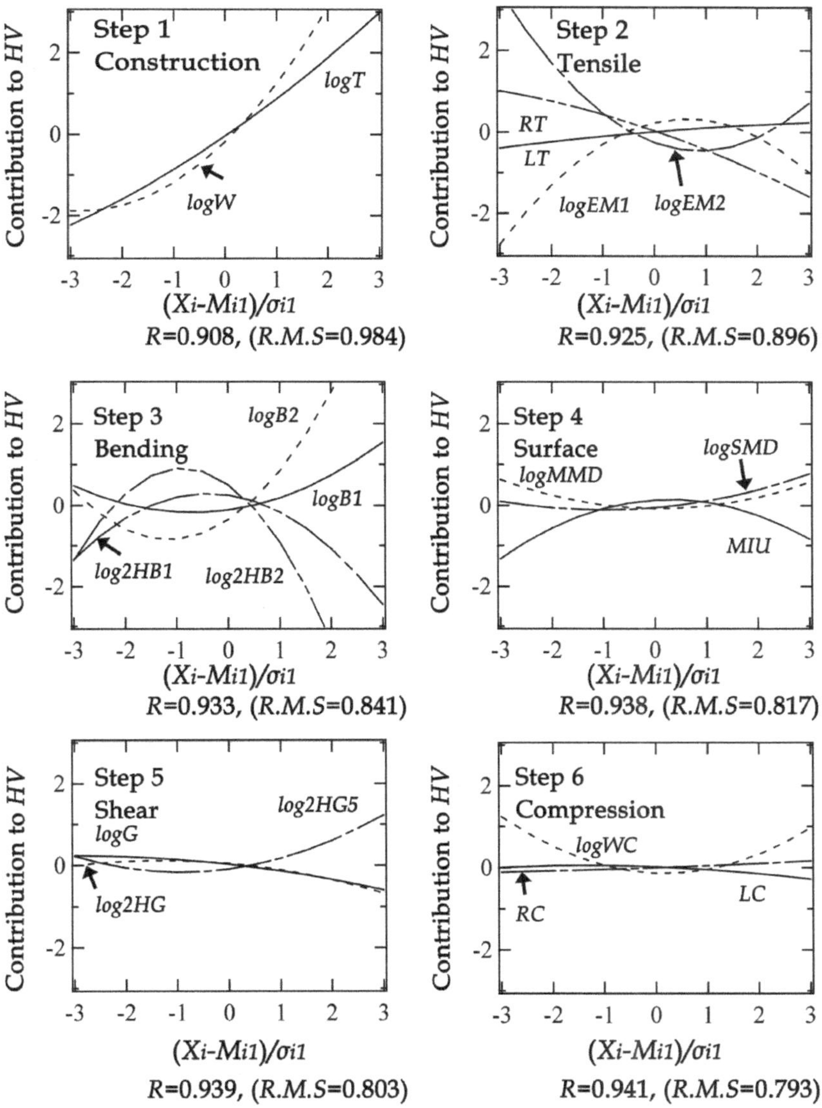

(Number of samples =139, number of subjects =4,
suffix *1* in the figure of tensile and bending properties: warp direction,
suffix *2* in the figure of tensile and bending properties: weft direction.)

Figure 9. The contribution of each mechanical property to the hand value *TEKASA* of Chirimen

5. Silhouette design of 'Chirimen' fabrics

5.1. Silhouette distribution of fabrics used in clothing in the west

The determination of silhouette design under standard conditions was also made for men's suiting (Kawabata, 1980); the results are shown in Fig. 10. The materials for men's suits were manufactured in Japan, and their sample values from 1980 were used for this analysis. Winter suits are either wool or blends, while those for summer use are also mainly wool or blends, with some linen or cotton. The center of gravity of the men's winter suiting is in almost the same area as that of ladies' suiting, but it is located away from the drape area, and is inclined in the direction of that for the *Hari* type.

Men's summer suiting is similar to the ladies' spring and summer suiting analyzed thus far (Inoue & Niwa, 2009), both being inclined in the direction of the *Hari* type. The difference between men's and ladies' suiting for spring and summer is that many ladies' silhouettes are located near the Drape type area (Inoue & Niwa, 2009), while men's silhouettes are away from the Drape type area. It was established that the silhouettes of men's suiting are distributed in the Tailored type area, in which silhouettes conform to the human's body shape, and in the direction of the *Hari* type area, in which silhouettes do not conform to the human's body shape.

Fig. 10 also shows the distribution of the silhouettes of ladies' medium thick suiting (Kawabata, 1980) and dress shirt fabrics (Matsudaira et al., 1984). These fabrics were manufactured in Japan; the sample values used for this analysis are from the 1980s. The materials were either wool or blends in ladies' medium thick suiting, and cotton, silk, polyester, wool or blends in dress shirt fabrics. Ladies' suiting is totally different from men's suiting, having a wide distribution, and being in the area where the beauty of each silhouette can be demonstrated. The silhouette range of dress shirt fabrics is between the Drape type area and the *Hari* type area, and there is a wide distribution.

As a result of clarifying the silhouette area of each fabric for each purpose via the above analysis, it was established that the range of clothing that humans are comfortable with is narrow.

5.2. Silhouette distribution of 'Chirimen' fabrics

The 271 silk Chirimen fabrics and 33 polyester Chirimen fabrics were divided into three optimum silhouette design groups on the basis of their mechanical properties. Fig. 10 shows the Z_1-Z_2 values plotted for each silhouette type of polyester Chirimen fabrics; those for silk Chirimen fabrics are shown in Fig. 11. Each symbol plotted represents a separate type of Chirimen fabric.

Chirimen fabrics are widely distributed in the range from Drape type to *Hari* type. Hitokoshi Chirimen fabrics are located towards the Drape type area, while Kodai Chirimen fabrics are inclined towards the Tailored type a little more than Hitokoshi Chirimen fabrics. Muji Isho Chirimen fabrics, meanwhile, are widely distributed in the range from *Hari* type to Drape type, although some are distributed in the range from Tailored type to Drape type. Rinzu Chirimen fabrics are also widely distributed in the range from *Hari* type to Drape type, but they tend to be further away from the center of gravity than the Muji Isho Chirimen fabrics. Chirimen fabrics are more widely distributed in the range from *Hari* type to Drape type than western fabrics used in clothing; their individual distribution areas have their own characteristic silhouettes depending on fabric type.

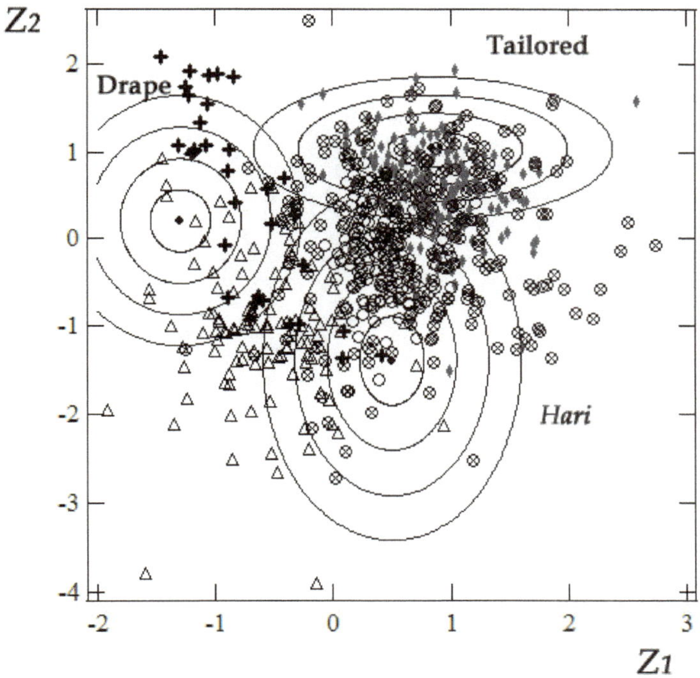

Men's winter suiting: ♦ (n=214), men's summer suiting : ο (n=156),
Women's suiting : ⊗ (n=220), dress shirt: Δ(n=116), polyester Chirimen: + (n=33)

Figure 10. Determination of optimum silhouette design for fabrics used in clothing in the west and for polyester Chirimen fabrics, based on measurements made under standard conditions

The 33 polyester Chirimen fabrics used have silhouette areas largely overlapping with those of Hitokoshi Chirimen fabrics, but, unlike silk Chirimen fabrics, some of the polyester ones have silhouette distributions farther away from the center of gravity of the Drape type, and others have distributions which are located at the center of gravity of the *Hari* type. From this it is clear that, when used for making the traditional Japanese kimono, polyester Chirimen fabrics may not perform at the level of quality achievable with silk Chirimen ones, since it will yield different silhouettes, and cannot create the same unique atmosphere as the silk Chirimen fabrics can. On the positive side, however, the fact that polyester Chirimen fabrics will yield different silhouettes also means that it is possible with polyester Chirimen to express silhouettes which cannot be created with silk Chirimen.

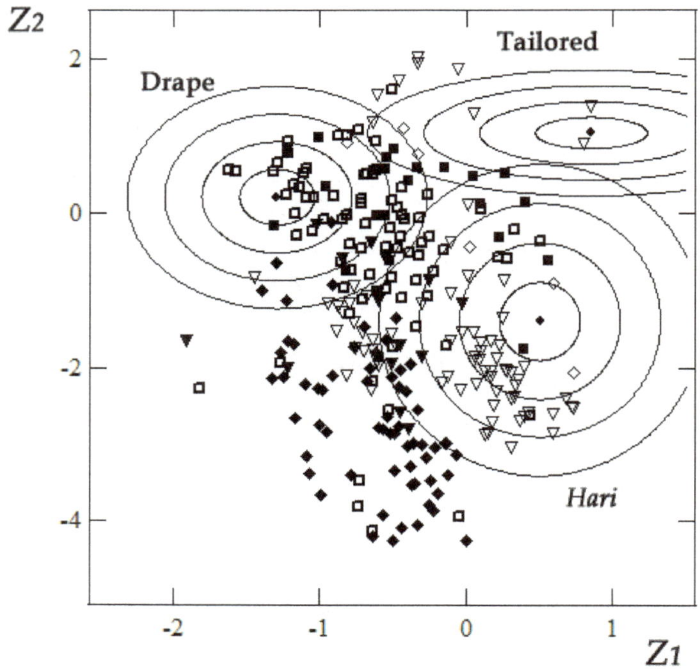

Hitokoshi Chirimen: □ (n=78), Muji Isho Chirimen: ▽(n=76)
Rinzu Chirimen: ◆ (n=63), Kawari Muji Chirimen: ▼(n=14)
Kodai Chirimen: ■ (n=26), Fuutsuu Chirimen: ◇(n=6)

Figure 11. Determination of optimum silhouette design for silk Chirimen fabrics based on measurements made under standard conditions

6. Conclusions

Japanese traditional 'Chirimen' fabrics are used for making kimonos, which have a fixed structure and are worn in very particular ways. These fabrics have also been used as dress fabrics in recent years. When the characteristics of the mechanical properties of various types of Chirimen were investigated to clarify differences in their hand value and appearance in clothing, the principal results were as follows.

1. Values for the weft direction of bending properties of all Chirimen groups, men and women's suit fabrics, and dress shirt fabrics were at the same level.

2. A significant feature of the mechanical parameters of each Chirimen group (excluding *SP* and *BS2* which are compound values of bending properties and shearing properties) was that they were in the range for ideal men's suiting.
3. Regarding hand value, '*KOSHI*' (stiffness) of Chirimen was found to be closely related to the bending properties, thickness and weight of the fabric, and '*TEKASA*' (hand quantity) of Chirimen was found to be closely related to the thickness and weight of the fabric.
4. When silhouettes were determined, the Chirimen fabrics were found to be widely distributed in the range from Drape type to Hari type, with their individual distribution areas having their own characteristic silhouettes depending on fabric type.
5. Differences in the silhouette designs of Chirimen fabrics and fabrics used in clothing in the West became clear.
6. The range of silhouette distribution for each kind of Chirimen fabric was clarified; this revealed that the silhouettes possible with Chirimen fabrics varied according to the type of Chirimen used.

It is hoped that the knowledge of the basic performance of Japanese traditional 'Chirimen' fabrics will help those in the industry recognize the importance of the mechanical properties of fabric, and further promote the designing of polyester 'Chirimen' fabrics.

Author details

Takako Inoue and Masako Niwa
Sugiyama Jogakuen University, Nara Women's University, Japan

7. References

Inoue, T. & Niwa, M. (2003). Objective evaluation of the quality of ladies' garment fabrics, (Paper translated from the Japanese ed.), *J. Text. Eng.*, Vol. 49, No.2, pp.33-45

Inoue, T. & Niwa, M. (2009). Objective evaluation of the quality of fabrics for ladies' Tailored-type jackets for spring and summer, *J. Text. Eng.*, Vol. 55, No.1, pp.1-11

Inoue, T. & Niwa, M. (2010a). Hand evaluation and formability of Japanese traditional 'Chirimen' fabric, *Int J Cloth Sci Tech*, Vol.22, No. 4, 234-247

Inoue, T. & Niwa, M. (2010b). Determination of optimum silhouette design for ladies' garment fabrics based on measurements made under the standard conditions for fabric mechanical properties, *Proceedings of the 39th Textile Research Symposium at IIT Delhi*, pp. 338-344

Inoue, T. & Niwa, M. (2011). Optimum silhouette design for traditional Japanese kimono fabrics, *Proceedings of Annual Meeting in Nishinomiya, Jpn. Res. Assn. Text. End-Uses.*, p.14

Inoue, T., Nakayama, A. & Niwa, M. (2010). Relationship between the warm/cool feeling of fabric and the subjective evaluation of the quality of ladies' knitted fabrics, *Int J Cloth Sci Tech*, Vol.22, No. 1, pp.7-15

Kawabata, S. & Niwa, M. (1989). Fabric performance in clothing and clothing manufacture, *J. Text. Inst.*, Vol. 80, No. 1, pp.19-50

Kawabata, S. (1972). Measuring system of fabric handle, *Sen-I To Kogyo (Fabric Science and Industry, Japan)*, 5, 277-285

Kawabata, S. (1980). *The standardization and analysis of hand evaluation second edition. HESC*, the Textile Machinery Society of Japan, Osaka, Japan

Kawabata, S. et al. (1996~1997). Development of Design Guidelines for Manufacturing Ideal Fabrics for Apparel Fabrics, *Research Project Grant-in Aid for Scientific Research (A)*, Project Number 08555237, p. 22

Kawabata, S., Ito, K. & Niwa, M. (1992). Tailoring process control, *J. Textile Inst.*, 83, 361-74

Kawabata, S., Niwa, M. & Yamashita, Y. (2002). Recent developments in the evaluation technology of fiber and textiles: Toward the engineered design of textile performance, *Journal of Applied Polymer Science*, Vol. 83, pp.687-702

Komatsu, K. & Niwa, M. (1981). Characteristics of physical properties of kimono fabric, *Research Journal of Living Science*, Vol.28, No.1, pp.235-243

Matsudaira, M., Kawabata, S. & Niwa, M. (1984). Measurements of mechanical properties of thin dress fabrics for hand evaluation, *J. Text. Mach. Soc. Japan (predecessor Journal of J. Text. Eng.)*, Vol.37, T49-T57

Nakae, K. (editor) (1993), *Encyclopedia of Dyeing*, Tairyusya, Tokyo

Niwa, M., Nakanishi, M., Ayada, M. & Kawabata, S. (1998). Optimum silhouette design for ladies' garments based on the mechanical properties of a fabric, *Textile Res. J.*, 69, 578-588

Tomiyama, H. & Ohno, C. (1967). *Japanese Traditonal Textile*, Tokuma Shoten, Tokyo

Yamamoto, R. (1960). *New Japanese Dressmaking (5th Edition)*, Kobun Shoin, Tokyo

Chapter 13

Compressive Stress Relaxation and Creep Properties of Synthetic Fiber and Regenerated Fiber Assemblies

Yoneda Morihiro and Nakajima Chie

Additional information is available at the end of the chapter

1. Introduction

In order to realize comfort environment for sleeping, material property of fiber which consists bedclothes is very important. Compression property of fiber assembly is needed to support human body comfortable, and heat and water transport property is needed to keep micro climate in bedclothes comfortable. In Japan, "futon" has been used for bedclothes, and use of "futon" is divided into mattress to support human body ("shiki-futon" in Japanese) and quilt to cover human body ("kake-futon" in Japanese). Futon consists of futon wadding made from fiber assembly and shell fabrics which covers futon wadding. Cotton fiber has been used for futon wadding use, and wool fiber increases its use for futon wadding recently. Therefore, in Japan, many studies concerning compression properties of futon wadding made from natural fiber such as cotton or wool has been conducted [1-5]. Recently, use of synthetic fiber including polyester for wadding increases because it is light and warm, easy to handle and easy to attach many kinds of functionalities. However, a study on compression properties of synthetic fiber assembly has not been conducted so much.

In this study, therefore, compression properties of fiber assembly made of synthetic and regenerated fiber for futon wadding will be investigated. The reason why we focus our study on compression properties is to explore feasibility of these materials as mattress use. In particular, compression viscoelastic properties such as stress relaxation and creep properties will be carried out in this study, because compression viscoelastic properties have not been carried out systematically in spite of its potential effect on sleeping comfort.

In this paper, results of repeated compression-recovery test, compression stress relaxation test and compression creep test of synthetic and regenerated staple fiber assembly are reported [6-9]. In addition, results of stress relaxation and creep behaviors of fiber assemblies are analyzed based on non-linear viscoelastic model.

2. Samples of fibers used

2.1. Fiber materials

Twelve kinds of staple fiber materials are used for experiment such as three kinds of polyester fiber with round section, three kinds of polyester fiber with heteromorphic section (w-shaped), one kind of polyester fiber with hollow section, three kinds of Cupra fiber and one kind of Lyocell fiber. Sample code is shown in Table1, and the details of samples are shown in Table 2. Web made from raw fiber material which is subjected to opening and carding process is used. Fiber assembly is conditioned in constant temperature and humidity room (20°C, 65%RH) over 24 hrs and served for experiment.

Sample code	Detail of sample
WPE-1	Polyester staple fiber with w-shaped heteromorphic section
WPE-2	
WPE-3	
RPE-1	Polyester staple fiber with round section
RPE-2	
RPE-3	
RPE-4	Polyester staple fiber with hollow section
PTT	Polytrimethyleneterephtalate staple fiber
CU-1	Cupra-ammonium (Cupra) staple fiber
CU-2	
CU-3	
LY	Lyocell staple fiber

Table 1. Sample code

2.2. Standard condition for the measurement

Important measurement condition in compression test is fiber volume fraction and maximum compression stress. Standard values for fiber volume fraction and maximum compression stress are estimated as follows. Standard value for fiber volume fraction is 0.025, of which value is estimated from standard size and wadding weight of futon commercially available in Japan. Standard value for maximum compression stress is 2352 Pa, of which value is based on the data that average pressure applied to futon by male adult is about 2.3 kPa. Measurement condition used in this study is determined based on these values of standard condition.

2.3. Sample preparation for fiber assembly

In the case of repeated compression-recovery test and compression stress relaxation test, fiber assembly sample is prepared by filling staple fiber into cylindrical cell made of Acrylic resin (inner diameter: 54mm, height: 100mm). In this case, setting of fiber volume fraction is important as one of the measurement conditions. Fiber volume fraction Φ_f in cylindrical cell is estimated as follows.

$$\Phi_f = 1.27 \text{ x } W/\rho d^2 h \tag{1}$$

where, W: sample weight (g), ρ: specific gravity of fiber (n.d.), d: inner diameter of cell (cm), h: height of cell (cm). Amount of staple fiber estimated by equation (1) is filled uniformly into cylindrical cell, and plastic disk of diameter 50mm is put on the top of fiber assembly. This cylindrical fiber assembly is served as samples to compression test.

Sample code	Fineness (dtex)	Fiber length (mm)	Fiber diameter (μm)	Specific gravity (n.d.)	Percentage of crimp (%)	Apparent young's modulus (kg/mm^2)	Bending rigidity (gf·cm^2)
WPE-1	1.4	38	24.22[*2]	1.38	14.80	318	5.37×10^{-5}[*4]
			6.27[*3]				2.41×10^{-7}[*5]
WPE-2	1.4	51	24.31[*2]	1.38	34.71	291	4.99×10^{-5}[*4]
			6.78[*3]				3.01×10^{-7}[*5]
WPE-3	2.2	B64[*1]	29.06[*2]	1.38	28.29	297	1.04×10^{-4}[*4]
			8.71[*3]				8.39×10^{-7}[*5]
RPE-1	1.3	38	12.76	1.38	26.70	412	5.36×10^{-6}
RPE-2	2.2	51	18.15	1.38	27.24	379	2.02×10^{-5}
RPE-3	6.6	51	29.17	1.38	33.33	262	9.31×10^{-5}
RPE-4	6.6	51	29.79	1.38	27.59	252	1.34×10^{-3}
PTT	1.7	51	15.02	1.38	15.43	154	3.85×10^{-6}
CU-1	1.4	38	11.84	1.50	8.79	645	6.22×10^{-6}
CU-2	1.4	51	13.73	1.50	15.06	639	1.11×10^{-5}
CU-3	2.2	76	15.46	1.50	20.51	475	1.33×10^{-5}
LY	7.0	64	33.56	1.50	14.62	359	2.23×10^{-4}

Table 2. Details of fiber samples

In the case of compression creep test, fiber assembly of which shape is rectangular prism (10cm square base) is used as sample. Fiber volume fraction Φ_f in rectangular prism is estimated as follows.

$$\Phi_f = W/\rho a^2 h \tag{2}$$

where, W: sample weight (g), ρ: specific gravity of fiber (n.d.), a: length of base (cm), h: height of prism (cm). Amount of staple fiber assembly estimated by equation (2) is served as samples to compression creep test after pre-processing.

3. Experimental measurements

3.1. Repeated compression-recovery test

Repeated compression-recovery behavior is measured using KES-G5 compression tester (Figure 1)(Kato Tech Co.)[6]. The movement of compression and recovery is applied to fiber assembly filled in cylindrical cell by metal plate with area 20cm^2 with constant rate, 1mm/sec. The displacement of metal plate is detected by potentiometer, and the movement is controlled by automatic control. Compression stress is measured by strain gauge attached to metal rod which is connected perpendicular to the metal plate. When compression stress reaches maximum value, Pmax, the compression movement is turned to recovery one. The test was carried out at four different level of Pmax, 784, 1176, 2352 and 3528 Pa. Number of repeating cycles of compression and recovery is 31 times. The output of electronic signal of the compression stress and the displacement of metal plate from amplifier is recorded by data logger with constant sampling time, 1 sec, and data processing is carried out using personal computer.

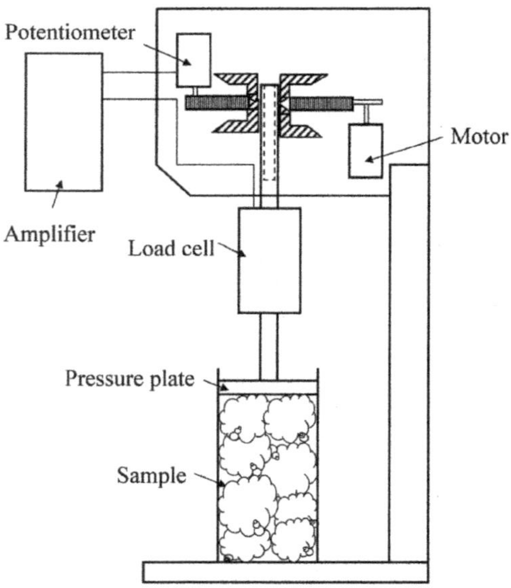

Figure 1. KES-G5 Compression Tester

3.2. Compression stress relaxation test

Compression stress relaxation behavior is measured using KES-G5 compression tester (Kato Tech Co.) on stress relaxation mode [7,8]. Compression displacement is applied to fiber assembly filled in cylindrical cell by metal plate with area 20 cm^2, and initial height of sample is fixed at 10cm. Compression stress at this point is regarded as 0. Starting from this point, metal plate is driven downward with speed 5mm/sec to the initial compression displacement, and fixed. Compression stress change at constant displacement is measured from 0 to 10^4 sec with time elapsed. The initial compression displacement is set at five different levels (1, 2, 3, 4 and 5 cm). The output signal of compression stress change is recorded using data logger with sampling time, 1 sec, and the data processing is carried out by personal computer.

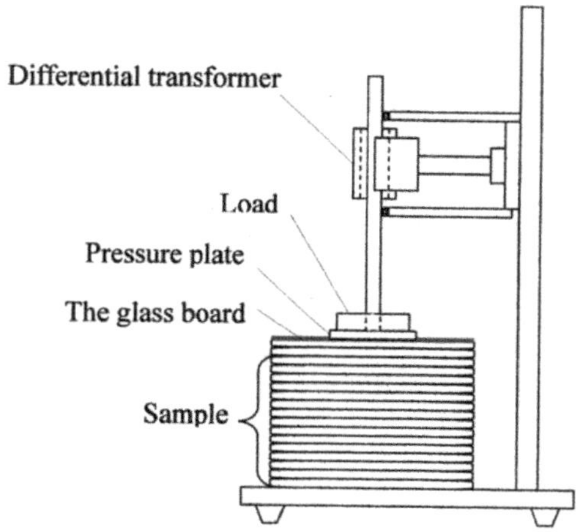

Figure 2. Compression Creep Tester

3.3. Compression creep test

Compression creep behavior is measured using Compression Creep Tester (Figure 2) (Kato Tech Co.)[9]. Metal plate of 10 cm square loaded with constant compression weight is allowed to move downward slowly, and let the metal plate touch to the top of fiber assembly of rectangular prism with 10 cm square. The time when metal plate touches top of sample is taken as 0sec, and measurement of creep starts. Creep deformation is detected by the displacement of metal rod connected perpendicular to metal plate using differential transformer. Creep test is carried out at two different compression load level, 1176 and 2156 Pa. The output signal of creep displacement is recorded by data logger with time elapsed, and data processing is carried out using personal computer. The measurement is carried out from t=0 to 10^4 sec. Compression creep rate, R_t (%) is defined as follows,

$$R_t = \varepsilon_t / h_0 \times 100\ (\%). \tag{3}$$

where, h_0: sample height at t=0 (sec), ε_t: compression creep displacement (mm) at time t (sec).

4. Results and discussion

4.1. Repeated compression-recovery test

4.1.1. Evaluation method

In general, evaluation of compression properties of fiber assembly (including fabrics) is conducted by characteristic parameters obtained from the measurement of compression-recovery curve as shown in Figure 3. In the case of KES evaluation system, for example, following parameters are used for the evaluation of compression properties [10].

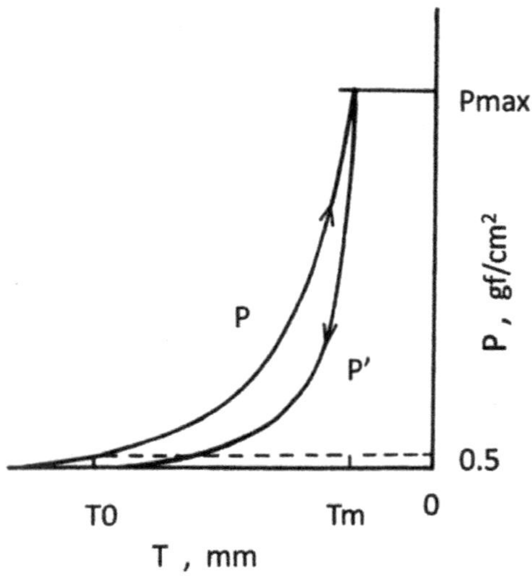

Figure 3. Compression-Recovery Curve

WC: Compressional energy obtained from compression curve P (gf.cm/cm^2)
LC: Linearity of compression curve (n.d.)
RC: Compressional resilience (%)
Each parameter is defined as follows.

$$WC = \int_{Tm}^{T0} PdT \tag{4}$$

$$LC = WC / WOC \tag{5}$$

$$RC = \left(WC' / WC\right) \text{ x } 100 \tag{6}$$

where, T: Thickness of sample (mm)
T_0: Thickness of sample at P=0.5gf/cm^2 (mm)

T_m: Thickness of sample at maximum pressure, P_m (mm).

$$WOC = P_m(T_0 - T_m)/2 \tag{7}$$

WC′: Recovery energy obtained from recovery curve P′

$$WC' = \int_{Tm}^{T0} P'dT \tag{8}$$

In the case of evaluating compression properties of fabrics, compression-recovery test is carried out only in one cycle. This is because that natural state of fabrics is generally defined clearly. In contrast, the natural state of three-dimensional fiber assembly is difficult to define, because it does not have specific shape. It is supposed that state of assemble is not uniform and residual stress is still remained for the bulk fiber sample filled in cylindrical cell. In this study, therefore, we try to trace the change of compression properties of sample by repeating a cycle of compression-recovery process for suitable number of cycles. After the investigation of property change with repeating cycles, we will decide in how many cycles are suitable for evaluating compression properties.

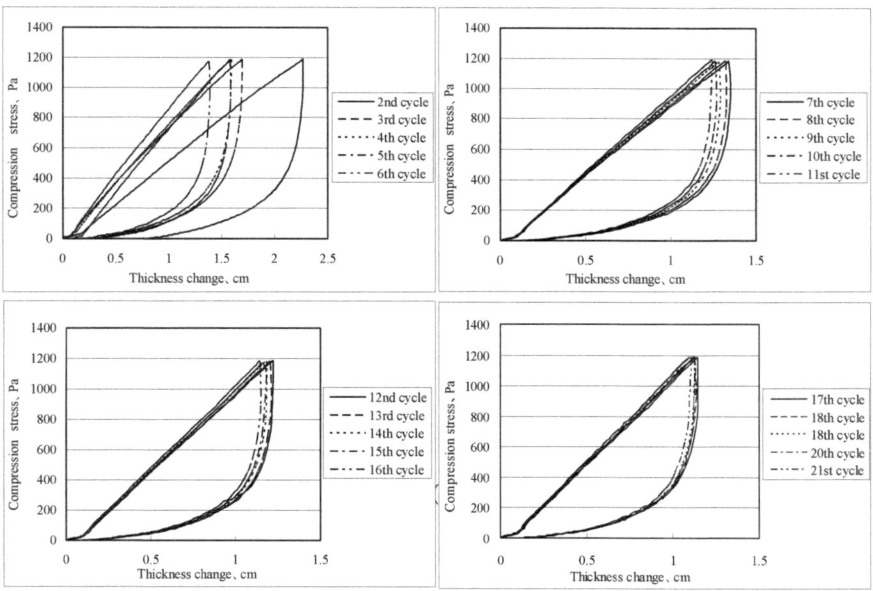

Figure 4. Compression-recovery curve for sample WPE-1

Figure 4 shows the compression-recovery curve in repeated compression test for the polyester staple fiber with heteromorphic section (sample: WPE-1). In this study, the measurement is carried out under the condition that initial fiber volume fraction is 0.0125 for three different levels of maximum compressional stress (784, 1176, 2352 Pa). Each graph includes five cycles of repetition. The data are shown from 2nd cycle to 21st cycle considering nonuniformity of fiber density and residual stress at 1st cycle. Each compression-recovery curve can be distinguished up to 6th cycle, and thereafter the curves have a tendency to overlap each other, and finally difference between each curve is not recognized from 22nd to 21st cycles.

In order to analyze the change of compression-recovery curve against repeating cycles, characteristic parameters, WC, WC′ and Tm−T0 per one cycle are obtained. Since WC′ and Tm−T0 have the same tendency as WC, results of WC for three samples are discussed in this section.

Behavior of compression energy in compression process, WC is shown in Figures 5,6,7 (sample: WPE-1, RPE-1, CU-1). Characteristic value, WC decreases with increasing repeating cycle, and the degree of decreasing is divided into three stages. At the first stage, from 2nd to 5th cycle, WC decreases rapidly with increasing number of cycles. At the second stage, from 6th to 20th cycles, decrease of WC becomes slowly. At the third stage, from 21st to 31st cycles, WC curve levels off and reaches equilibrium state.

Figures 5,6,7 show results for three different levels of maximum compression stress, Pm, 784, 1176 and 2352 Pa, respectively. From the graph, it is clear that maximum compression stress, Pm influences the relationship between samples in magnitude of WC. When Pm is 784 Pa, WC of PET fiber with heteromorphic section and Cupra fiber is same level, and WC of PET fiber with round section is smaller compared to the others. When Pm is 1176 Pa, WC of Cupra fiber is largest for all cycles, and PET fiber with heteromorphic section follows, and PET fiber with round section is smallest. When Pm is 2352 Pa, magnitude of WC for three samples becomes of the same level for all cycles.

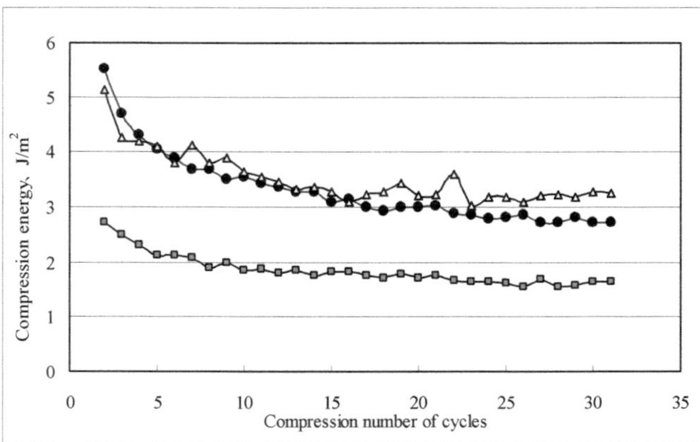

Figure 5. WC vs. Compression number of cycles (784Pa)
Sample: ●:WPE-1, ■: RPE-1, △:CU-1

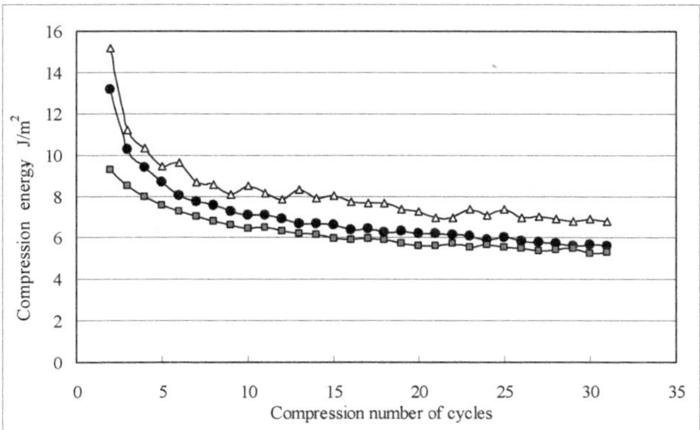

Figure 6. WC vs. Compression number of cycles (1176Pa)

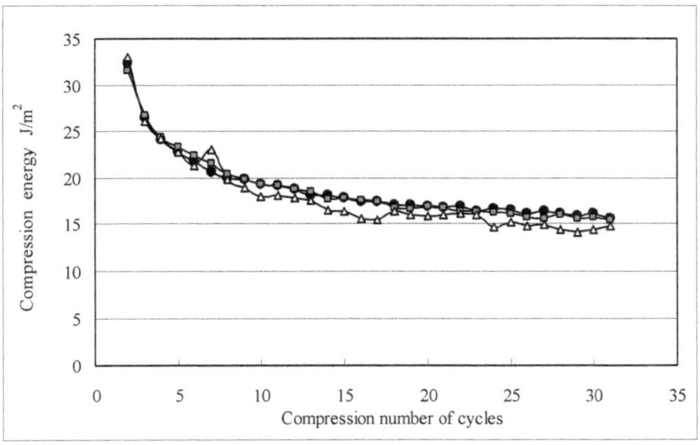

Figure 7. WC vs. Compression number of cycles (2352Pa)

The results show that, in repeated compression-recovery test for fiber assembly, the difference in mechanical parameters between samples appears in some cases or does not in other cases depending on maximum pressure condition. Therefore, in order to characterize the feature in compression properties of fiber assembly, suitable sampling condition in maximum compression pressure must be selected in carrying out compression test.

4.1.2. Analysis by linearizing method

It is confirmed that the feature of compression properties for each sample appears at the 6th cycle of repeated compression-recovery test under the condition of maximum compression pressure, Pm 1176 Pa and fiber volume fraction 0.0125 as discussed in the last section. In this study, the shape of the 6th cycle of repeated compression-recovery curve at the condition mentioned above is analyzed based on linearized method proposed by Kawabata [11].

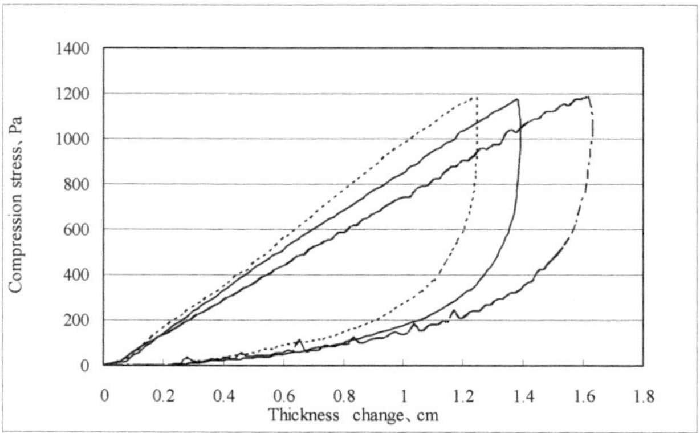

Figure 8. Compression-recovery curve at 6th cycle under the condition that volume fraction is 0.0125 and maximum compression stress is 1176 Pa for three samples.
Straight line: WPE-1
Dotted line: RPE-1
Dots and dashes: CU-1

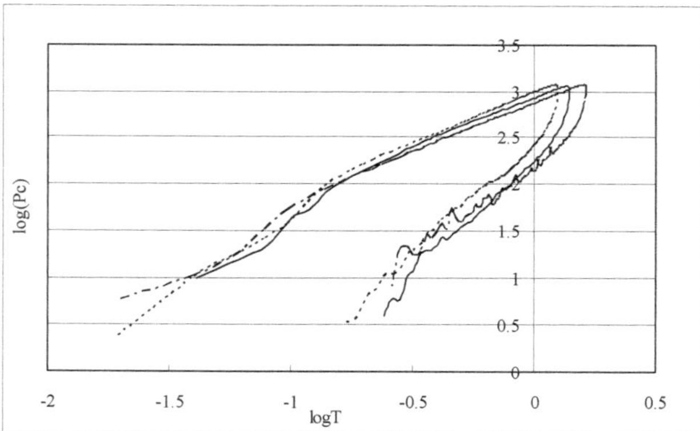

Figure 9. A plot of the relationship between the logarithm of compression stress and the logarithm of thickness change at 6th cycle.
Straight line: WPE-1
Dotted line: RPE-1
Dots and dashes: CU-1

The compression-recovery curves for three samples (WPE-1, RPE-1, CU-1) are shown in Figure 8. The ordinate is compression stress, Pc (Pa), and the abscissa is compression displacement, T (cm). A plot of logarithm of Pc against logarithm of T is shown in Figure 9. It is confirmed that the relationship between log Pc and log T is almost linear, and the line has inflexion point both in compression and recovery process. In other words, the relationship between log Pc and log T is divided into two stages at a certain point. The range in which linear relationship holds is summarized in Table 3. The range in which linear relationship at lower stress holds is called as lower stress region, and that of higher stress is called as higher stress region. The relationship between log Pc and log T in compression and recovery process is expressed in the following way.

	Compression process		Recovery process	
Code	Low stress area	High stress area	Low stress area	High stress area
WPE-1	0~81.8	81.8~1176	0~27.4	27.4~1176
RPE-1	0~63.7	63.7~1176	0~20.6	20.6~1176
CU-1	0~70.1	70.1~1176	0~30.0	30.0~1176

Table 3. The range where logPc-logT shows linear relationship (unit:Pa)

Parameter	Code	Compression process		Recovery process	
		Low stress area	High stress area	Low stress area	High stress area
n	WPE-1	1.752	1.104	3.553	2.487
	RPE-1	1.727	1.146	2.756	2.213
	CU-1	1.436	1.063	3.097	1.862
A	WPE-1	2275	865	664	182
	RPE-1	2313	984	470	272
	CU-1	1299	740	622	160

Table 4. Parameters n and A

$$\log Pc = \log A + n \log T \tag{9}$$

where, A and n are constants determined by intersection and slope of the curves shown in Figure 9. The A and n depend on mechanical properties of sample, measurement condition and stress region. The values of A and n for three samples and compression and recovery process are summarized in Table 4. Taking equation (9) into account, relationship between compression stress, Pc (Pa) and compression displacement, T (cm) is expressed as follows,

$$Pc = A\,T^{n}. \tag{10}$$

The feature of compression-recovery curve can be characterized by parameters A and n.

The effect of properties of fiber on the behavior of compression-recovery curve is investigated. The relationship between parameters A, n and fiber property is different for compression or recovery process and for lower stress region or higher stress region. In the compression process, A and n values depend on such parameter as number of crimp, percentage of crimp and apparent Young's modulus. In the recovery process, A and n values depend on bending rigidity in addition to parameters mentioned above. This fact may support the supposition that the driving force in the recovery process may be bending recovery energy of fiber element.

4.2. Compression stress relaxation test

4.2.1. Residual stress ratio and approximation equation

In this section, functional form of compression stress relaxation behavior for fiber assembly is discussed.

Residual stress ratio $S\gamma$ is defined as follows,

$$S_Y = \sigma / \sigma_0, \tag{11}$$

where, σ: stress at time t (Pa), σ_0: initial stress (stress at t=0) (Pa).

Results of measurement of residual stress ratio for all samples are shown in Figures 10～12. Figure 10 shows the results of PET with heteromorphic section and PTT, Figure 11 shows that of PET with round and hollow sections, and Figure 12 shows that of Cupra and Lyocell fibers. In each figure, compression displacement is 5 cm. Ordinate is residual stress ratio S_Y (n.d.) in linear scale, and abscissa is natural logarithm of time t (s). As shown in figure, the curve can be approximated by linear function above $t=10^2$ sec for all the samples. Therefore, the curve of residual stress ratio S_Y for $t>10^2$ sec can be expressed approximately as follows,

$$S_Y = K(a_0 - \ln t) \tag{12}$$

where, t: time (s), K, a_0: constants determined by sample type and measurement condition (n.d.). K and a_0 were determined by least square method based on measurement curve. K values of regenerated fiber assembly are greater than that of synthetic fiber assembly for all displacement conditions. While K values of synthetic fiber range from 0.0075 to 0.0221, K values of regenerated fiber assembly range from 0.0354 to 0.0438. The smaller K value is, the slower compression relaxation speed becomes. It is supposed that synthetic fiber assembly has high sustainability of elasticity and high resistance to compression judging from its small K values. Summarizing these results, evaluation of compression relaxation by K value is very important from the viewpoint of performance in use such as sleeping comfort. It is conjectured that K value is influence by fiber material properties such as number of crimp, crimp percentage, apparent Young's modulus, bending rigidity, inter-fiber friction and viscoelastic properties.

4.2.2. Functional expression for compression stress relaxation of regenerated fiber assembly

In this section, functional expression for compression stress relaxation of regenerated fiber assembly which holds all time regions in this measurement ($0<t<10^4$ sec) was discussed.

Compression stress relaxation phenomena for regenerated fiber assembly including Cupra fiber is expressed in the following way [2].

$$S_Y = K \ln\left[\coth\{1/2\left(2e^{-a_0}t + B\right)\}\right] \tag{13}$$

where, K and a_0 are constants determined from equation (2) (n.d.). Constant B can be obtained by substituting t=0 and $\sigma=\sigma_0$ into equation (13). Functional expression (13) is derived from the nonlinear two-element viscoelastic model as shown in Figure 13 [12]. This model consists of an elastic element and a Non-Newtonian viscous element. Its viscous behavior is expressed as follows,

$$d\varepsilon / dt = K \sinh(\alpha f) \tag{14}$$

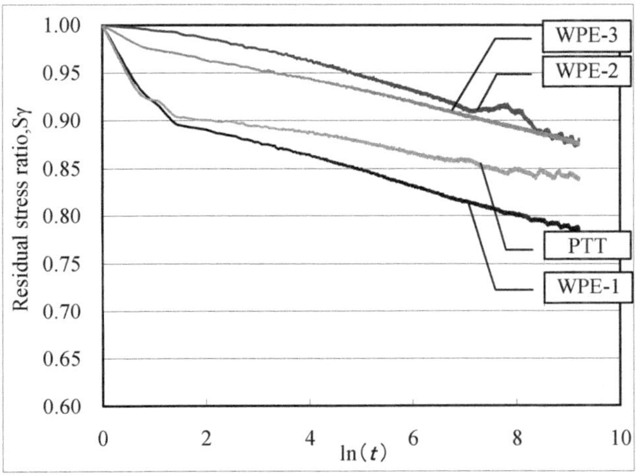

Figure 10. Compression relaxation curve (WPE-1,2,3, PTT, disp.=5cm)

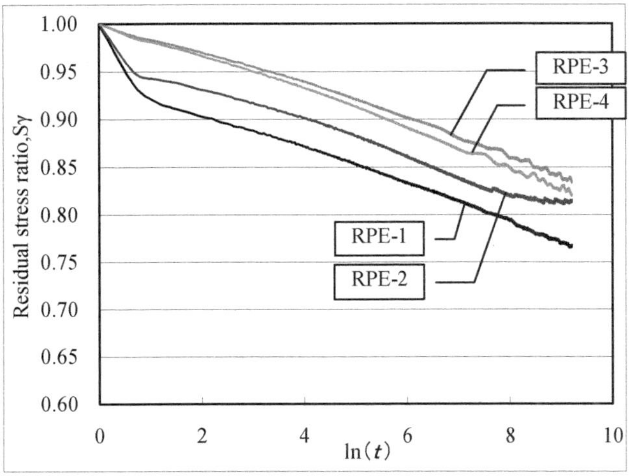

Figure 11. Compression relaxation curve (RPE-1,2,3,4, disp.=5cm)

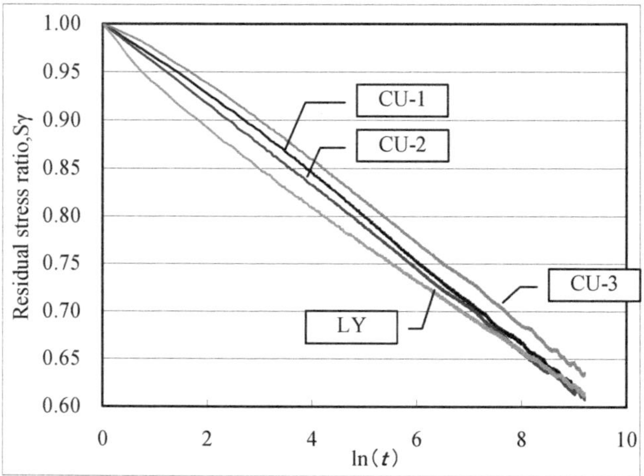

Figure 12. Compression relaxation curve (CU-1,2,3,LY, disp.= 5cm)

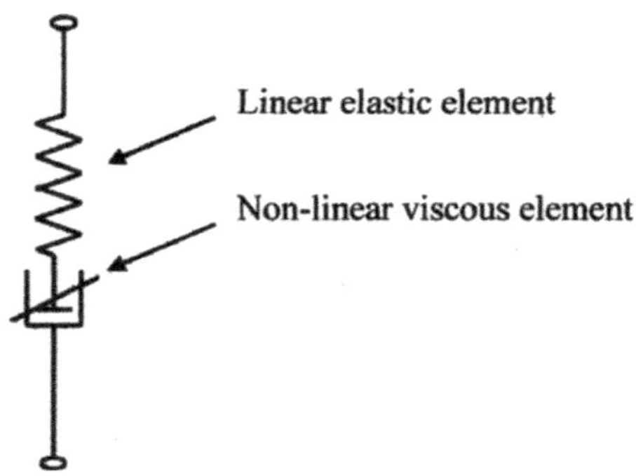

Figure 13. Nonlinear two element model

where, $d\varepsilon/dt$: strain rate, f: force, and K, α: constants concerning non-linear viscous element. The behavior of equation (14) when f $\rightarrow 0$ agrees with that of Newtonian viscosity.

The comparison between calculated curves by equation (13) and measurement curves is shown in Figures 14 and 15. Figure 14 shows the result of Lyocell staple fiber assembly (7dtex, 64mm) (LY) for initial displacement 2 cm. Figure 15 shows the result of Cupra staple fiber assembly (CU-2) for initial displacement 2 cm. In each figure, calculated and measurement curves have good agreement for whole time region. It was known that equation (13) holds for compression stress relaxation phenomena of cotton fiber assembly by Nogai et al [2]. It is confirmed that equation (13) also holds for stress relaxation of cellulosic regenerated fiber assembly such as Cupra and Lyocell fibers in this study.

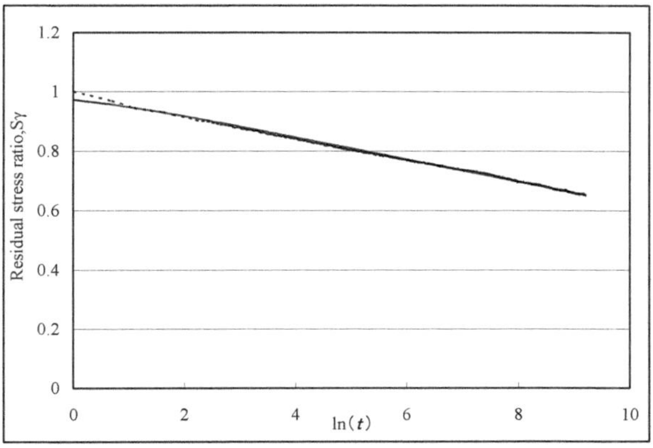

Figure 14. Comparison between experimental and calculated curves (LY) straight line: calculation, dotted line: experimental

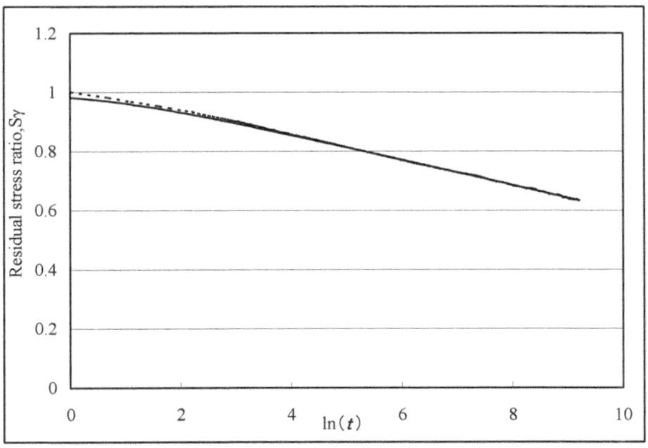

Figure 15. Comparison between experimental and calculated curves (CU-2) straight lime: calculation, dotted line: experimental

4.2.3. Functional expression for compression stress relaxation of synthetic fiber assembly

For synthetic fiber assembly, stress relaxation phenomena for whole time region ($0<t<10^4$ sec) cannot be expressed by equation (13). This arises from the reason that coth type solution cannot be obtained because K values for synthetic fibers are smaller compared to regenerated fibers.

Empirical equation to express compression stress relaxation behavior for synthetic staple fiber assembly is investigated. It is confirmed that linear relationship holds for all samples by bi-logarithmic plot of stress-time curve. Therefore, the empirical equation is obtained as follows,

$$\sigma = \beta t^{-\alpha}. \tag{15}$$

where, σ: stress at time t (Pa), t: time (s), and α,β: constants determined by sample type and measurement condition. Agreement of compression stress relaxation curve for synthetic staple fiber assembly with empirical equation (15) is investigated. An example of α and β values for PET with heteromorphic section (1.4dtex, 51mm) and fiber volume fraction 0.0139 (WPE-2) is substituted into equation (15). Following equation is obtained.

$$\sigma = 4.68\, t^{-0.0164} \tag{16}$$

Comparison between calculated and measurement curves is shown in Figure 16. The agreement of curves calculated by equation (16) with measured curve (WPE-2) is good. As shown in Figure 16, calculated curves are in agreement with measurement curves very well for PET with heteromorphic section, PET with round section, PET with hollow section and PTT fibers. (It is hard to distinguish each curve, because the calculated and measured curves show good agreement.)

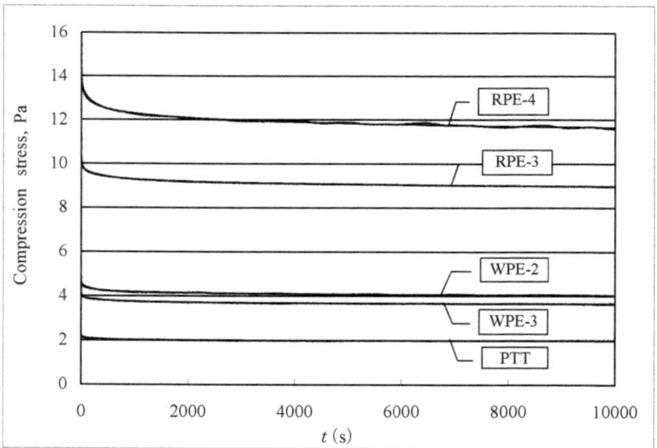

Figure 16. Comparison between experimental and calculated curves using eq.(15)

The relationship between α, β values and fiber volume fraction is shown in Figures 17 and 18, respectively. α value is almost constant against fiber volume fraction. β values increase with increasing fiber volume fraction. β values of PET with round and hollow sections of which count is high and having a tendency to be large.

In the case of synthetic staple fiber assembly, stress relaxation function has a form of decreasing power function with respect to time. This type of relaxation function does not have corresponding viscoelastic model. This is an empirical equation which holds when magnitude and rate in stress relaxation are very small (i.e. elastic deformation).

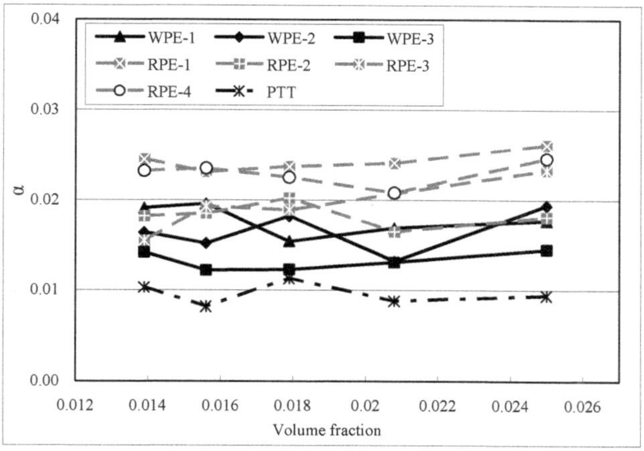

Figure 17. Relationship between α and volume fraction

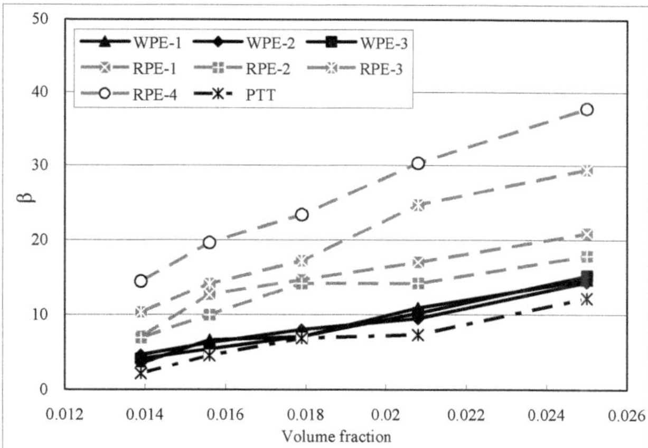

Figure 18. Relationship between β and volume fraction

4.3. Compression creep test

4.3.1. Results of compression creep test

Compression creep test for twelve samples were carried out under low stress condition (1176 Pa) and high stress condition (2156 Pa). Examples of the results for low stress condition are shown in Figures 19~21. Figure 19 shows the result of PET with heteromorphic section and PTT, Figure 20 shows that of PET with round section and hollow section, and Figure 21 shows that of Cupra and Lyocell fibers. The ordinate is compression creep displacement ε_t(mm), and the abscissa is natural logarithm of time t(s). As seen from the figure, compression creep curve is almost linear against logarithm of time from 10 to 10^4 sec. As for type of fiber material, compression creep displacement of regenerated fiber assembly such as Cupra and Lyocell fibers is greater than that of synthetic fiber assembly.

Compression creep displacement ratio Rt(%) for different fiber material is investigated. Compression creep displacement ratio Rt at $t=10^4$ sec of regenerated fiber assembly is greater than that of synthetic fiber assembly.

Rt at 10^4 sec for low stress condition is as follows. PET with heteromorphic section (WPE-1, 2, 3) : 0.5 ~ 1.2%, PET with round section (RPE-1, 2, 3) : 1.1~3.2%, PET with hollow section (RPE-4): 1.5%, PTT: 0.8%, Cupra (CU-1, 2, 3): 7.0~8.7% and Lyocell (LY): 10.1%.

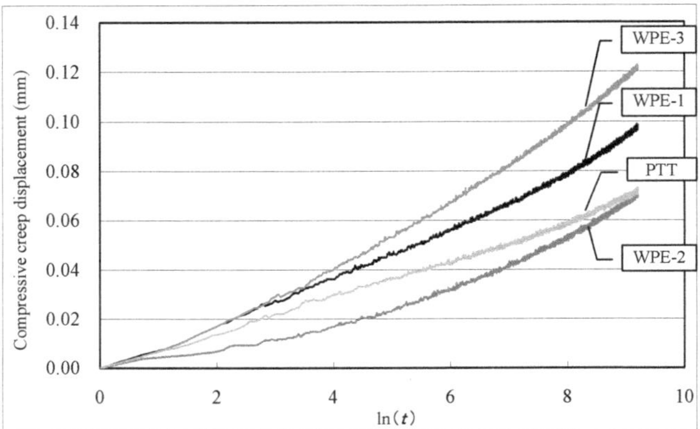

Figure 19. Compressive displacement curve (WPE-1,2,3, PTT)

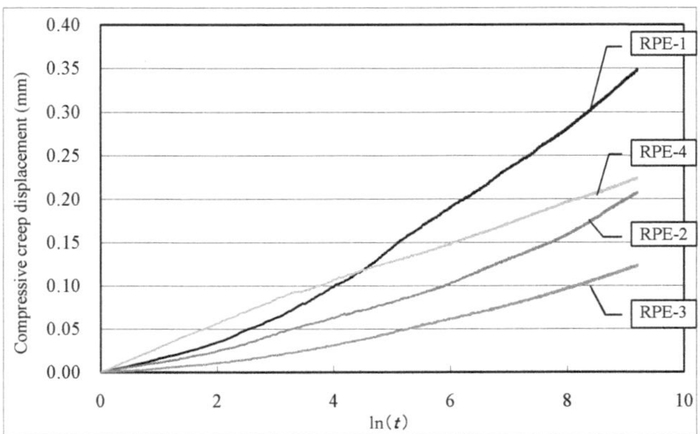

Figure 20. Compressive displacement curve (RPE-1,2,3,4)

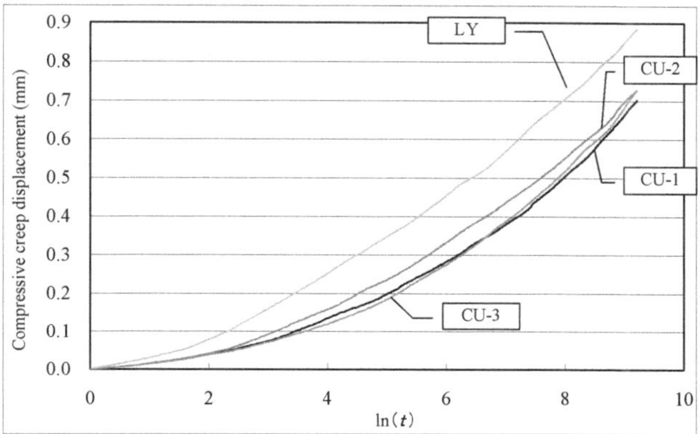

Figure 21. Compressive displacement curve (CU-1,2,3, LY)

Rt at 10^4 sec for high stress condition is as follows. PET with heteromorphic section (WPE-1, 2, 3) : 1.2 ~ 2.2%, PET with round section (RPE-1, 2, 3) : 2.6~4.2%, PET with hollow section (RPE-4): 4.1%, PTT: 1.1%, Cupra (CU-1, 2, 3): 4.9~9.7% and Lyocell (LY): 9.1%.

Creep displacement ratio Rt of samples that have same fineness and same fiber length (1.3 or 1.4 dtex and 38mm) are compared. Rt of Cupra (CU-1) is largest, PET with round section (RPE-1) follows, and PET with heteromorphic section (WPE-1) is smallest. Creep displacement ratio Rt of samples that have same fineness and same fiber length (6.6 dtex and 55mm) are also compared. Rt of PET with round section (RPE-3) is larger than that of PET with hollow section (RPE-4).

In the next place, compression stress dependence of creep displacement is examined. Figure 22 shows Rt values at $t=10^3$ sec for low stress condition (1176 Pa) and high stress condition (2156 Pa). Samples of which Rt at high stress condition are larger than that of low stress condition are as follows; WPE-1, WPE-2, RPE-1, RPE-3 RPE-4 and CU-1. Samples of which stress level dependence is not observed are as follows; WPE-3, RPE-2 and PTT. Samples of which Rt at high stress condition is lower than that of low stress condition are as follows; CU-2, CU-3 and LY. As a result, the behavior of stress level dependence shows different types for different samples. It is conjectured that this phenomena may be arisen from nonlinear viscoelastic behavior in compression creep of fiber assembly.

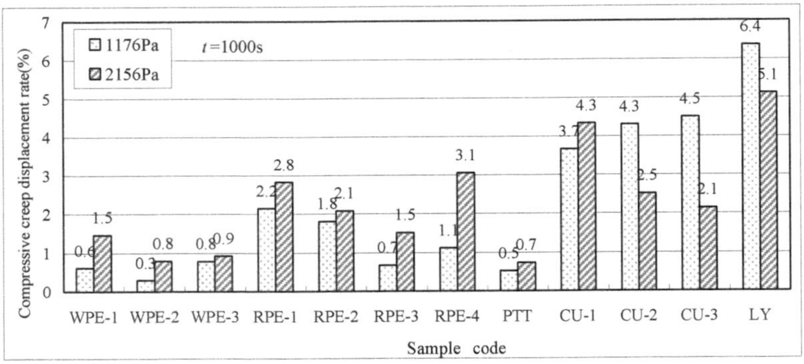

Figure 22. Compressive displacement rate at t=1000s

4.3.2. *Functional expression for compression creep behavior*

In this section, the equation for compression creep behavior is discussed. Nogai and Narumi[3] showed that following equation holds good for compression creep behavior of cotton fiber mass.

$$\varepsilon_t = Y_0 \ln\left(vt + 1\right) \tag{17}$$

where, εt: compression displacement (mm), Y_0(mm) and v(s^{-1}) are constants determined by material properties, measurement condition and compression stress. Equation (17) is derived from non-linear three element viscoelastic model including Eyring viscous element (Figure 23) [12]. In this study, equation (17) is applied to analyze compression creep curve of synthetic and regenerated fiber assembly.

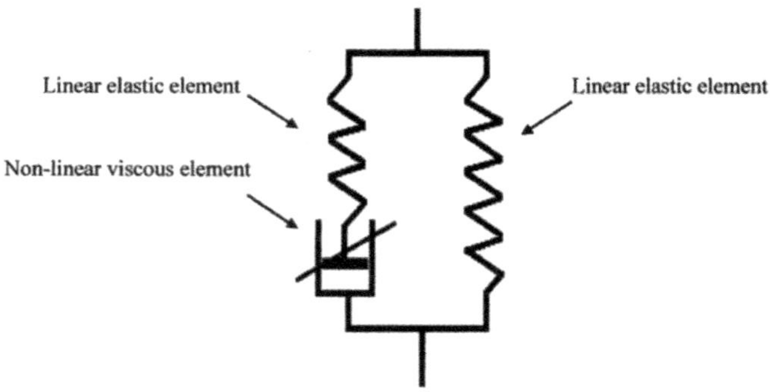

Figure 23. Non-linear three element model

Parameters Y_0 and v are derived from measured curves of compression creep, and the calculated curve is obtained by substituting Y_0, v values into equation (17). The comparison between calculated and measured curves is shown in Figures 24~26. Figure 24 shows the result of PET with heteromorphic section (WPE-1), Figure 25 shows that of PET with hollow section (RPE-4), and Figure 26 shows that of Cupra fiber (CU-2). The ordinate in Figures 24~26 is compression displacement ε_t (mm), and the abscissa is time t (s). In each graph, the curves for low stress condition (1176 Pa) and high stress condition are included. Y_0 and v values are obtained as follows. Measured value of ε_t at $t=10^3$ s and $t=10^4$ s are substituted into equation (17), and Y_0 and v are obtained by solving simultaneous equations for $t=10^3$ s and $t=10^4$ s.

As seen from Figures 24~26, the agreement between calculated and measured curves is very well for all samples used in this experiment. Therefore, it is concluded that compression creep behavior of synthetic and regenerated fiber assembly can be well explained by non-linear three element viscoelastic model.

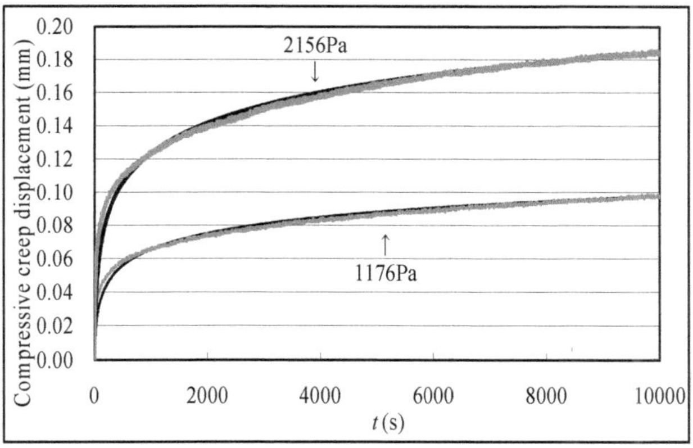

Figure 24. Comparison between experimental and calculated curves using eq(17) (WPE-1)

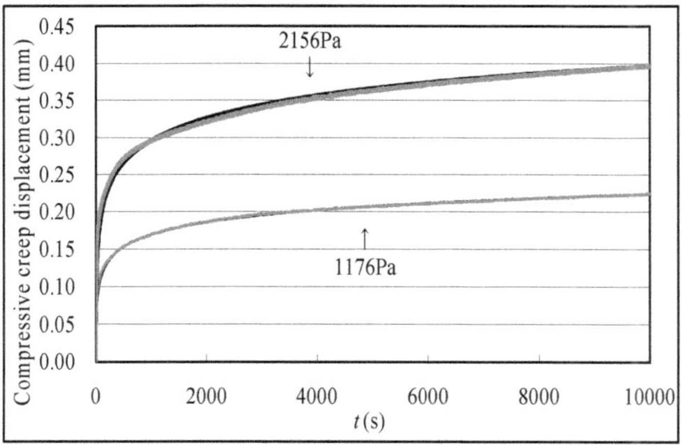

Figure 25. Comparison between experimental and calculated curves using eq(17) (RPE-4)

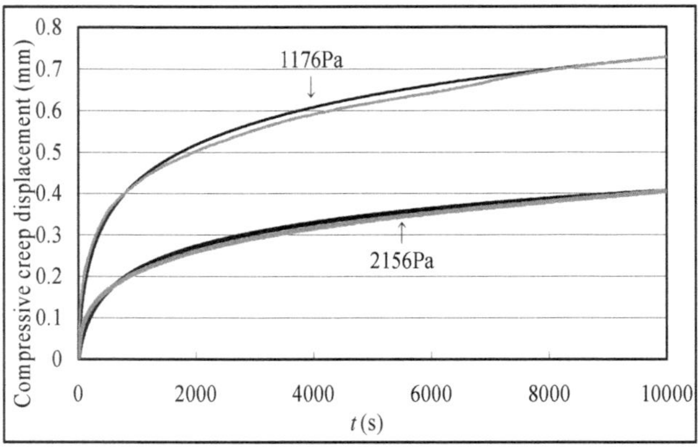

Figure 26. Comparison between experimental and calculated curves using eq(17) (CU-2)

In order to find the relationship between Y_0 ,v values and material properties of fiber, linear regression analysis is carried out. Y_0 has good correlation with apparent Young's modulus, number of crimp and crimp ratio. In particular, correlation between Y_0 and apparent Young's modulus is strong (R=0.8). v has good correlation with bending rigidity (R=0.95~ 0.98).

5. Conclusion

In this study, investigation on compression properties of three-dimensional fiber assembly for futon wadding use is presented. Twelve kinds of synthetic staple fiber assembly and regenerated staple fiber assembly were investigated. KES-G5 compression tester (Kato Tech Co.) is used for the measurement. Repeated compression-recovery test, compression stress relaxation test and compression creep test are carried out. Compression stress relaxation and creep properties of fiber assembly are analyzed based on viscoelastic model. Results are obtained as follows.

i. Repeated Compression-Recovery Test

In order to characterize compression properties of fiber assembly, compression-recovery test with constant strain rate is carried out, and characteristic parameters derived from measured curves are used for evaluation of compression properties. As for the compression measurement of three-dimensional fiber assembly, attention must be paid because fiber assembly does not have specific shape in natural state. To find sampling conditions suitable for characterizing fiber properties, repeated compression-recovery test was carried out for three different levels of maximum compression stress. As a result of sampling, compression-recovery curve at 6th cycles under maximum stress 1176 Pa is selected for the analysis. Linearizing method is applied to the analysis to characterize compression-recovery curve. Functional expression ($P_c=AT^n$) and characteristic parameters A and n are obtained.

ii. Compression Stress Relaxation Test

Two different types of expression for compression stress relaxation function are obtained, approximate expression which holds for long term region and strict expression which holds for whole time region, respectively. Approximate expression of stress relaxation which holds when $t>10^2$ sec is obtained as follows, $S_Y=K(a_0 - \ln t)$. Stress relaxation rate K of regenerated fiber assembly is larger than that of synthetic fiber assembly. Strict equation of stress relaxation function for whole time region ($0<t<10^4$ sec) has two different expressions for regenerated and synthetic fiber assemblies, respectively. In the case of regenerated fiber assembly, strict equation is expressed as follows, $S_Y = K \ln[\coth\{1/2(2e^{-a0}t + B)\}]$. Stress relaxation function of coth type is derived from nonlinear two element model including Eyring viscous element. In the case of synthetic fiber assembly, strict equation is expressed as follows, $\sigma = \beta t^{-\alpha}$.

iii. Compression Creep Test

The degree of creep deformation is estimated by creep displacement ratio, Rt (%) at definite time. Rt of regenerated fiber assembly is larger than that of synthetic fiber assembly. The equation for creep displacement, εt is obtained as follows, $\varepsilon t = Y_0 \ln(vt + 1)$. Creep compliance of ln type is derived from nonlinear three element model including Eyring viscous element.

The results obtained in this study will be the basic information to design and evaluate fiber materials for futon wadding use with regard to sleeping comfort. Analysis of heat and water transport properties of futon wadding concerning micro climate will be also needed for sleeping thermal comfort.

Author details

Yoneda Morihiro*
Faculty of Human Life and Environment, Nara Women's University, Nara, Japan

Nakajima Chie
Graduate School of Human Culture, Nara Women's University, Nara, Japan

6. References

[1] Nogai T, Narumi Y (1972) Senki Ronbunshu , 25, 180-188
[2] Nogai T, Narumi Y, Tanaka K (1974) Senki Ronbunshu , 27, 177-182
[3] Nogai T, Narumi Y, Tanaka K (1975) Senki Ronbunshu , 28, 77-82
[4] Yokura H, Sukigara S, Niwa M (1995) Jpn Res Assn Text End-Uses, 36, 594-601
[5] Yokura H, Sukigara S, Niwa M (1996) Jpn Res Assn Text End-Uses , 37, 535-543
[6] Nakajima C, Yoneda M, Itoh Y(2010) J Text Eng , 56, 29-38
[7] Nakajima C, Yoneda M, Itoh Y(2010) J Text Eng , 56, 39-46

* Corresponding Author

[8] Nakajima C, Yoneda M(2010) J Text Eng , 56, 139-146
[9] Nakajima C, Yoneda M, Nishioka S, Miyazaki A(2010) J Text Eng , 56, 195-202
[10] Kawabata S (1980) The Standardization and Analysis of Hand Evaluation 2nd ed, The Textile Machinery Society of Japan, Chap.4
[11] Kawabata S (1986) Senki Ronbunshu , 39, 169-173
[12] Morton ME, Hearle JWS (1975) Physical Properties of Textile Fibers, The Textile Institute, Chap.18

Section 4

Electrical Applications

Chapter 14

Improvement of the Electrochemical Behavior of 2024-T3 Alclad Using Polyester Coatings

Willian Aperador Chaparro

Additional information is available at the end of the chapter

1. Introduction

Organic coatings based on synthetic resins provide excellent corrosion protection after making a good cleaning and surface preparation, followed by a suitable pretreatment depending on the type of substrate to be protected (Atta, 2006, Atta, 2007; Sathyanarayan, 2008 .) Such protection systems can be applied in a single layer or multilayer, as in the case of coatings for aircraft industry, which are generally constituted by a pretreatment, a primer and a topcoat. Several studies concerning innovation and optimization of new alternative coatings corrosion protective aerospace application have been developed (Reynolds, 1997), researches have been oriented in two directions: first, alternative coatings to traditional surface treatment (pretreatment) , being the most representative case, the coatings by chemical oxidation from zirconium and titanium oxides (Sanchez-Amaya, 2007; Osborne, 2001; Neuder, 2003), and second, alternative coatings of traditional organic primers from resins and primers without any incorporation of cromatants pigments as corrosion inhibitors (Twite, 1998).

One of the disadvantages of the polyester-styrene copolymers has always been their tendency to shrinkage, which is due to the internal contraction of the molecule when the polymerization occurs. This has been improved with the addition of glass fibers or flakes and other fillings. However, substantial contraction can occur when a large area is coated with polyester resin and fiberglass, as in the case of the bottom of a tank or other structure. Therefore, the contraction should be a parameter to be considered during application (Munger, 1984; Groover 1997).

The adaptation of these protective corrosion systems requires a test to evaluate the performance of the material in realistic exposure conditions through or during or under diverse environments, and even more, to establish an estimate of the life span of the protection system. Salt Fog Spray Test (SFST) described by ASTM B117 is one of the most widespread testing methods (ASTM B117, 2008). In this evaluation process, SFST, a particular coatings of Hetron 197-3 and polyester are exposed to a salt spray to imitate the artificial marine environment and exaggerated conditions of temperature and relative humidity. These accelerated aggressive atmospheric environmental conditions can be considered to evaluate the behavior of the corrosion protective coating system and to assess its efficiency in terms of durability.

The aim of this work is the implementation of protective coating Hetron 197-3 polyester type applied on substrates of aluminum alloy Alclad 2024-T3, which is used as barriers to corrosion. This type of substrate is widely used in manufacturing the battery compartment of T-41 Aircraft. The coatings were exposed to 0 and 2000 hours in aggressive atmospheric conditions (acidified saline medium). The degree of deterioration was compared in protection systems based synthetic resin (Hetron 197-3) and polyester. The morphology surface analysis was made using Scanning Electron Microscopy (SEM) to observe the damage generated by exposure to aggressive media. The electrochemical performance of the coatings was evaluated through Electrochemical Impedance Spectroscopy (EIS) and Tafel polarization curves, polarization curves and Nyquist plots indicate that antacid polyester resin coating have the lowest value of current and corrosion rate with the system traditionally used in the aviation industry (Plating Process (Alclad), Conversion Coatings for Aluminum Alloys, chemical oxidation)

2. Experimental details and results

2.1. Material, experimental processes and characterization techniques

The coverings fuselage of the T41 aircraft as well as internal areas prone to corrosion, such as the battery compartment, are made of aluminum alloy 2024-T3 sheet whose chemical composition is shown in Table 1, plated with alloy 1230 in accordance with international specifications of ASTM B209 (ASTM, 2007) (Table 2), and corrosion protective coatings, paint base according to military standards for aviation (surface pretreatment, primer and finish paint). This protection system consists of a surface treatment, commercially called Alodine 5700 that generates a chromium-free conversion coating in a chemical solution composed of 80% of an inorganic component of oxyfluoride species of titanium, zirconium and silicon (H_2TiF_6+$ZrO(OH)(CO_3)_{0.5}$+SiO_2). Also owns 20% of an organic component poly-4-vinyl-phenol and amines, thus promoting an inorganic protective coating by conversion from an organometallic complex of zirconates, and a good bonding surface for subsequent coating organic, and second, a primer surface with epoxy primer cromatante base, being its main component an epoxy resin diluted under strontium chromate pigments as corrosion protection, and finally a topcoat (polyurethane) (Twite, 1998)

The protection system based on synthetic resin comprises the following stages of implementation: The pretreatment was performed chemically, initiating with a process of cleaning and degreasing with an acid detergent, based on a phosphoric acid at a temperature of 46 °C for 10 minutes, then etching was performed with NaOH at a concentration of 50 g/l dissolved in distilled water for 1 minute at a temperature of 52 °C, finally generated anchor treatment with Alodine 5700 was applied in a thin layer of polyester resin Polisec 115, of 50 microns per flush, then homogenizing the coating over the entire area of application. The resin had a curing time of 20 minutes. To finish the process was applied a coating polyester resin compound antacid Hetron 197-3, which joined as a glass flake filler by 20% (weight/weight) with respect to the total weight and 2% (weight /weight) of catalyst on the total weight. Two systems were obtained, H1 = 1layer (100 μm), H2 = 2 layers (200 μm). These systems were compared with commonly used coating which is a combination of Alodine 5700, epoxy primer and polyurethane paint, this coating was called PP.

Chemical composition of the substrate (AA 2024) approx. 91 %Al										
									Others	
% Weight	Si	Fe	Cu	Mn	Mg	Cr	Zn	Ti	unspecified others	Total
Minimum	-	-	3.8	0.30	1.2	-	-	-	-	-
Maximum	0.50	0.50	4.9	0,9	1.8	0.10	0.25	0.15	0.05	0.15

Table 1. Chemical composition of the substrate alloy (Al 2024)

Chemical composition of the clad (AA1230) approx.99 % Al									
								Others	
% Weight	Si +Fe	Cu	Mn	Mg	Cr	Zn	Ti	unspecified others	Total
	0.7	0.1	0.05	0.05	-	0.1	0.03	0.03	0.15

Table 2. Chemical composition of the clad alloy (Al 1230)

The specimens were subjected to ultrasonic cleaning process in an acetone bath and dried prior to trial. The coatings were evaluated in an acidified salt spray according to ASTM B117 (ASTM B 117, 2008) norms. The electrochemical characterization was performed on a Gamry equipment model PCI-4 using the techniques of Tafel polarization curves and electrochemical impedance spectroscopy at room temperature, using a cell consisting of a typical 3-electrode arrangement, the working electrode with an exposed area of 1 cm^2, a reference electrode of Ag / AgCl and as counter electrode a platinum wire, in a solution of NaCl to 3.5 weight.% (ASTM B 209, 2007). Measurements of Tafel polarization curves were obtained at a scan rate of 0.125 mV / s in a voltage range of -200 mV to +1200 mV vs. Ecorr. Nyquist diagrams were obtained by frequency sweeps in the range from 0.001 Hz to 100 kHz with sinusoidal signal amplitude of 10 mV (Aperador, 2010). A scanning electron microscope (SEM) of high resolution (JEOL JSM - 6490LV) was used to measure the thicknesses in cross section after a pre-coating by gold.

2.2. Scanning electron microscopy

The microstructural analysis obtained by scanning electron microscopy (Figure 1), together with EDS analysis used in chemical analysis (Table 3), confirms the formation of conversion coating on Alclad substrate surface (aluminum alloy 1230) . The surface morphology observed by SEM (Figure 2) clearly illustrates the roughness generated by chemical oxidation and the characteristic values recorded EDS chemical composition of elements such as Zr, Ti, Al, F and O, proper treatment with Alodine 5700.The composition of 35% in C, verified the presence of an additional organic binder, which promotes chemical affinity with the organic coatings applied on plating Al1230.

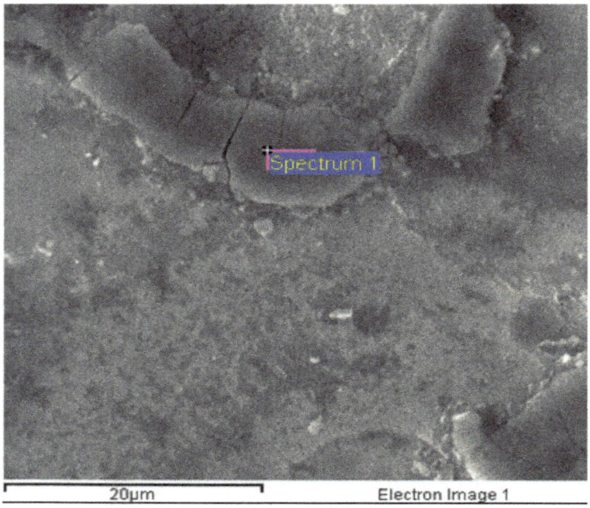

Figure 1. SEM micrograph which shows the region that performed the analysis of chemical composition by EDS to the surface of the substrate (AA 1230) prepared with Alodine 5700.

Element	Weight%	Atomic%
C	35.35	52.77
O	13.02	14.59
F	2.28	2.15
Na	0.24	0.18
Al	43.23	28.73
Si	0.22	0.14
Cl	0.11	0.05
K	0.20	0.09
Ti	1.18	0.44
Fe	0.10	0.03
Cu	0.22	0.06
Zr	3.87	0.76
Total	**100.00**	

Table 3. Chemical composition obtained by EDS for the surface of the substrate (AA 1230) prepared with Alodine 5700. The spectrum is shown in Figure 3.

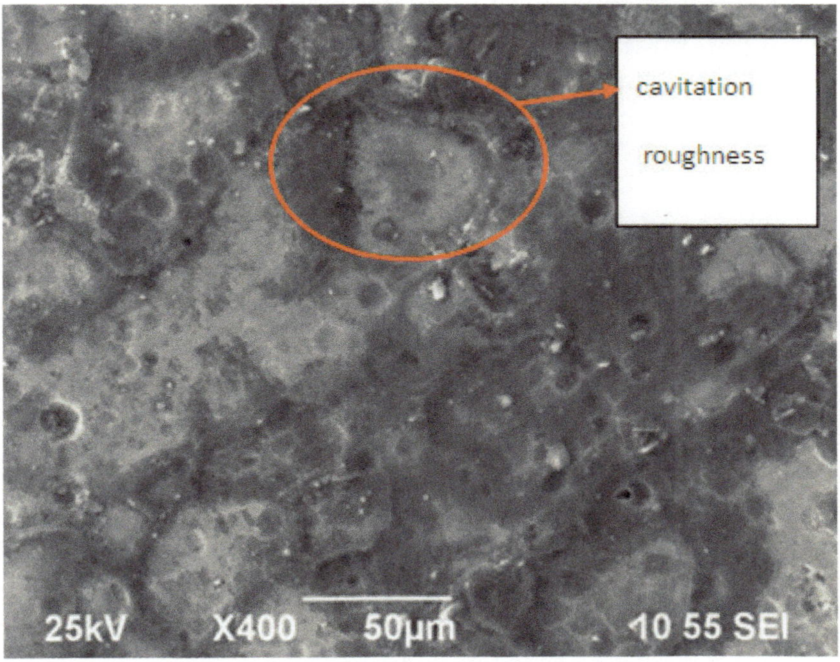

Figure 2. SEM micrograph of the substrate under chemical oxidation with Alodine 5700.

A scanning electron microcopy was used to determine the thickness of the coating, besides the conditions of the various layers and their thicknesses in the base polyester resin coating and Hetron 197-3 (Figures 3 and 4). However, for coating H2 can be seen an increase in the thickness of the antacid resin layer of about 50 microns after having been subjected to 2000 hours of exposure in the saline chamber, this is due to the permeability of the layer of polyester resin and a moisture retention in the molecular structure of the layer. In the micrographs of the H2 coating, it can be checked the continuous arrangement and parallel to the surface of the glass flakes simulating the shape of a surface veil and intensify the chemical barrier coating. Likewise there is a good cohesion of the layer to the substrate along the entire interface even after 2000 hours of salt spray exposure, corroborating a good anchor pattern (between the coating and substrate) despite the high viscosity of the resin.

2.3. Electrochemistry

Figure 5 shows the Tafel polarization curves for the coating polyurethane and polyester antacid resin (AH1, AH2 and PP). The polarization curves were obtained at 0 hours and 2000 hours. Found values of the anodic and cathodic slopes in each case, which are recorded in Table 4 along with values of current density, corrosion potential and corrosion rate for each of the cases were studied.

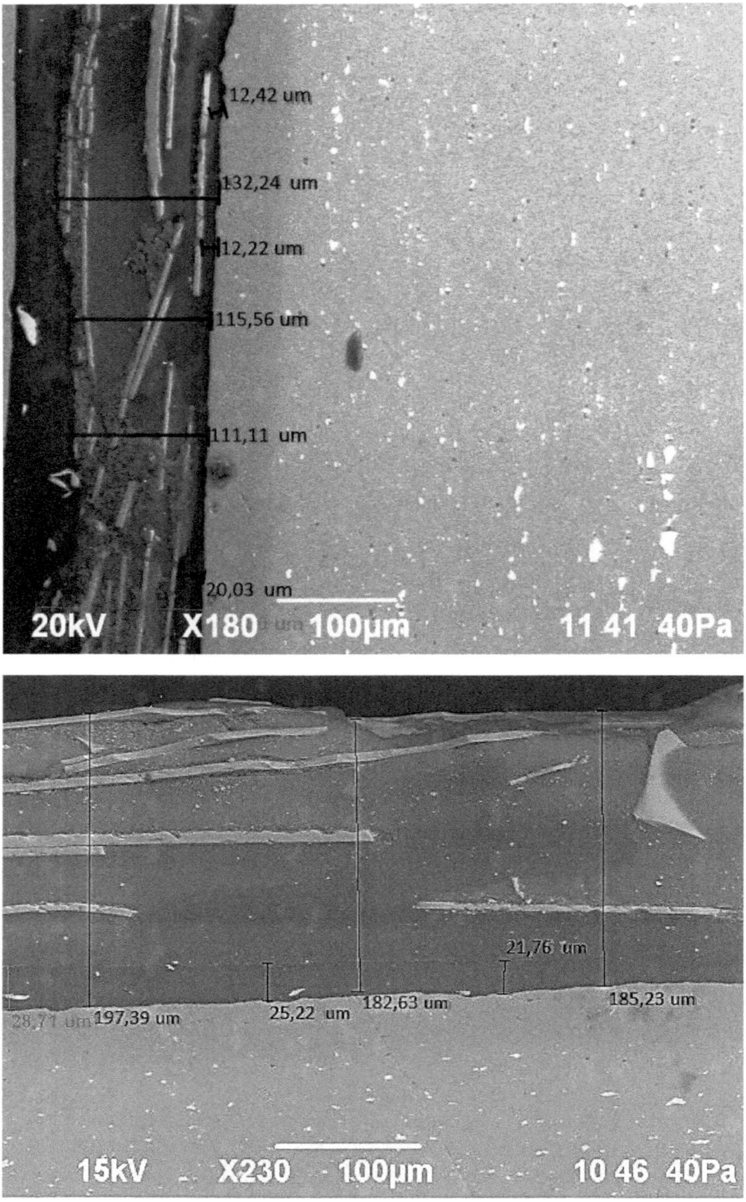

Figure 3. SEM micrographs of the H2 coating after exposure to salt chamber for (a) 0 hours and (b) 2000 hours.

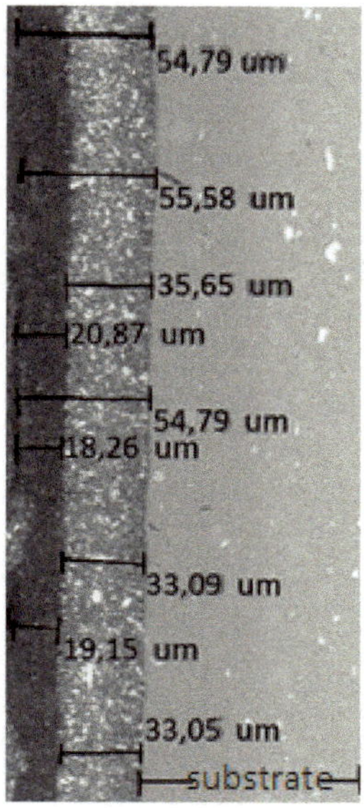

Figure 4. SEM micrograph of the PP coating after exposure to salt chamber for 2000 hours

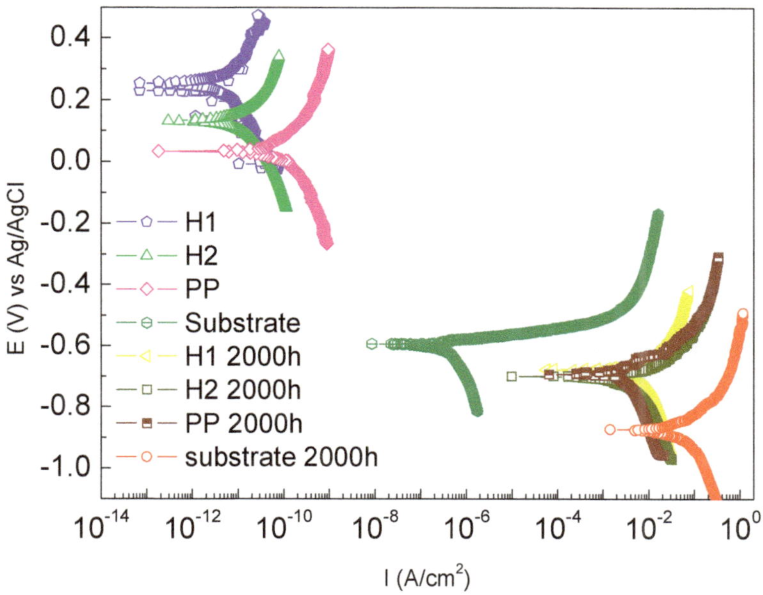

Figure 5. Tafel polarization curves of the substrate and coating types PP and H, evaluated in the salt chamber at 0 and 2000 hours.

The corrosion rate indicates a good performance of the coatings compared to the substrate, for 0 hours of exposure in the salt chamber. The coatings which have a lower density value of corrosion are called H with 1 and 2 layers, followed by those used by the military aviation industry. Subsequently, measures of Tafel polarization curves for 2000 hours of exposure to saline-acidified were performed. One can observe the increase in corrosion rate of the coating and the substrate showing the destructive effect of the medium exposed. Also it distinguishes the protective effect of the coatings compared to the substrate under these conditions. The lowest values of the corrosion rate were reported at (2.31×10^{-9} μmy) for PP coating. However, the coatings implemented in this study show a similar behavior to PP, the Nyquist plots corresponding to the substrate and coatings polyurethane type and synthetic resin exposed to 0 hours and 2000 hours.

	Substrate	H1	H2	PP	Substrate 2000h	H1 2000h	H2 2000h	PP 2000h
Anodic beta (V/decade)	$28{,}70 \times 10^{-3}$	145×10^{-3}	30×10^{-3}	387×10^{-3}	$497{,}1 \times 10^{-3}$	$79{,}80 \times 10^{-3}$	$510{,}1 \times 10^{-3}$	$239{,}7 \times 10^{-3}$
Cathodic beta (V/decade)	263×10^{-3}	245×10^{-3}	77×10^{-3}	389×10^{-3}	737,2	$178{,}9 \times 10^{-3}$	$475{,}0 \times 10^{-3}$	$371{,}5 \times 10^{-3}$
Corrosion current (A/cm^2)	370×10^{-9}	121×10^{-12}	3.31×10^{-12}	210×10^{-12}	$38{,}30 \times 10^{-6}$	$1{,}790 \times 10^{-6}$	$13{,}30 \times 10^{-6}$	$40{,}90 \times 10^{-9}$
Corrosion potential (mV vs Ag / AgCl)	-594	133.4	250	37,10	-686	-710	-740	-750
Corrosion rate (μmy)	4.03×10^{-6}	2.66×10^{-11}	3.60×10^{-11}	2.29×10^{-9}	443.8	20,71	15.41	$1{,}86 \times 10^{-2}$

Table 4. Electrochemical parameters generated from the Tafel polarization curves.

Table 4 and Fig.5 showing the decrease in the corrosion potential when the specimens have been subjected to half-acidified saline for 2000 hours, indicating active corrosion generated. PP type coatings and H, generated similar value of potential corrosion. By comparing this value with substrate, a protective effect against degradation phenomena of the coatings was observed. Density and corrosion rate were decreased substantially by applying the protective film, the corrosion rate was decreases by five orders of magnitude. As a result of exposing the specimens to salt spray for 2000 hours, there was an increase in density and corrosion rate in all tested cases, being most pronounced for the substrate. However, it still maintains the protective effect of the coatings corroborating these results with the values obtained for the corrosion potentials.

The formation of the impedance spectra in the Nyquist plot for both the PP probe, and for the H1 and H2, are graphed consisting of two semicircles (fig 4). These systems are represented by an equivalent circuit shown in Figure 7.

The equivalent circuit of Figure 7, is applicable to coatings exposed to salt chamber in 0 hours and 2000 hours. The difference is set to the magnitude of the polarization resistance values (Rct). The parameter Rs represents the resistance of the electrolyte; in this case the evaluation in saline (NaCl) acidified which corresponds to the resistance between the working electrode and reference electrode. In case of PP and H coatings, its magnitude was small, it is in the order of $1x10^5$ Ω. Cc is the capacitance of the coating layer and R_{po} is the resistance of the solution-coating interface. In the case of base paint systems, refers to coating conformed by the system of the primer epoxy Alodine 5700 and topcoat, and in the case of system H refers to the system formed by Alodine 5700, antacid resin primer and Hetron197-3 , where the increase in Cc is consistent with a measure of the level of coating degradation.

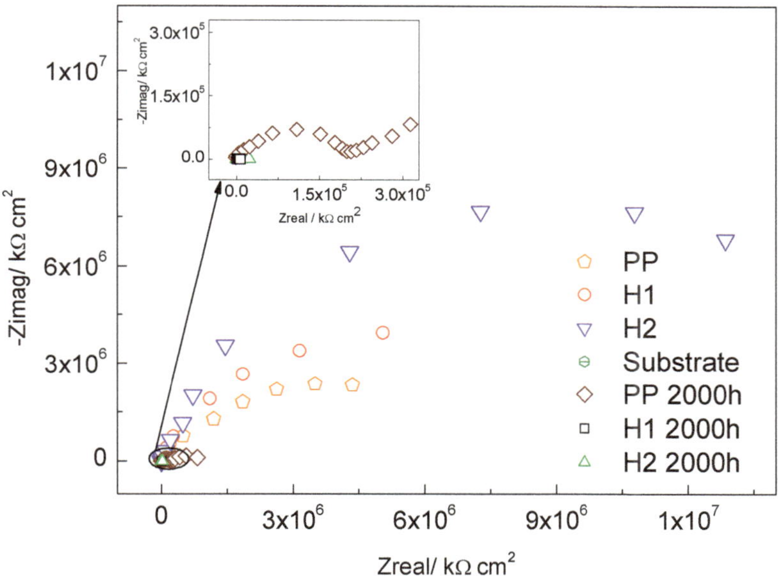

Figure 6. Nyquist plots of the substrate and coating types PP and H, evaluated in the salt chamber at 0 hours and 2000.

The decreasing in R_{po} is corresponding with increasing of water absorption or permeability thereof. Differentiate the contribution of each of the layers to the total resistance to corrosion may be performed by EIS, it follows that the first semicircle the Nyquist plots corresponding to the resistive contribution of the organic coatings (epoxy primer and synthetic resins), because the input resistive the conversion coating is relatively small compared to the organic coatings and therefore it is not clearly recorded in the input impedance in the system of corrosion protection. At 0 hours of exposure, the R_{po} value of the H coating with 1 and 2 layers is close to $1x10^9$ Ω and the PP coating exhibits $1x10^8$ Ω for the PP, however after 2000 hours of exposure, the R_{po} of the H and PP coatings decreased to $1x10^6$ Ω, also the value of PP is higher. This indicates that the permeability is increased dramatically in the H coating compared to the PP coating, and is also reflected in the decrease in polarization resistance and corrosion kinetics because when the coating is permeated, the ionic species are reaching the coating-substrate interface and reduced the polarization resistance and increased the corrosion rate, this behavior is similar to that found in the Tafel polarization curves. The Cdl is the capacitance of the double layer while its increase is an indicator of coating adhesion loss (Osborne, 2001). The Rct represents the resistance to the charge transfer (polarization resistance) associated with the kinetics of the corrosion process. This parameter is inversely proportional to the corrosion rate and the oxidation occurrence area (Reynolds, 1997). Therefore, coating systems have a good performance as a corrosion protection system with respect to the substrate, due to high Rpo and Rct values and low values of Cc and Cdl for 0 hours to 2000 hours of exposure in the mist chamber.

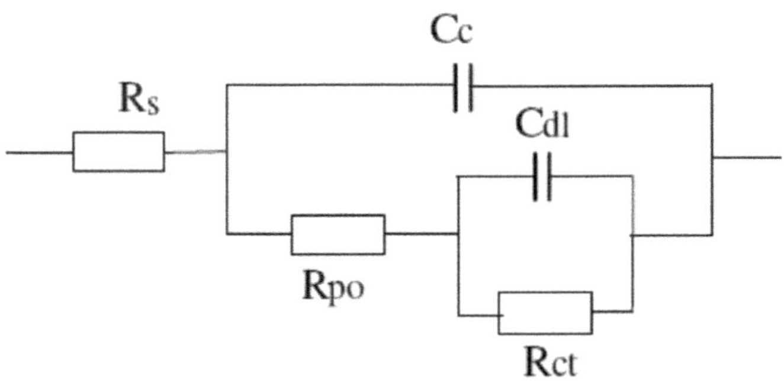

Figure 7. Equivalent circuit diagrams used to explain Nyquist coatings for H1, H2 and PP.

3. Conclusions

According to the Tafel polarization curves and Nyquist plots, the coatings of polyester antacid resin has the lowest value of current and rate corrosion with the traditionally system used in the aviation industry without exposure to mist chamber.

After 2000 hours of mist chamber exposure, the system of the PP has a smaller decrease in the corrosion current comparing to H system, which is due to the higher moisture permeability of the polyester resin.

The effect of thickness on the coating base polyester antacid resin is not significant (corrosion rates of H1 and H2 are 0.817 and 0.608 respectively) as to the results of corrosion rate when there is an increased thickness around 100 μm, passing from one layer (1 layer) to 200 μm (2 layers) approximately.

Characterization techniques by SEM confirmed that after 2000 hours of exposure in saline acidified chamber, the adhesion and coating substrate interface, in the systems of the PP and H haven't visible deterioration.

Synthetic resin based coatings can be used in areas not subjected to impact, unless they are reinforced with glass fiber fabrics, which potencialice impact resistance and coating integrity guarantee and the mechanism of protection by chemical barrier.

The mist chamber during 2000 hours of operation, was under the conditions laid down by ASTM B117, with respect to the variables of temperature in the exhibition area, condensate collection, relative humidity and thus the reliability of results.

The coating AH1 presents an added value in terms of their relative low cost and readily available in the market, therefore, is of great interest to continue the use of such configurations in synthetic resin coatings.

Author details

Willian Aperador Chaparro
Ingeniería Mecatrónica, Universidad Militar Nueva Granada, Bogotá, Colombia

Acknowledgement

This research was supported by Universidad Militar Nueva Granada.

4. References

Airframe, Hanbook. *Airframe & Powerplant Mechanics* (1999), U.S Departament of transportation, Federal Aviation Administration.

Aperador, W., Vera, E. and Vargas, A. (2010) "Estudio de la resistencia a la corrosión electroquímica de electro-recubrimientos níquel/cobre obtenidos por corriente pulsante" Ingeniería & Desarrollo. Universidad del Norte. 27(2): 48-61.

ASM, Hanbook. *Metallography and Microstructures.* (2005). ASM international the materials information society.

ASTM B 117 – 09 (2008). Standard, Standard Practice for Operating Salt Spray (Fog) Apparatus, West Conshohocken, PA, American Society for Testing and Materials.

ASTM B 209 – 07 (2007) Standard, Standard Specification for Aluminum and Aluminum-Alloy Sheet and Plate, West Conshohocken, PA, American Society for Testing and Materials.

Atta, A, M. and Elsaeed, A,M. (2006). "New Vinyl Ester Resins Based on Rosin for Coating Applications". Reactive & Functional Polymers, 66(6): 1596–1608.

Atta, A, M. and Nassar, I.F. (2007). "Unsaturated polyester resins based on rosin maleic anhydride adduct as corrosion protections of steel" Reactive & Functional Polymers, 67(4): 617–626.

Atta, A,M. and Elsaeed, A,M. (2007). "Unsaturated Polyester Resins Based on Rosin Maleic Anhydride Adduct as Corrosion Protections of Steel". Reactive & Functional Polymers, 67(3): 549–563

Bierwagen,G.P. and Tallman, D.E. (2001) "Choice and measurement of crucial aircraft coatings system properties" Progress in Organic Coatings, 41(2): 201–216.

Buchheit, R.G. and Guan,H. (2003). "Active corrosion protection and corrosion sensing in chromate-free organic coatings" Progress in Organic Coatings, 47(4): 174–182.

ESTCP, reporte de. *Non-Chromate Aluminium Pretratments*. (2004). Environmental Security Techology Certification Program.

Groover M.P, Pearson Prentice Hall. 1997.Dr H, Lijain, "Aircraft corrosion, Agard - Corrosion protection schemes for aircraft structures: some examples for the corrosion behavior of al-alloys"; Germany.

Munger C. G, Corrosion Prevention by Protective Coatings, NACE; Houston, Texas, 1984, cap. 3, 4, 5.

Neuder, H. and Sizemore, C. "Chrome-Free Single-Step In-Situ Phosphatizing Coatings on a Ti-6A1-4V Titanium Alloy" Progress in organic coatings, 47(3): 307 -311.

Oñate, A. E. *Las Aeronaves y sus Materiales. Tecnología Aeronáutica*. (1991). Madrid, España, Paraninfo. ISBN: 0-89100-058-5.

Osborne, J. H. (2001). "Inorganic/organic hybrid coatings for aircraft aluminum alloy substrates" Progress in Organic Coatings, 41(4): 217-225.

Reynolds, L.B. and Twite, R. (1997). "Preliminary evaluation of the anticorrosive properties of aircraft coatings by electrochemical methods" Progress in Organic Coatings, 32(3): 31-34.

Sanchez-Amaya, J.M. and Osuna, R.M. (2007). "Monitoring the degradation of a high solids epoxy coating by means of EIS and EN" Progress in Organic Coatings, 60 (4): 248–254.

Sathiyanarayanan, S., and AZIM, S. (2008). "A new corrosion protection coating with polyaniline–TiO2 composite for steel" Electrochimica Acta, 53(6): 2087–2094.

Twite, R.L. and Bierwagen, G. P. (1998) "Review of alternatives to chromate for corrosion protection of aluminum aerospace alloys", Progress in organic coatings, 33(2): 91-100.

Chapter 15

Electric Breakdown Model for Super-Thin Polyester Foil

Haiyang Wang and Zhengzhong Zeng

Additional information is available at the end of the chapter

1. Introduction

Polyester has been more and more widely applied to the field of electrical insulation because of their excellent electrical and mechanical properties. One of the commonest concerns in industry is the reliability of materials. Therefore, the study of breakdown of solid dielectrics is extremely important in industrial applications. The increasing demanding for polyester films requires that the dielectric strength of polyester with various film thicknesses should be measured accurately. Lack of accurate data on dielectric strength could lead to design shortcomings: excessive insulation could lead to more expensive equipments, while inadequate insulation could lead to premature failure. The data of breakdown strength for polyester from several hundred microns to several millimeters can be directly obtained from the handbook of dielectric materials. However, there are few data of breakdown strength in micro-meter regime.

The dielectric strength of solid dielectrics is affected by many factors, such as ambient temperature, humidity, duration of test, impurities or structural defects, whether a.c., d.c. or impulse voltages are being used, pressure applied to these electrodes etc. The mechanism of breakdown is complex in the case of solids, and varies depending on the time of application of voltage as shown Fig. 1 (Naidu & Kamaraju, 1995; Pai & Zhang, 1995). The various mechanisms are intrinsic breakdown (Callen, 1949; Hippel & Alger, 1949; O'Dwyer, 1964; Seeger & Teller, 1939; Seitz, 1948; Whitehead, 1951), electromechanical breakdown (Laghari & Sarjeant, 1992; Stark & Garton, 1955), breakdown due to treeing and tracking, thermal breakdown, electrochemical breakdown (Sawa, 1986).

1. Intrinsic Breakdown -- it is generally considered to be due to electrons in the insulator gain sufficient energy from the applied field to cross the forbidden energy gap from valence to the conduction band. Intrinsic breakdown is usually accomplished in the order of 10^{-8}s (Seeger & Teller, 1939).

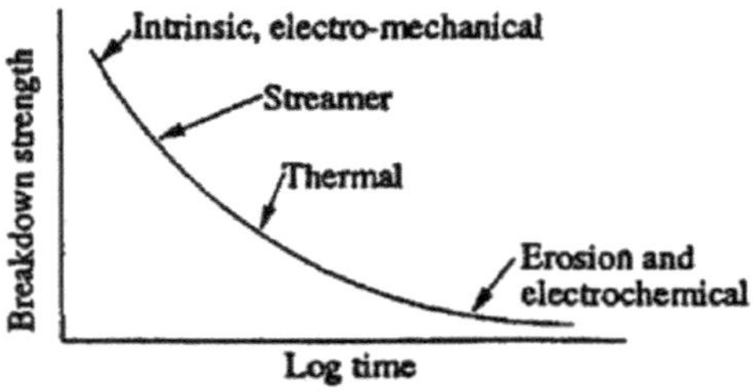

Figure 1. Variation of breakdown strength with time after application of voltage

2. Streamer Breakdown -- the mechanism is conceptually similar to the streamer mechanism in gases, namely when the avalanche exceeds certain critical size breakdown will occur. Transit time is usually short.
3. Thermal breakdown -- when an insulator is stressed electrically, because of currents and dielectric losses due to polarization, heart is continuously generated within the insulator. If the rate of heat generation exceeds the rate of heat losses, then the insulator will undergo thermal breakdown. Such breakdown takes place usually in a much slower time scale.
4. Erosion Breakdown -- insulators of contain voids or cavities within the material. These cavities are usually filled with gas or liquid medium of lower breakdown strength than the solid insulator. Accordingly, under normal working stress of the insulator the voltage across the cavities may exceed the breakdown value hence initiating breakdown in the cavities.
5. Breakdown due to tracking and other causes -- Tracking is the formation of some permanent conducting paths, usually carbon, along the insulator surface due to degradation of the insulator or other causes.
6. Electromechanical breakdown – Electromechanical breakdown occurs in dielectric foils of low Young's modulus and easy deforming and it is because of the force of attraction between the surface charges.
7. Treeing breakdown -- treeing occurs due to the erosion of material at the tips of the spark. Erosion results in the roughening of the surfaces, and hence becomes a source of dirt and contamination. This causes increased conductivity resulting either in the formation of conducting path bridging the electrodes or in a mechanical failure of the dielectric.

Electrical breakdown testing of polymers for insulator applications has long been a subject of interest. In order to improve insulator performance, it is essential to understand the mechanisms of electrical breakdown in solid dielectrics. However, unlike the case for gases, electrical breakdown and conduction mechanisms in polymeric solids are less understood. In solid dielectrics, the electrical transport phenomena include currents due to orientation and interfacial polarization in addition to electronic and ionic charge carriers (Barber et al., 2009). On account of the time of dielectric breakdown, this chapter mainly presents intrinsic breakdown and electromechanical breakdown.

2. Intrinsic breakdown

When an electric field is applied to an insulator, electrons gain sufficient energy from the applied field to cross the forbidden energy gap from the valence band to conduction band. When this process is repeated, more and more electrons become available in the conduction band, eventually leading to breakdown. Intrinsic breakdown usually occurs in the order of 10-8s. The initial density of free electrons is assumed to be large, and electron-electron collisions occur. The critical criterion of intrinsic breakdown is initial electrons formed by impact ionization. It consists of Hippel's low-energy criterion and Frohlich's high-energy criterion (Callen, 1949; Frohlich, 1939; Hippel & Alger, 1949; O'Dwyer, 1964; Seeger & Teller, 1939; Seitz, 1948; Whitehead, 1951).

2.1. Hippel low-energy criterion

Hippel low-energy criterion assumed that the energy E breakdown occurs when the applied field is high enough to accelerate electrons of any given energy in the conduction band to sufficient energy to further ionize the ions of the crystal by collision, thus leading to an exponential increase in the number of electrons in the conduction band. It may be easily seen that the applied field necessary to accelerate (on the average) an electron of given energy has a maximum when considered as a function of electron energy. In order to accelerate an electron, the applied field must compensate the energy lost by this electron to the lattice; but slow electrons have insufficient energy to excite the vibrational modes of the lattice, whereas very fast electrons interact for too short a time with the ions which they pass to transfer energy to them with appreciable probability. The breakdown strength, according to von Hippel's "low energy" criterion, therefore corresponds to the maximum, and an electron brought into the conduction band with low energy is immediately accelerated to the ionization energy and produces an additional conduction electron, the process thus building up in the form of an avalanche.

2.2. Frohlich high-energy criterion

Frohlich high-energy criterion is based on that an applied field sufficient only to accelerate electrons already having the ionization energy may lead to an instability and to electric breakdown (Callen, 1949; Cooper, 1966).

Eq. (1) shows the critical breakdown strength of intrinsic electron breakdown (Callen, 1949). For low energy breakdown theory $u = u_c \approx 4\hbar\omega$, and for high-energy breakdown theory, $u = u_I$.

$$E_{\mathrm{B}} = \sqrt{\frac{m^{*}B}{e^{2}\tau}} \tag{1}$$

Eq. (1) may be written as

$$E_{\mathrm{B}} = E_0\sqrt{\frac{\hbar\omega}{C}\cdot\frac{1}{\tau}\cdot\frac{B}{C}} \tag{2}$$

Where

$$E_0 = \frac{\sqrt{2}\pi m^* e e^{*2}}{Ma^3 \hbar\omega} \tag{3}$$

B is the average rate of energy loss for a given electronic energy, and is expressed by Eq. (4), where u is the average effect of collisions on the electrons of energy. C is expressed by Eq. (5). τ is the relaxation time, e is charge of electron, e^* is the effective charge per ion, m^* is the effective mass of the electron, M is the reduced mass of the ions, a is the interionic distance, ω is the natural angular frequency and ω_t is the natural angular frequency of the transverse mode.

$$B(u) = \mathrm{C}\left(\frac{\hbar\omega}{u}\right)^{1/2}\left[(\overline{n}+1)\ln\frac{1+(1-\hbar\omega/u)^{1/2}}{1-(1-\hbar\omega/u)^{1/2}} - \overline{n}\ln\frac{1+(1+\hbar\omega/u)^{1/2}}{1-(1+\hbar\omega/u)^{1/2}}\right] \tag{4}$$

$$\mathrm{C} = \frac{\pi(2m^*)^{1/2}e^2}{Ma^3(\hbar\omega)^{1/2}} \tag{5}$$

$\overline{n}$ is the average probability of scattering of an electron and is given by

$$\overline{n} = \frac{1}{e^{\hbar\omega/kT} - 1} \tag{6}$$

Where k is Boltamann's constant =8.622×10^{-5} eV/°C.

$$\omega^2 = \frac{2\pi e^{*2}}{Ma^3}\left(\frac{\varepsilon_\mathrm{r}\varepsilon_\mathrm{op}}{\varepsilon_\mathrm{op} - \varepsilon_\mathrm{r}}\right) \tag{7}$$

$$\frac{\omega^2}{\omega_t^2} = \frac{\varepsilon_\mathrm{r}}{\varepsilon_\mathrm{op}} \tag{8}$$

$$\frac{1}{\tau} = \frac{B}{2\hbar\omega}\sqrt{\frac{\hbar\omega}{u}} + \frac{\mathrm{C}}{\hbar\omega}\sqrt{\frac{\hbar\omega}{u}}\left[(\overline{n}+1)\sqrt{1-\frac{\hbar\omega}{u}} + \overline{n}\sqrt{1+\frac{\hbar\omega}{u}}\right] \tag{9}$$

Where e^* is expressed in terms of ω by Eq. (7) and ω is expressed in terms of ω_t by Eq. (8). We obtain

$$E_0 = \frac{2^{3/2}\pi^2 m^* e}{\hbar^2}\hbar\omega_t \frac{\varepsilon_\mathrm{r} - \varepsilon_\mathrm{op}}{(\varepsilon_\mathrm{r}\varepsilon_\mathrm{op}^3)^{1/2}} = 1.34\times 10^8(\hbar\omega_t)\frac{\varepsilon_\mathrm{r} - \varepsilon_\mathrm{op}}{(\varepsilon_\mathrm{r}\varepsilon_\mathrm{op}^3)^{1/2}} \tag{10}$$

$$E_\mathrm{BH} = \frac{E_0}{\mathrm{C}}\sqrt{\frac{\hbar\omega B(T,u)}{\tau}} \quad \left(u = u_c \approx 4\hbar\omega\right) \tag{11}$$

$$E_{BF} = \frac{E_0}{C}\sqrt{\frac{\hbar\omega B(T,u)}{\tau}} \quad (u = u_I) \tag{12}$$

Where ε_r and ε_{op} are the relative and optical values of the dielectric constant, T is the temperature of the lattice in K.

Material	ε_r	ε_{op}	a/ 10^{-10}m	M/ 10^{-20}g	E_{BH} / (kV•mm^{-1})	E_{BF} / (kV•mm^{-1})	E_E / (kV•mm^{-1})
PE	2.4	2.28	2.52	2.8	79	17	20
polycarbonate	3	2.5	20.8	40	255	54	90
polyester	3.2	10.7	3	301	65	65	180

Table 1. Breakdown strength for polymer

In Table 1 the pertinent breakdown strength data for polymer are shown according to Eq. (10) and Eq. (11). The electron energy is 0.1~0.2 eV for Hippel's criterion and 5 eV for Frohlich's criterion. The theoretical values are compared with the experimental breakdown strengths in mm regime. The experimental breakdown strength E_E which is given by Ref. (Wang et al. 1992) in mm regime is less than the breakdown strength of Hippel's criterion E_{BH} and greater than the breakdown strength of Forhlich's criterion E_{BF}. The critical breakdown strength of intrinsic breakdown model is calculated on the basis of Hippel's low-energy criterion and Frohlich's high-energy criterion (Wang et al., 2008).

3. Electromechanical breakdown

3.1. Electromechanical breakdown

Electromechanical breakdown is likely to happen in dielectric foils of low Young's modulus and easy deforming. When a dielectric foil is subjected to a high voltage, an electrostatic compressive force which will cause deformation of the dielectric foil is produced. If the electric stress exceeds the mechanical compressive strength of the dielectric foil, breakdown of the dielectric foil takes place. The electromechanical breakdown model is shown in Fig. 2. If the thickness of the specimen is d_0 and is compressed to d under an applied voltage U, then the electrically developed compressive stress is in equilibrium

$$\varepsilon_0\varepsilon_r \frac{U^2}{2d^2} = Y\ln\left(\frac{d_0}{d}\right) \tag{13}$$

Where Y is the Young's modulus.

A plot of thickness as a function of voltage on normalized scales is given in Fig. 3. It is obviously seen that there is a critical voltage above which the thickness goes to zero. This occurs when $d[d(U)]/dU = -\infty$ or $d(U)/d[d(U)] = 0$. Therefore, the critical breakdown strength of electromechanical breakdown is (Laghari & Sarjeant, 1992; Stark & Garton, 1955)

$$E_{EM} = \frac{U}{d_0} = 0.6\left(\frac{Y}{\varepsilon_0\varepsilon_r}\right)^{1/2} \tag{14}$$

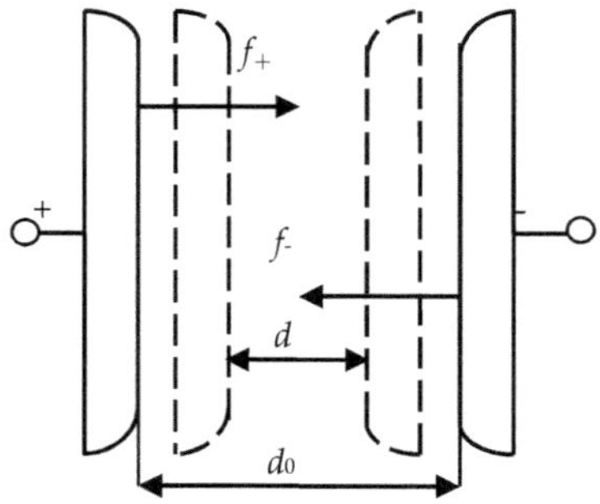

Figure 2. Electromechanical breakdown model

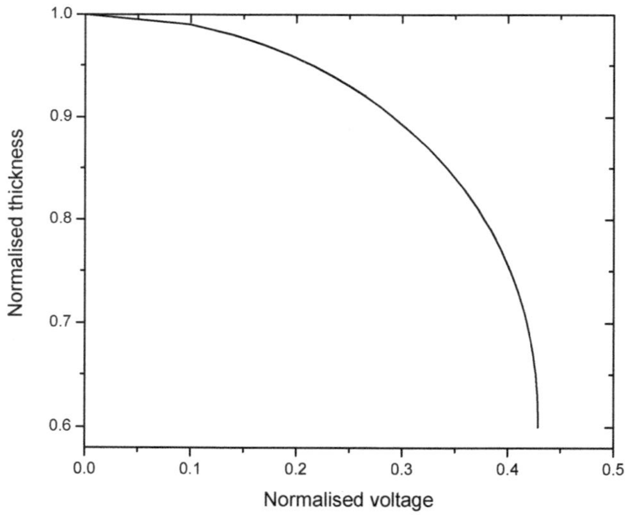

Figure 3. Thickness as function of voltage for electromechanical breakdown

The electromechanical breakdown model is rather unrealistic as it assumed that the dielectric material somehow disappeared to an infinitesimal thickness at $U \geq U_{em}$. Therefore, it may ignore several factors, such as possible earlier instability at microscopic areas of stress concentration, the dependence of Y on time and stress and plastic flow. Stack and Garton pointed out that the stress concentration at microscopic areas is instable, which may result in a localized thinning and breakdown. Blok and LeGrand considered that local regions subjected to higher-than-average electric fields and experienced a shear stress which tended to form an indentation. The indentation gives rise to an even more inhomogeneous field, produces a sharp depression and eventually causes the material to flow radially away (Blok & LeGrand, 1969).

3.2. Electrofracture breakdown

Zeller et al (Zeller et al., 1984; Zeller & Schneider, 1984) proposed electrofracture as an aging mechanism in which the growth of a filamentary crack through a dielectric was caused by electrostatic forces. The model is based on the concepts of fracture mechanics in which the change in electrostatic energy at growth is equal to the mechanical energy required for partial discharge channel formation.

In the electrofracture mechanism the local field at the crack tip is considered to be insufficiently high for the strain energy density W_{em} contribution to be insignificant, i.e. electrostatic energy density $W_{es} \gg W_{em}$. So that the criterion for electrofracture is:

$$W_{es} \gg W_p \tag{15}$$

The electrofracture breakdown strength at the end of the crack tip is given by

$$E_{\mathrm{EF}} = \left[\frac{2Y}{\varepsilon_0 \varepsilon_r} \right]^{1/2} \tag{16}$$

E_{EF} is the local field at the tip of the crack whereas the electromechanical breakdown field is the applied field.

3.3. Filamentary electromechanical breakdown

Fothergill proposed the filamentary electromechanical breakdown in which a filamentary-shaped crack propagates through a dielectric releasing both electrostatic energy and electromechanical strain energy stored in the material due to the applied electric field (Fothergill, 1991, 1992). It operates at higher local fields.

The filamentary electromechanical breakdown model assumed electric field at the filament tip that are sufficiently high for $W_{em} > W_{es}$, so that the appropriate criterion is:

$$W_{em} > W_s + W_p \tag{17}$$

The filamentary electromechanical breakdown strength is defined as

$$E_{\mathrm{FEM}} = \left(\frac{8Y(2G + Yr_f)}{\varepsilon_0^2 \varepsilon_r^2 r_f} \right)^{1/4} \tag{18}$$

In general, $2G \gg Yr_f$, so that eq. (18) is written as (Hikita et al., 1987)

$$E_{\mathrm{FEM}} = \left(\frac{16GY}{\varepsilon_0^2 \varepsilon_r^2 r_f} \right)^{1/4} \tag{19}$$

A modified relationship between applied voltage and electric field at the tip of a filament is given by Eichhorn (Eichhorn, 1977) as:

$$E = \frac{V(1 + r_f / d)^{1/2}}{r_f \ \operatorname{arctan}h((1 + r_f / d)^{-1/2})} \tag{20}$$

Where d is the distance from the filament to the counter electrode, and is assumed to be the full thickness of the specimen in the worst-case estimate. r_f is the tubular crack of radius and depends on what initiates the filament. If it is assumed of $E = U / r_f$, the radius of the filament can be predicted from eq. (20). Solving eq. (20) give $r_f = 0.29\mu m$ which is quite reasonable.

Both the electrofracture and filamentary electromechanical breakdown mechanisms assume a crack propagation process. The limiting speed of a crack is equal to that of a longitudinal elastic wave in the material.

According to the Young's modulus of different materials and Eq. (14), Eq. (16) and Eq. (19), the critical strengths of polyester, polyimide, polycarbonate, PE and polypropylene are obtained. The results are shown in Table 2, from which it is apparent that the critical strength of electric-mechanical breakdown is higher than the strength given by Ref. (Wang et al. 1992) in the mm regime. Therefore, the above results support the electric-mechanical breakdown mechanism for relatively thinner dielectric foils.

Material	Y/ (MN•m^{-2})	G/ (J•m^{-2})	ε_r	E_{EM} / (kV•mm^{-1})	E_{EF} / (kV•mm^{-1})	E_{FEM} / (kV•mm^{-1})
polyester	35	3600	3.2	667	1554	633
polyimide	28	6000	3.5	571	1330	542
polycarbonate	30	3000	9.2	638	1486	606
PE	9.2	3800	2.3	410	955	389
polypropylene	16	7900	2.2	544	1267	516

Table 2. Critical strength of materials

4. Recent development in polymer breakdown theory

Many studies have been published on high-field electrical conduction and breakdown of insulating polymers and there are also many good reviews and books which deal with this subject (Bartnikas & Eichhorn, 1983; Ieda, 1980, 1984; O'Dwyer, 1973). However, the dielectric breakdown processes of polymers are complicated due to their highly complex structures. Therefore, much work has been devoted to clarify the correlation between the breakdown characteristics and the properties inherent to polymers. Theories on the breakdown mechanisms are usually divided into three categories: electronic, thermal (Hap & Raju, 2006) and mechanical processes.

The different degrees of ordering of macromolecules and the presence of regions of varying densities gives rise to different conductivities under an applied field. A non-uniform voltage drop occurs within the polymer when a current begins to flow, leading to variable distribution of electric field. A detailed analysis of breakdown processes in polymers has been discussed by Ieda et al (Ieda, 1980). Artbauer proposed the basic idea of Free Volume Theory (Artbauer, 1996), which was taken in explaining the breakdown phenomena of polymer around the glass transition temperature. The amorphous phase in polymers with unoccupied volume consisting of holes of molecular order presents a free path for the electrons to accelerate under an external field and gain energy. Job et al. improved dielectric strength of polymers by chemical modifications. They have increased the breakdown strength of polyethylene terephthalate (PET) films by *insitu* polymerization of a layer of polyaniline (PANI). A 30% increase in dielectric strength was observed when the non-conductive PANI filled the voids of PET. And they proposed the breakdown was the electron avalanche process (Job et al., 2003). Breakdown in polymers with saturated bonds, such as polyethylene and polypropylene, can not be explained by the electron avalanche mechanism as they have a short free path length. As a result, the electrons cannot gain sufficient energy by acceleration to cause avalanche. In these polymers, electrical breakdown is the last step in a process of polymer degradation that results in the formation of conducting channels. The formation of these conducting channels depends on the intensity and time of the applied external field. The lifetime of these polymers decreases exponentially with applied electric field (Zakrevski et al., 2003). Overall, a general consensus has emerged that the reduction in dielectric breakdown strength of polymers is closely related to the dissociation (bond scission) of polymer molecules under an applied external field. The dissociation process is considered to be the initiation of polymeric breakdown. However, the exact mechanism of dissociation is unclear. One approach to explain this phenomenon is the non-uniform distribution of electric field intensity within a polymer due to electrode defects and inherent differences within the structure of polymer (Zakrevski et al., 2003).

The dielectric breakdown field of polyimide (PI) thin films has been studied across the temperature range from 25 to 400 °C under dc conditions by Diaham et al. it appeared that both the breakdown field value and β–scale parameters decrease with increasing electrode area following the area extrapolation law. It was found that the β–parameter has presented an unexpected increase with increasing the thickness thanks to percolation theories (Diaham et al., 2010).

Peruani et al described the dielectric breakdown model in conductor-loaded composites. Conducting particles are distributed at random in the insulating matrix, and the dielectric breakdown propagates according to new rules to take into account electrical properties and particle size. Dielectric breakdown patterns are characterized by their fractal dimension and the parameters of the Weibull distribution. Studies are carried out as a function of the fraction of conducting in homogeneities. The fractal dimension of electrical trees approaches the fractal dimension of a percolation cluster when the fraction of conducting particles approximates the percolation limit (Peruani et al., 2003). DammigQuiña et al developed a capacitive dielectric breakdown model (CDBM) of electrical tree growth to simulate dielectric breakdown in polymer materials. Dependence of the fractal dimension on the applied voltage, the breakdown process has its thermodynamic origin in the interplay between energy injection and space charge formation processes. The model is capable of predicting electrical trees with fractal dimensions and qualitatively reproducing the temporal evolution of the breakdown process (DammigQuiña et al., 2008).

Kim and Shi investigated the dielectric strength of an interlevel low relative permittivity dielectric for various film thicknesses and temperatures by using I-V measurements with metal-insulator-semiconductor (MIS) structures. It is found that the dielectric breakdown mechanism is electromechanical in origin for relatively thick films (thickness >500 nm). The dielectric strength is proportional to the square root of Young's modulus of the films (Kim & Shi, 2001).

5. Experimental

5.1. Polyester foil

The polyester foil studied here is compound of biaxially-oriented polyethylene terephthalate, as shown in Fig. 4. Biaxially-oriented Polyethylene terephthalate, commonly abbreviated BOPET, is made from stretched polyethylene terephthalate (PET) and used for its high tensile strength, chemical and dimensional stability, transparency, gas and aroma barrier properties and electrical insulation (Sarker et al., 2001).

The end properties of the BOPET are direct consequence of their super molecular structure. The final structure of the BOPET depends mainly by the melt spinning conditions and by the subsequent heat-mechanical modifications, too. The desirable super molecular structure and the needed final morphologies and properties of the PET can be obtained by controlling of the forming process parameters as well as the subsequent treatment. For example, BOPET can be metallized by vapor deposition of a thin film of evaporated aluminum, gold, or other metal onto it. Metallized BOPET film, along with other plastic films, is used as a dielectric in foil capacitor. Four film thicknesses of 9μm, 12μm, 15μm and 18 μm were selected to study the breakdown strength.

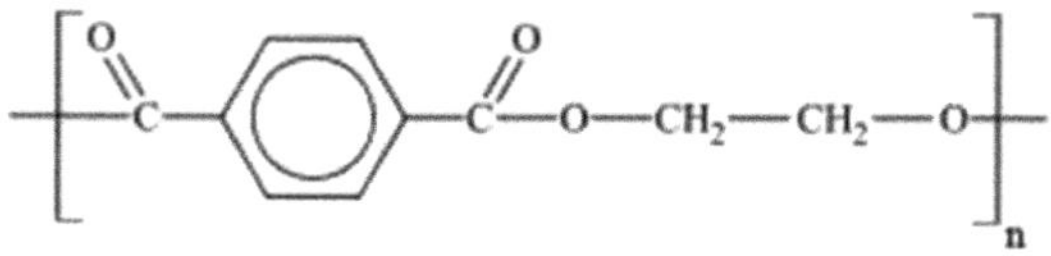

Figure 4. Chemical structure of the Polyethylene terephthalate

5.2. Experimental setup

Dielectric breakdown field measurements have been performed using a dc voltage supplied by a 15 kV voltage source connected to the sample. The electrical schematic diagram is shown in Fig. 5. In Fig. 5, 1, 2 and 6 are transmission lines. 3 is the polymer-foil switch. 4, 5 are the charging resistance. The breakdown field was identified when the polymer-foil switch is self-breakdown, and output a rectangular pulse. The maximum voltage supplied was also recorded using a divider connected to the generator and was considered as the breakdown voltage (Wang et al., 2004, 2006).

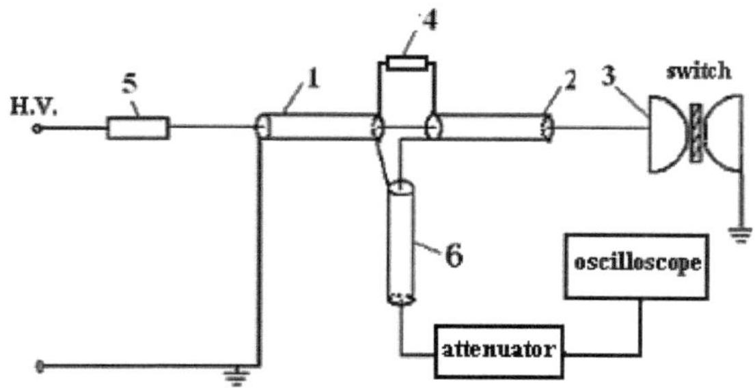

1,2,6 transmission line; 3 polymer-foil switch; 4,5 charging resistance

Figure 5. Electrical schematic diagram of experiment

Fig. 6 is the experimental arrangement of polyester-foil switch. The high voltage electrode (HE) and ground electrode (GE) are planes made of copper, and between HE and GE is the polyester foil, whose thickness varies among 9μm, 12μm, 5μm and 18μm. The self-breakdown voltage is attained by a resistive divider. The measurement is accomplished by a 500MHz digital storage oscilloscope.

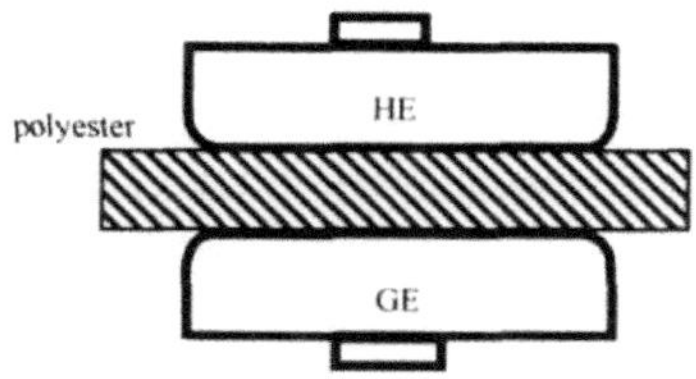

Figure 6. Experimental setup

6. Statistical analysis

Although a number of samples of the same material may be tested, each sample might exhibit breakdown at a different voltage due to random structural and measurement differences. The breakdown voltage can be considered as a random variable, which necessitates a statistical analysis of the breakdown data. Electrical breakdown can be considered as a system failure under the condition where a weak link in the system fails. The Weibull distribution, which is a type of extreme value distribution, is the most common for such applications (Weibull, 1951; Wu & Vollertsen, 2002).

$$F(x) = 1 - e^{-(\frac{x-\gamma}{\alpha})^{\beta}} \tag{21}$$

Where $F(x)$ is the cumulative probability of failure, α is the scale parameter (V/cm) corresponding to a probability of failure $F(\alpha)$ =63.2%, β is the shape parameter. A high β-value is related to a low scattering of the data. The location (or threshold) parameter γ has been set to zero.

The experimental data have been ranking using the median rank approximation given by (Chauvet & Laurent, 1993; Fothergill, 1990):

$$F(i,n) = \frac{i - 0.3}{n + 0.4} \qquad (22)$$

Where, i and n are the rank of a failed sample and the total number of tested samples, respectively. For plotting the Weibull distribution law, the transformation of equ. (21) into equ. (23) has been realized:

$$\log_{10}\left[\log_e\left(\frac{1}{1 - F(x)}\right)\right] = \beta\left[\log_{10}(x) - \log_{10}(\alpha)\right] \qquad (23)$$

The Weibull parameters have been extracted considering a confidence interval of 90 %. Both the maximum likelihood and least square fit methods have been also applied leading to similar α– and β–parameter values (Diaham et al., 2010; Laihonen et al., 2007a, 2007b).

7. Results and discussion

Experiments were performed to measure the self-breakdown voltage of 10μm level foil. Fig. 7 and Fig. 8 show self-breakdown voltage and deviation thickness-varied polyester foils under positive voltage, respectively. The dashed lines through the data points represent the average self-breakdown voltage of the foils. From Fig. 7, it is clearly seen that the average self-breakdown voltage under positive voltage for 9μm, 12μm, 15μm, 18μm foils are 4.2, 5.7, 7.2 and 9.3 kV respectively, and the corresponding standard deviations are 0.6, 0.5, 0.8 and 0.9 kV. As shown in Fig. 8, the breakdown strength for 9, 12, 15, 18μm foils under positive voltage are 467, 475, 480 and 517kV/mm, and the corresponding standard deviations are 67, 42, 51 and 50 kV/mm, respectively.

Compared to the dielectric strength of thickness in mm regime, it can obviously that the breakdown strength for 10μm level foil is far greater than that breakdown strength in mm regime. That is named 'film strengthenization effect'. Nevertheless, as far as the breakdown strength for those four kinds of thickness foils are concerned, it exhibits a slight increase with increasing thickness for our results. It seems to be contrary to the film strengthenization effect. The reasons for this phenomenon are surface roughness conditions of electrodes, thickness deviation, defects and treatment processing, and so on. This 'curious' behavior also can be interpreted with the percolation theory. It based on a priori supposed that when the probability to find defects into the bulk increases, an increase in the scattering of the breakdown and a decrease in the shape parameter value ought to be observed (Diaham et al., 2010; Essam, 1980; Helgee&Bjellheim, 1991).

Fig. 9 and Fig. 10 show self-breakdown voltage and breakdown strength for thickness-varied polyester foils under negative voltage, respectively. The average self-breakdown voltage under negative voltage for 9, 12, 15, 18μm foils are 3.8, 5.4, 7.4, and 9.0kV respectively, and the corresponding standard deviations are 0.5, 0.7, 0.4 and 0.9 kV. The breakdown strength for 9, 12, 15, 18μm foils are 444, 450, 490 and 500kV/mm, and the corresponding standard deviations are 53, 58, 28 and 49 kV/mm. It is also found that the breakdown voltage and dielectric strength of polyester film under negative voltage is smaller than that one under positive voltage. In other words, the dielectric strength of polyester film is evidently dependent of the voltage polarity. It can be interpreted by the space charge effect in relation to polyester solid structure (Ieda, 1987).

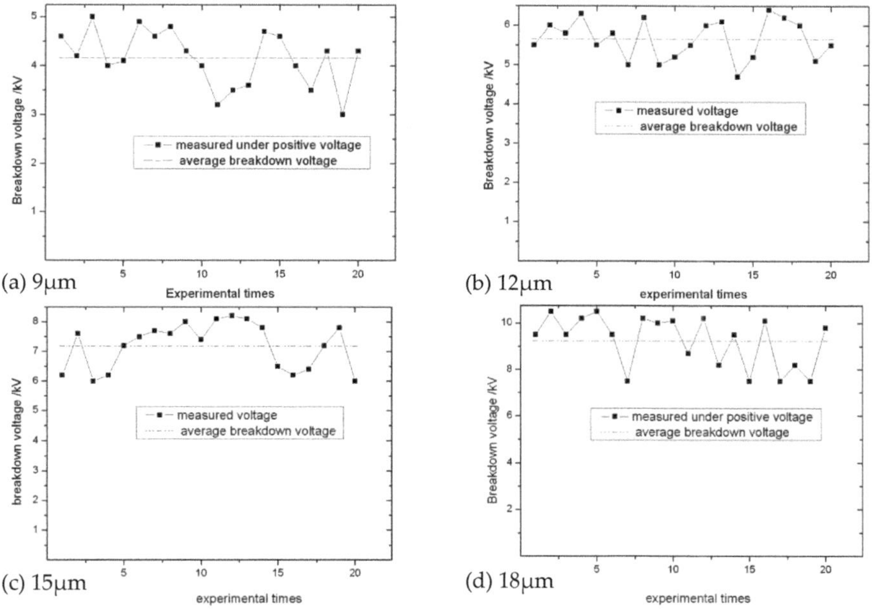

Figure 7. Self-breakdown voltage of thickness-varied polyester-foils under positive voltage

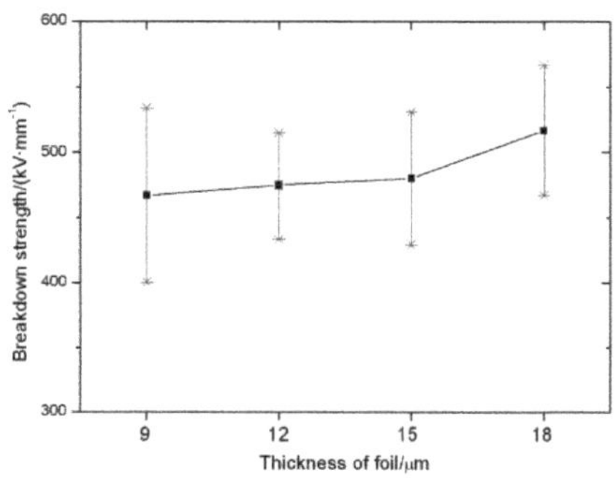

Figure 8. Breakdown strength for thickness-varied polyester-foil under positive voltage

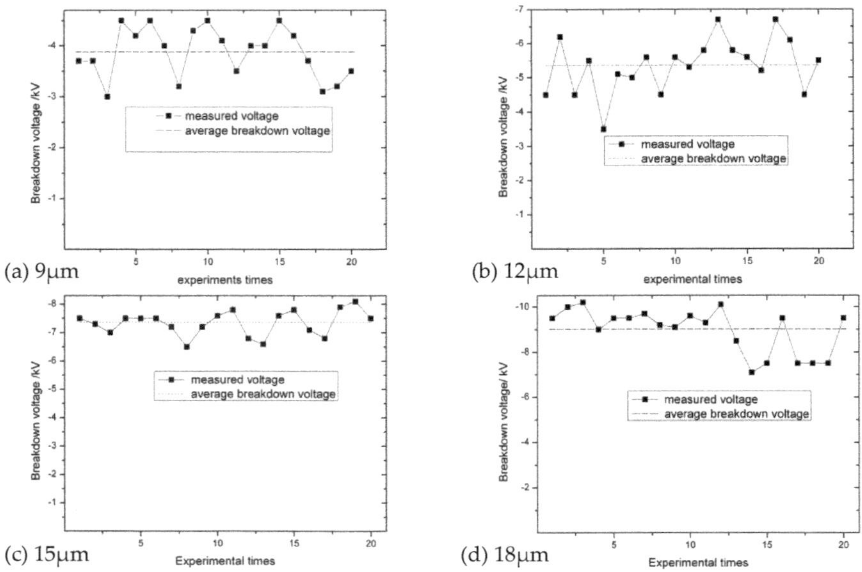

Figure 9. Self-breakdown voltage of thickness-varied polyester-foils under negative voltage

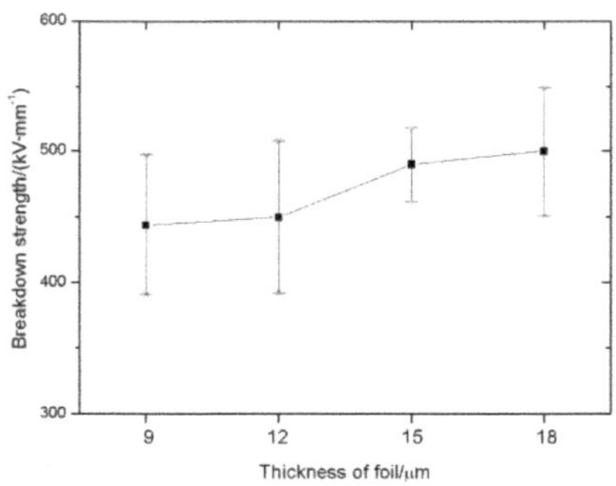

Figure 10. Breakdown strength for thickness-varied polyester-foil under negative voltage

The influence of the thickness on the Weibull parameters was also investigated. Fig. 11 shows the α–parameter values in a plot versus the polyester film thickness. The solid line and dash line represent the linear fits. This result is not in good agreement with typical value found in the literature for polymers and for such a litter variation of the thickness (Diaham et al., 2010; Helgee&Bjellheim, 1991).

In comparison with the theoretical results, the experimental results show that the critical breakdown strength of the polyester film in the intrinsic breakdown model is far lower than the experimental results and the critical breakdown strength of the polyester film in the electromechanical breakdown model and filamentary electromechanical breakdown model almost agrees with the experimental data of the polyester film. Therefore, it is concluded that the dielectric breakdown mechanism for thickness in micro-meter range is dominantly controlled by the electromechanical breakdown mechanism.

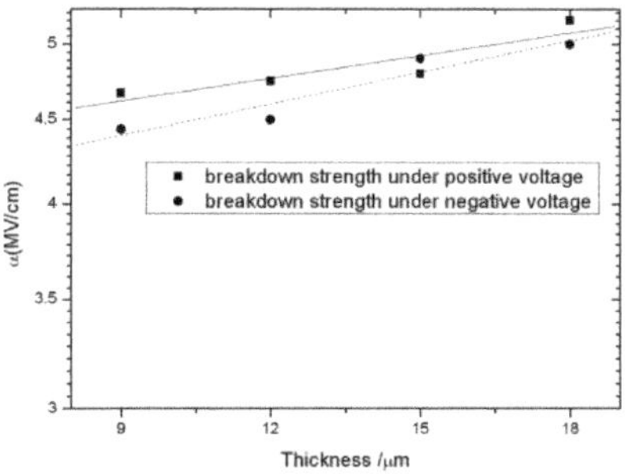

Figure 11. Changes in the scale parameter versus the polyester-foil thickness

8. Conclusion

The theoretical and experimental results of dielectric strength of polyester films in micro-meter regime are presented. Theoretical analysis was carried out based on the intrinsic breakdown model and the electromechanical breakdown model. The critical breakdown strength of intrinsic breakdown model is estimated on the basis of Hippel's low-energy criterion and Frohlich's high-energy criterion. The electromechanical breakdown model, electrofracture and filamentary electromechanical breakdown model are also introduced. The thickness dependence of dielectric strength of the polyester films was also investigated in a plane-plane electrode arrangement with different polar DC voltage. The thickness of the polyester films varied among 9μm, 12μm, 15μm and 18μm. Experimental results show that there is greatly increasing on the dielectric strength in micro-meter regime, compared to the dielectric strength of thickness in mm regime. Nevertheless, it exhibits a slight decrease with increasing thickness for our results. This 'curious' behavior is discussed on the basis of the percolation theory. It is also found that the dielectric strength of polyester film under negative voltage is smaller than that one under positive voltage. In other words, the dielectric strength of polyester film is evidently dependent of the voltage polarity. It can be interpreted by the space charge effect in relation to polyester solid structure. The comparison between the theoretical analysis and the experimental results is also presented. The results show that the critical breakdown strength of the polyester film in the intrinsic breakdown model is far lower than the experimental results, and the critical breakdown strength of the polyester film in the electromechanical breakdown model almost agrees with the experimental data of the polyester film. Therefore, it is concluded that the dielectric breakdown mechanism for thickness in micro-meter range is dominantly controlled by the electromechanical breakdown mechanism.

Author details

Haiyang Wang and Zhengzhong Zeng
Northwest Institute of Nuclear Technology, Xi'an, People's Republic of China

9. References

Artbauer, J. (1996). Electric Strength of Polymers. *J. Phys. D: Appl. Phys.*, Vol. 29, No. 2, (February 1996), pp. 446-457, ISSN 0022-3727.

Barber, P.; Balasubramanian, S.; Anguchamy, Y.; Gong, S. S.; Wibowo, A.; Gao, H. S.; Ploehn, H. J. & Zur Loye, H. -C. (2009). Polymer Composite and Nanocomposite Dielectric Materials for Pulse Power Energy Storage. *Materials*, Vol. 2, No. 4, pp. 1697-1733, (October 2009), ISSN 1996-1944.

Bartnikas, R. & Eichhorn, R. M. (1983). *Engineering Dielectrics. Volume IIA—Electrical Properties of Solid Insulating Materials: Molecular Structure and Electrical Behavior*, ASTM Publication, ISSN 978-0-8031-0228-6, Baltimore.

Blok, J. & LeGrand, D. G. (1969). Dielectric Breakdown of Polymer Films. *J. Appl. Phys.*, Vol. 40, No. 1, (January 1969), pp. 288-293, ISSN 0021-8979.

Callen, H. B. (1949). Electric Breakdown in Ionic Crystals. *Phys. Rev.*, Vol. 76, No. 9, (November 1949), pp. 1394-1402, ISSN 0031-899X.

Chauvet, C., & Laurent, C. (1993). Weibull Statistics in Short term Dielectric Breakdown of Thin Polyethylene Films. *IEEE Trans. Electr. Insul.*, Vol. 28, No. 1, (February 1993), pp. 18-29, ISSN 0018-9367.

Cooper, R. (1966). The Electric Strength of Solid Dielectrics. *Br. J. Appl. Phys.*, Vol. 17, No. 2, (February 1966), pp. 149-161, ISSN 0508-3443.

DammigQuiña, P. L.; Herrera, L.; Irurzun, I. M. & Mola, E. E. (2008). A Capacitive Model for Dielectric Breakdown in Polymer Materials. *Computational Materials Science*, Vol. 44, No. 2, (December 2008), pp. 330-338, ISSN 0927-0256.

Diaham, S.; Zelmat, S.; Locatelli, M. L.; Dinculescu, S.; Decup, M. & Lebey, T. (2010). Dielectric Breakdown of Polyimide Films: Area, Thickness and Temperature Dependence. *IEEE Trans. Dielectr. Electr. Insul.*, Vol. 17, No. 1, (February 2010), pp. 18-27, ISSN 1070-9878.

Eichhorn, R. M. (1977). Treeing in Solid Extruded Electrical Insulation. *IEEE Trans. Electr. Insul.*, Vol. EI-12, No. 1, (February 1977), pp. 2-18, ISSN 0018-9367.

Essam, J. W. (1980). Percolation theory. *Rep. Prog. Phys.*, Vol. 43, No. 7, (July 1980), pp. 53-134, ISSN 0034-4885.

Fothergill, J. C. (1990). Estimating the Cumulative Probability of Failure Data Points to be Plotted on Weibull and other Probalility Paper. *IEEE Trans. Electr. Insul.*, Vol. 25, No. 3, (June 1990), pp. 489-492, ISSN 0018-9367.

Fothergill, J. C. (1991). Filamentary Electromechanical Breakdown. *IEEE Trans. Electr. Insul.*, Vol. 26, No. 6, (December 1991), pp. 1124-1130, ISSN 0018-9367.

Fothergill, J. C. (1992). Filamentary Electromechanical Breakdown in Polymers. *Proceedings of the 4th International Conference on Conduction and Breakdown in Solid Dielectrics*, pp. 323-327, ISSN 0-7803-0129-3, Sestri Levante, Italy, June 22-25, 1992.

Frohlich, H. (1939). Dielectric Breakdown in Ionic Crystlas. *Phys. Rev.*, Vol. 56, No. 4, (August 1939), pp. 394-352, ISSN 0031-899X.

Hap, S. H. & Raju, G. G. (2006). DC Breakdown Characteristics of High Temperature Polymer Films. *IEEE Trans. Dielectr. Electr. Insul.*, Vol. 13, No. 4, (August 2006), pp. 917-926, ISSN 1070-9878.

Helgee, B. & Bjellheim, P. (1991). Electric Breakdown Strength of Aromatic Polymers: Dependence on Film Thickness and Chemical Structure. *IEEE Trans. Electr. Insul.*, Vol. 26, No. 4, (December 1991), pp. 1147-1152, ISSN 0018-9367.

Hikita, M.; Kanno, I; Ieda, M.; Ishino, I.; Doi, S. & Sawa, G. (1987). Electrical Breakdown of Ethylene Copolymers. *IEEE Trans. Electr. Insul.*, Vol. EI-22, No. 4, (April 1987), pp. 175-179, ISSN 0018-9367.

Hippel, A. V. & Alger, R. S. (1949). Breakdown of Ionic Crystals by Electron Avalanches. *Phys. Rev.*, Vol. 76, No. 1, (July 1949),pp. 127-133, ISSN 0031-899X.

Ieda, M. (1980). Dielectric Breakdown Processof Polymer. *IEEE Trans. Electr. Insul.*, Vol. EI-15, No. 3, (June 1980), pp. 206-217, ISSN 0018-9367.

Ieda, M. (1984). Electrical Conduction and Carrier Traps in Polymeric Materials. *IEEE Trans. Electr. Insul.*, Vol. EI-19, No. 3, (June 1984), pp. 162-179, ISSN 0018-9367.

Ieda, M. (1987). Carrier Injection, Space Charge and Electrical Breakdown in Insulating Polymers. *IEEE Trans. Electr. Insul.*, Vol. EI-22, No. 3, (June 1984), pp. 261-267, ISSN 0018-9367.

Job, A E; Alves, N.; Zanin, M.; Ueki, M. M.; Mattoso, L. H.; Teruya, M. Y. & Giacometti, J. A. (2003). Increasing the Dielectric Breakdown Strength of Poly (ethylene terephthalate) Films Using a Coated Polyaniline Layer. *J. Phys. D: Appl. Phys.*, Vol. 36, No. 12, (June 2003), pp. 1414-1418, ISSN 0022-3727.

Kim H K, Shi F G (2001). Thickness Dependent Dielectric Strength of a Low-permittivity Dielectric Film. *IEEE Trans. Dielectr. Electr. Insul.*, Vol. 8, No. 2, (June 2001), pp. 248-253, ISSN 1070-9878.

Laghari, J. R. & Sarjeant, W. J. (1992). Energy Storage Pulsed Power Capacitor Technology. *IEEE Trans. Power Electr.*, Vol. 7, No. 1, (June 2001), pp. 251-257, ISSN 0885-8993.

Laihonen, S. J.; Gafvert, U.; Schutte, T. & Gedde, U. W. (2007a). DC Breakdown Strength of Polypropylene Films: Area Dependence and Statistical Behavior. *IEEE Trans. Dielectr. Electr. Insul.*, Vol. 14, No. 2, (April 2007), pp. 275-286, ISSN 1070-9878.

Laihonen, S. J.; Gustafsson, A.; Schutte, T. & Gedde, U. W. (2007b). Area Dependence of Breakdown Strength of Polymer Films: Automatic Measurement Method. *IEEE Trans. Dielectr. Electr. Insul.*, Vol. 14, No. 2, (April 2007), pp. 263-274, ISSN 1070-9878.

Naidu, M. S. & Kamaraju, V. (1995). *High Voltage Engineering* (Second ed). McGraw-Hill Companies, Inc., ISBN 0-07-462286-2, New York, USA.

O'Dwyer, J. J. (1964). *Theory of Dielectric Breakdown in Solids*. Clarendon Press., ISBN 101-209-528 Oxford, U.K.

O'Dwyer, J. J. (1973). *The Theory of Electrical Conduction and Breakdown in Solid Dielectrics*. Calrendon Press., ISBN 0198513321, Oxford, U.K.

Pai, S. T., Zhang, Q. (1995). *Introduction to High Power Pulse Technology*. World Scientific Publishing Co. Pte. Ltd., ISBN 978-981-02-1714-3, Singapore.

Peruani, F.; Solovey, G.; Irurzun, I. M.; Mola, E. E.; Marzocca, A. & Vicente, J. L. (2003). Dielectric Breakdown Model for Composite Materials. *Phys. Rev. E*, Vol. 67, No. 6, (June 2003), pp. 1-6, ISSN 1550-2376.

Sarker, A. K.; Kimura, K; Yokoyama, F. & Yamashita, Y. (2001). Control of the Length of Aromatic Polyester Whiskers. *High Perform. Polym.*, Vol. 13, No. 2, (June 2001), pp. S351-S364, ISSN 0954-0083.

Sawa, G. (1986). Dielectric Breakdown in Solid Dielectrics. *IEEE Trans. Electr. Insul.*, Vol. EI-21, No. 6, (December 1986), pp. 841-846, ISSN 0018-9367.

Seeger, J. & Teller, E. (1939). Remarks on the Dielectric Breakdown. *Phys. Rev.*, Vol. 56, No. 4, (August 1939), pp. 352-354, ISSN 0031-899X.

Seitz, F. (1948). On the Mobility of Electrons in Pure Non-polar Insulators. *Phys. Rev.*, Vol. 73, No. 6, (March 1939), pp. 549-564, ISSN 0031-899X.

Stark, K. H. & Garton, C. G. (1955). Electric Strength of Irradiated Polythene. *Nature*, Vol. 176, (December 1955), pp. 1225-1226, ISSN 0028-0836.

Wang Z. D.; Chen, Z. T. & Wu, B. C. (1992). *Electrical Insulation Handbook.* (Second ed), China Machine Press, ISBN 7-111-00712-3, Beijing.

Wang, H. Y.; Ma, L. Y. & Zeng, Z. Z. (2004). Trigger Characteristics of Polymer-foil Switch with 1ns Delay and Sub-ns Jitter. *High Power Laser and Particle Beams,* Vol. 16, No. 12, (December 2004), pp. 1626-1628, ISSN 1001-4322.

Wang, H. Y.; Ma, L. Y. & Zeng, Z. Z. (2006). Sub-ns Jitter, 1ns Risetime and 10kV Rectangular Pulse Trigger Generator. *Plasma Science and Technology,* Vol. 8, No. 5, (September 2006), pp. 600-601, ISSN 1009-0630.

Wang, H. Y.; Ma, L. Y. & Zeng, Z. Z. (2008). Electric Breakdown Model for Dielectric Strength of 10μm Level Polyester Foil. *High Power Laser and Particle Beams,* Vol. 20, No. 10, (October 2008), pp. 1749-1752, ISSN 1001-4322.

Weibull, W. (1951). A Statistical Distribution Function of Wide Applicability. *J. Appl. Mechanics,* Vol. 9, (September 1951), pp. 293-297, ISSN 1528-9036.

Whitehead, S. (1951). *Dielectric Breakdown of Solids.* Oxford University Press., London and New York, 1951.

Wu, E. Y. & Vollertsen, R. P. (2002). On theWeibull Shape Factor of IntrinsicBreakdown of Dielectric Films and Its Accurate Experimental Determination —PartI: Theory, Methodology, Experimental Techniques. *IEEE Trans. Electron Devices,* Vol. 49, No. 12, (December 2002), pp. 2131-2140, ISSN 0018-9383.

Zakrevski, V. A.; Sudar, N. T.; Zappo, A. & Dubitsky, Y. A. (2003). Mechanism of Electrical Degradation and Breakdown of Insulating Polymers. *J. Appl. Phys.,* Vol. 93, No. 4, (February 2003), pp. 2135-2139, ISSN 0021-8979.

Zeller, H. R. & Schneider, W. R. (1984). Electrofracture Mechanics of Dielectric Aging. *J. Appl. Phys.,* Vol. 56, No. 2, (July 1984), pp. 455-159, ISSN 0021-8979.

Zeller, H. R.; Hibma, T. & Pfluger, P. (1984). Electrofracture Mechanics of Dielectric Aging. pp. 85-88, *1984 Annual Report Conf. Electr. Insul. and Dielectric Phenom,* ISSN 0084-9162, National Research Council (U.S.), Claymont, DE, October 21-25, 1984.

Permissions

The contributors of this book come from diverse backgrounds, making this book a truly international effort. This book will bring forth new frontiers with its revolutionizing research information and detailed analysis of the nascent developments around the world.

We would like to thank Hosam El-Din M. Saleh, PhD, for lending his expertise to make the book truly unique. He has played a crucial role in the development of this book. Without his invaluable contribution this book wouldn't have been possible. He has made vital efforts to compile up to date information on the varied aspects of this subject to make this book a valuable addition to the collection of many professionals and students.

This book was conceptualized with the vision of imparting up-to-date information and advanced data in this field. To ensure the same, a matchless editorial board was set up. Every individual on the board went through rigorous rounds of assessment to prove their worth. After which they invested a large part of their time researching and compiling the most relevant data for our readers. Conferences and sessions were held from time to time between the editorial board and the contributing authors to present the data in the most comprehensible form. The editorial team has worked tirelessly to provide valuable and valid information to help people across the globe.

Every chapter published in this book has been scrutinized by our experts. Their significance has been extensively debated. The topics covered herein carry significant findings which will fuel the growth of the discipline. They may even be implemented as practical applications or may be referred to as a beginning point for another development. Chapters in this book were first published by InTech; hereby published with permission under the Creative Commons Attribution License or equivalent.

The editorial board has been involved in producing this book since its inception. They have spent rigorous hours researching and exploring the diverse topics which have resulted in the successful publishing of this book. They have passed on their knowledge of decades through this book. To expedite this challenging task, the publisher supported the team at every step. A small team of assistant editors was also appointed to further simplify the editing procedure and attain best results for the readers.

Our editorial team has been hand-picked from every corner of the world. Their multi-ethnicity adds dynamic inputs to the discussions which result in innovative outcomes. These outcomes are then further discussed with the researchers and contributors who give their valuable feedback and opinion regarding the same. The feedback is then collaborated with the researches and they are edited in a comprehensive manner to aid the understanding of the subject.

Apart from the editorial board, the designing team has also invested a significant amount of their time in understanding the subject and creating the most relevant covers. They scrutinized every image to scout for the most suitable representation of the subject and create an appropriate cover for the book.

The publishing team has been involved in this book since its early stages. They were actively engaged in every process, be it collecting the data, connecting with the contributors or procuring relevant information. The team has been an ardent support to the editorial, designing and production team. Their endless efforts to recruit the best for this project, has resulted in the accomplishment of this book. They are a veteran in the field of academics and their pool of knowledge is as vast as their experience in printing. Their expertise and guidance has proved useful at every step. Their uncompromising quality standards have made this book an exceptional effort. Their encouragement from time to time has been an inspiration for everyone.

The publisher and the editorial board hope that this book will prove to be a valuable piece of knowledge for researchers, students, practitioners and scholars across the globe.

List of Contributors

Martin Koller
Graz University of Technology; Institute of Biotechnology & Biochemical Engineering, Graz, Austria
ARENA Arbeitsgemeinschaft für Ressourcenschonende & Nachhaltige Technologien, Graz, Austria

Anna Salerno, Alexander Muhr and Angelika Reiterer
Graz University of Technology, Institute of Biotechnology & Biochemical Engineering, Graz, Austria

Emo Chiellini
Laboratory of Bioactive Polymeric Materials for Biomedical and Environmental Applications via Vecchia Livornese, Loc. S. Piero a Grado (PI), University of Pisa, Department of Chemistry and Industrial Chemistry, Pisa, Italy

Sergio Casella
DAFNAE, Department of Agronomy Food Natural Resources Animals and Environment, Università degli Studi di Padova, Legnaro, Padova, Italy

Predrag Horvat
Department of Biochemical Engineering, Faculty of Food Technology and Biotechnology, University of Zagreb, Croatia

Gerhart Braunegg
ARENA Arbeitsgemeinschaft für Ressourcenschonende & Nachhaltige Technologien, Graz, Austria

Hossein Mighani
Uni of Golestan, Iran

S.Venkatachalam, Shilpa G. Nayak, Jayprakash V. Labde, Prashant R. Gharal, Krishna Rao and Anil K. Kelkar
Reliance Technology Group, Reliance Industries Limited, B4, MIDC Industrial Area, Patalganga, Raigad District, Maharashtra State, India

Sadia Sultan
Atta-ur-Rhaman Institute for Natural Product Discovery, Faculty of Pharmacy, University Tecknologi MARA, Puncak Alam, Selangor, Malaysia

Mashiur Rahman
Department of Textile Sciences, Faculty of Human Ecology, University of Manitoba, Winnipeg, Manitoba, Canada

Salar Bagherpour
Islamic Azad University, Department of Materials Science and Engineering, Najafabad-Branch, Iran

Ibrahim Refaay Ahmed
Faculty of engineering, Beni-Suif University, Egypt

Ali Waheed Yousry
Faculty of engineering, Minia University, Egypt

Bharat Dholakiya
Applied Chemistry Department, Sardar Vallabhbhai National Institute of Technology (SVNIT), Gujarat, India

Ghi Ho Tae
Department of Structural Engineering, Seoul National University of Science & Technology, Korea

Eun Soo Choi
Department of Civil Engineering, Hongik University, Korea

Hosam El-Din Saleh, Talat Bayoumi and Samir Eskander
Radioisotope Department, Nuclear Research Center, Atomic Energy Authority, Egypt

Ewa Skrzetuska, Wiesława Urbaniak – Domagała, Barbara Lipp-Symonowicz and Izabella Krucińska
Lodz University of Technology, Departament of Material and Commodity Sciences and Textile Metrology, Poland

Takako Inoue and Masako Niwa
Sugiyama Jogakuen University, Nara Women's University, Japan

Yoneda Morihiro
Faculty of Human Life and Environment, Nara Women's University, Nara, Japan

Nakajima Chie
Graduate School of Human Culture, Nara Women's University, Nara, Japan

Willian Aperador Chaparro
Ingeniería Mecatrónica, Universidad Militar Nueva Granada, Bogotá, Colombia

Haiyang Wang and Zhengzhong Zeng
Northwest Institute of Nuclear Technology, Xi'an, People's Republic of China